Environmental
MAGNETISM

Environmental MAGNETISM

ROY THOMPSON

Department of Geophysics, University of Edinburgh

and

FRANK OLDFIELD

Department of Geography, University of Liverpool

London
ALLEN & UNWIN
Boston   Sydney

Allen & Unwin (Publishers) Ltd,
40 Museum Street, London WC1A 1LU, UK

Allen & Unwin (Publishers) Ltd,
Park Lane, Hemel Hempstead, Herts HP2 4TE, UK

Allen & Unwin Inc.,
8 Winchester Place, Winchester, Mass. 01890, USA

Allen & Unwin (Australia) Ltd,
8 Napier Street, North Sydney, NSW 2060, Australia

First published in 1986

British Library Cataloguing in Publication Data

Thompson, Roy
   Environmental magnetism.
1. Magnetism, Terrestrial
I. Title   II. Oldfield, Frank
333.7 QC815.2
ISBN 0-04-538003-1

Library of Congress Cataloging in Publication Data

Thompson, Roy, 1948–
   Environmental magnetism.
Includes bibliographies and index.
1. Magnetic measurements. 2. Magnetism, Terrestrial.
I. Oldfield, Frank. II. Title.
QC820.T57 1986 538'.7 85-22885
ISBN 0-04-538003-1

Set in 10 on 11 pt Ehrhardt by Fotographics (Bedford) Ltd
and printed in Great Britain by William Clowes Limited, Beccles, Suffolk

Preface

The scientist will be forced, in the unenthusiastic words of one of my scientific colleagues, 'to slosh about in the primordial ooze known as inter-disciplinary studies'.

John Passmore
Man's responsibility for nature

The present text has arisen from some thirteen years of collaboration between the two authors. During that period, upwards of a dozen postgraduates in Edinburgh, the New University of Ulster and Liverpool have been closely involved in exploring many of the applications of magnetic measurements described in the second half of the book. Much of the text is based on their work, both published and unpublished. A great deal of the work summarised reflects extensive co-operation not only between the authors and among their postgraduate groups, but also involving colleagues in geology, geography, ecology, hydrology, meteorology, glaciology, archaeology, limnology, oceanography, chemistry and physics. It is from this wide range of collaborative, interdisciplinary work that the concept of the book has arisen. If indeed, as we hope to have shown, magnetic measurements have a growing rôle to play in many diverse and important areas of the environmental sciences, then there is a case for trying to bring together and present in an organised way, illustrations of as many applications as possible. The text also includes enough grounding in principles and techniques to encourage the interested research worker in any of the potential 'user' disciplines to understand the methods and results more fully as well as to approach the practical use of the techniques with more background, insight and confidence.

Progress in empirical science often grows from advances in our perception, appraisal and creative use of order in natural systems. Out of this can come enhanced insight into processes, structures and systems interactions on all temporal and spatial scales and at all integrative levels from subatomic to cosmic. In the environment, elements of order are often difficult to appraise and analyse, not only because of intrinsic complexity, but as a consequence of our lack of techniques, instrumentation and suitable methodologies. Magnetic properties, whether natural or induced, reflect forms of order which, in recent years, have become dramatically more accessible to a growing range of instruments and techniques. Developments in electronics and computing have been vital elements in this trend, complementing and, at times, stimulating growing interest in and concern for the environmental studies to which magnetic measurements can contribute. Thus methodological advance, problem definition and hypothesis development have been mutually interactive at all stages in the work, each aspect giving the lead from time to time.

Much of the second half of the book has arisen out of work stimulated by the late F. J. H. Mackereth of the Freshwater Biological Association Laboratories in the English Lake District. His studies of the Flandrian sediments of Windermere showed that the stable natural remanent magnetisation locked into them gave a consistent record of secular variation which, once calibrated by ^{14}C dating and through comparison with Observatory records and archaeomagnetic evidence, could provide a chronological framework for palaeolimnological studies. At the same time, studies there and elsewhere were showing that lake sediment chronologies for the last few millenia, based on ^{14}C dates alone, almost always involved large errors arising from the inwash of old particulate carbon from persistent organic residues in terrestrial soils as a result of erosion following human activity. In the Lough Neagh sediments, which were the subject of a research project arising from serious aquatic pollution problems in the late 1960s, the 'old particulate carbon' error led to stratigraphically inverted ^{14}C dates over the past 2000 years. Consequently, much of the initial palaeolimnological work designed to reconstruct the historical background for assessing the present problems of cultural eutrophication in the lake was vitiated by lack of a chronology of sedimentation for the period of maximum human impact. The possibility of identifying the 1820 AD westerly declination

maximum in the recent sediments prompted the application of Mackereth's palaeomagnetic approach to Lough Neagh. Not only did this exercise provide a suite of cores with stable, consistent and repeatable secular variation from a second lake (Thompson 1973), it played a crucial rôle in the development of the chronology of sedimentation (O'Sullivan et al. 1973) and led to a growing interest in the mineral magnetic properties of the sediment (Thompson et al. 1975). The realisation that the magnetic susceptibility variations provided a basis for rapid correlation and were a response to forest clearing and agriculture prompted first, a much wider range of catchment studies, and subsequently the extension of the approach to marine sediments and an interest in the magnetic properties of atmospheric particulates. Thus in a very real and direct sense, much of the work outlined in the second half of the book is a sequel to and consequence of the pioneering work of John Mackereth.

F. Oldfield
R. Thompson

Acknowledgements

We are grateful to the following individuals and organisations who have kindly given permission for the reproduction of copyright material (figure numbers in parenthesis):

Figure 2.8 reproduced from Stoner & Wohlfarth, *Phil Trans R. Soc.* **A240**, by permission of the Royal Society and E. P. Wohlfarth; Figure 2.9 reproduced from Butler & Banerjee, *J. Geophys. Res.* **80**, 4049–58, by permission of the AGU and R. F. Butler, copyright American Geophysical Union; Springer-Verlag (3.4); M. Fuller (4.12); N. Peddie and the United States Geological Survey (5.3); Terra Scientific Publishing Co. and N. Peddie (5.3, 5.9); Figure 5.4 reproduced from *Earth* (F. Press & R. Siever) by permission of W. H. Freeman and Company, © 1982 W. H. Freeman; M. J. Aitken and Oxford University Press (5.7); M. W. McElhinny (5.9); Figure 6.7 reproduced from Goree & Fuller, *Geophys. Space Phys.* **14**, 591–608, by permission of the AGU, copyright American Geophysical Union; the Editor, *Archaeometry* (6.10); W. Junk (10.6, 10.15, 10.16); Figure 13.5 reproduced from McDougall *et al., Geol Soc. Am. Bull.* **88**. 1–5, by permission of the Geological Society of America and I. McDougall; K. Kobayashi (13.6); T. D. Allan (13.7); Figure 13.8 reproduced from Phillips & Forsyth, *Geol Soc. Am. Bull.* **83**, 1579–600, by permission of the Geological Society of America, J. W. Phillips and D. W. Forsyth; P-L. Kalliomaki (15.2); North Holland Publishing Co. and S. J. Williamson (15.3).

The broad range of topics and disciplines covered in this book has been made possible by the help, advice and stimulation we have received from many friends and associates over the past two decades. Numerous colleagues have generously provided hard won, well characterised field samples for us to indulge our magnetic whims on; others have unselfishly supplied us with their unpublished data; while others have patiently donated their time in order to guide us through the intricacies of their own disciplines and to reduce our confusions and misapprehensions. To all these friends, companies and colleagues we are extremely grateful. Sections of text have been read and much improved by the constructive comments of Rick Battarbee, Jan Bloemendal, David Collinson, John Dearing, Ed Deevey, Ken Gregory, Roy Gill, Andy Hunt, Jen Jones, Norman Hamilton, Barbara Maher, Malcolm Newson, Bill O'Reilly, Simon Robinson, Don Tarling and Des Walling. Our technical helpers and assistants have developed and maintained our field and laboratory equipment despite (intermittent) abuses while new or rushed experiments have been attempted, and have striven to sustain high levels of 'quality control' over the hundreds of thousands of magnetic measurements made at Edinburgh and Liverpool during the course of the past years. Our students in particular have generated much of the data described in the book, and provoked and vitalised new lines of enquiry, and to them special thanks are due.

We also wish to thank Janet Hurst for her typing and mastery of the Liverpool University Faculty of Social and Environmental Studies word processor, and the staff of the drawing office in the Liverpool University Geography Department for their help in preparing the diagrams of the book.

Finally, we owe a great debt to our wives, not only for foregoing scrambling and skiing weekends in the Scottish mountains or days in the garden, but also for an acumen in spotting split infinitives and 'sentences that are not sentences'. Above all, we thank them for their continued support and encouragement. We dedicate this book to Mary and Christine.

Contents

CONTENTS

List of tables

[1]
Introduction

Its purpose is to be suggestive to him rather than didactic; to put
him in the way of intelligently observing for himself.

Archibald Geikie 1879
Outlines of field geology
London: Macmillan

Such is the scope and versatility of magnetic studies that the same apparatus can be used, within the span of a single day, to probe the orienting behaviour of living organisms, trace the origin of suspended stream sediments, provide estimates of particulate pollution, and explore, through measurements of palaeo-magnetic intensities and pole positions, the dynamics of the Earth's fluid core. This very diversity presents us with a formidable challenge at the outset, the more so since in most of the disciplines touched on, few workers are familiar with either the principles of magnetism involved or with the methods of study employed.

The text of this book seeks to introduce research workers and advanced students to the range of applications of magnetic measurements in their own and related disciplines, as well as to provide a theoretical and practical background should this be required. Chapters 2–5 are concerned mostly with the properties, definitions, concepts, principles and theories which are used in later chapters. In so far as possible, the level of explanation and discussion chosen has been designed to allow non-physicists to understand and appreciate the magnetic properties which are inherent or may be induced in natural materials, and which form the basis for the wide range of applications illustrated later. For any interested readers unfamiliar with basic concepts in physics, we hope in the early chapters to raise their knowledge of magnetism, at least in an intuitive or qualitative sense,

to the minimum level needed for making use of either the techniques directly or the results arising from their use.

Chapter 2 describes the various types of magnetic behaviour characteristic of natural materials. These categories of behaviour can be identified from relatively simple experiments which involve recording the response of specimens to changes in variables such as magnetic field and temperature. Many of the responses can be related to the properties of crystal structure and to the effects of crystal size and shape. These properties, along with others related to the concentration of magnetic minerals, form the basis for the magnetic differentiation of soils, sediments, peats and dusts (Chs 8–12 & 16) as well as for the property of natural magnetic remanence upon which magneto-stratigraphy depends (Chs 13 & 14). Chapter 3 concentrates on naturally occurring minerals and sets out their different magnetic characteristics in terms of the magnetic properties and categories of behaviour outlined in Chapter 2. Chapter 4 deals with the magnetic parameters which we actually measure and use in the later chapters of the book in more detail. Some of these parameters are related to properties of the Earth's magnetic field, past or present (4.3.1), and others are quite independent of this and are instead a function of magnetic mineralogy, grain size and shape (4.3.2–4.8). These magnetic parameters can be diagnostic of mineral type and origin; they are sensitive to chemical and thermal transformations and can reflect the ambient magnetic field at certain critical stages in processes such as sediment deposition, rock cooling or crystal growth. Used in combination, they provide a powerful new approach to the study of environmental systems. Chapter 5 considers the Earth's magnetic field at the present day and in the past. The definition of terms and concepts, the brief outline of sources of evidence and the introduction to the nature of variation in the Earth's field on a wide range of timescales all form a link with Chapters 13 and 14 dealing with magnetostratigraphy. Chapter 6 is devoted to methods of measurement and hence to instrumentation. Its main sections are designed to provide a practical guide to available techniques and appropriate equipment, largely for the benefit of prospective practitioners inexperienced in magnetic measurements. Chapters 2–6 as a whole serve both as an introduction to essential aspects of theory and practice and as a reference text once magnetic measurements and the appraisal of results are under way.

Chapter 7 introduces in general terms those environmental systems which involve the movement of material, including magnetic minerals. It develops the linking concept of particulate flux and fore-shadows the use of magnetic minerals to explore, quantify and characterise this flux. In addition, some general points are noted regarding methods of sampling and sample preparation. In Chapters 8–12, the magnetic characterisation of particulate flux is considered in the full range of environmental contexts studied so far. Each chapter first considers the nature of mineral particles in general and magnetic minerals in particular in the environmental system considered, then briefly outlines their origin, movement and transformation. Suitable methods of sampling and measurement particular to the problems and processes involved are also outlined. Finally, case studies are summarised and these are designed to illustrate the applications of mineral magnetic studies, discuss areas of particular interest or future relevance, and identify crucial problems and uncertainties for further study. The range of mineral

magnetic studies completed so far varies very much with the environmental system under consideration and this is reflected in the chapters themselves. Some are based on a wide range of empirical studies (e.g. Ch. 10) and others are necessarily more speculative (e.g. Ch. 11). Whereas Chapters 8–12 deal with 'artificial' *mineral* magnetic properties and their uses, Chapters 13 and 14 deal with natural remanent magnetisation and attempt to summarise the full range of palaeo-magnetic measurements in the many environmental contexts where they have been studied. Chapter 15, on Biomagnetism, is concerned with the way in which animals make use of iron minerals, with emphasis on detecting the Earth's magnetic field as a directional aid. Chapter 16 is an integrative study based on the Rhode River, a tidal arm of Chesapeake Bay. It provides, amongst other things, an opportunity to develop more fully the themes of systems linkage and the scope for integrated magnetic studies especially within watershed-based interdisciplinary research programmes. The final chapter attempts a preliminary appraisal of the future rôle of magnetic measurements in environmental science.

Depending on initial training, experience and motivation, the reader may either begin with the early Chapters (2–6) or any one of the later ones (7–16) most closely related to his or her academic interests and research programme. We hope that for the geophysicist/palaeomagnetist approaching the book, the later chapters will encourage an interest in the wide range of only partly explored and potentially exciting environmental applications of magnetic measurements touched on. Perhaps in this way more scholars with the appropriate geophysical background and expertise will become interested in collaboration with non-geophysicists in other disciplines. Equally, we hope that the non-geophysicist, led to the book by an interest in particular areas of application, will feel stimulated to seek out opportunities to develop and extend the contribution of magnetic measurements to studies in his own field, through collaboration and a growing direct personal involvement in the techniques and the results obtained.

[2]
Magnetic properties of solids

Couples and electrons have their moments.

Graffito,
James Clerk Maxwell Building,
Edinburgh University

2.1 Introduction

To the man-in-the-street, matter is often thought to be either magnetic or non-magnetic. An ordinary magnet attracts magnetic material, e.g. iron filings, pins, lodestone, whereas non-magnetic material, e.g. wood, chalk, is not attracted to the magnet. In fact, all materials show some reaction to a magnetic field though in the case of conventionally 'non-magnetic' materials the reaction will be very weak. A powerful electromagnet and sensitive measuring instrument are needed to demonstrate these weak reactions.

Later chapters of this book will be mainly concerned with the more strongly magnetic materials that can, under a range of conditions, retain some of their magnetism after removal from a field. These materials are broadly referred to as **ferromagnetic** or **ferrimagnetic**. Other materials, which do not retain their magnetism are called either **paramagnetic** or **diamagnetic** according to whether they are respectively pulled into or pushed out of regions of strong magnetic field. Occasionally these relatively weak magnetic properties can dominate the magnetic behaviour of natural samples.

At the atomic scale magnetic fields arise from the motion of electrons. Two possible electron motions may be imagined within an atom; the orbital rotation of an electron about an atom's nucleus, and the spin motion of an electron about its own axis. Both these whirling motions or currents in an atom produce a magnetic field, but in the natural iron oxide minerals, with which later chapters will be largely concerned, the spin moments are completely dominant.

2.2 Basic magnetic properties

2.2.1 Diamagnetism

Diamagnetism is a fundamental magnetic property. It is extremely weak compared with other magnetic effects and so it tends to be swamped by all other types of magnetic behaviour. Diamagnetism arises from the interaction of an applied magnetic field with the orbital motion of electrons and it results in a very weak negative **magnetisation**. The magnetisation is lost as soon as the magnetic field is removed. Strong magnetic fields tend to repel diamagnetic materials. The spin magnetic moments of electrons do not contribute to the magnetisation of diamagnets as all the electron spin motions are paired and cancel each other out.

Diamagnetism is for all practical purposes independent of temperature. Many common natural minerals, such as quartz, feldspar, calcite and water, exhibit diamagnetic behaviour.

3

2.2.2 Paramagnetism

Paramagnetic behaviour can occur when individual atoms, ions or molecules possess a permanent elementary magnetic dipole moment. Such magnetic dipoles tend to align themselves parallel with the direction of any applied field and to cause a weak positive magnetisation. However, the magnetisation of a paramagnet is lost once the field is removed because of thermal effects. In an applied field, paramagnetic materials behave in the opposite way to diamagnetic materials and tend to be attracted to regions of strong field. Many natural minerals, e.g. olivine, pyroxene, garnet, biotite, and carbonates of iron and manganese, are paramagnetic. The incompletely filled inner electron shells of Mn^{2+}, Fe^{2+} and Fe^{3+} ions are generally responsible for the paramagnetic behaviour of natural materials as they have unpaired electrons with free spin magnetic moments.

When a field is applied to a paramagnetic substance the spin magnetic moments tend to order and to orientate parallel to the applied field direction. However, the magnetic energies involved are small and thermal agitation constantly attempts to break down the magnetic ordering. A balance is reached between these two competing processes of thermal randomising and magnetic ordering. The magnetic moment which depends on this balance is thus a function of both the applied field and the absolute temperature. The magnetisation of a paramagnetic substance is very weak compared with that of a ferromagnet, but paramagnetic effects are in turn dominant over diamagnetic effects.

2.2.3 Ferromagnetism

Ferromagnetic materials such as iron are characterised by the way in which their magnetic properties change dramatically at a particular critical temperature, called the **Curie temperature**. Below the Curie temperature a ferromagnetic material can carry a strong remanent magnetisation, but above the Curie temperature, its ferromagnetic ordering is broken down by thermal energy and it behaves as a paramagnet.

The **remanent magnetisation** of ferromagnetic materials results from the phenomenon of spontaneous magnetisation – that is a magnetisation which exists even in the absence of a magnetic field. Spontaneous magnetisations arise from the group-magnetic phenomenon of exchange interactions. In these interactions all elementary magnetic moments of neighbouring electrons are aligned parallel with one another by quantum mechanical effects.

Another important property of ferromagnetic materials is that their net magnetic moment is much greater than that of paramagnetic and diamagnetic materials. This strong magnetic moment of ferromagnets arises because the magnetic exchange interactions between neighbouring atoms are so powerful that they are able to align the ferromagnetic atomic moments despite the continual disturbance of thermal agitation.

2.2.4 Ferrimagnetism and antiferromagnetism

The main natural magnetic minerals we shall be dealing with are special variants of ferromagnets known as ferrimagnets and imperfect **antiferromagnets**. Ferrimagnetic and antiferromagnetic behaviour in natural materials arises from ordering of the spin magnetic moments of electrons in the incompletely filled 3d shells of first transition series elements, particularly iron and manganese, by exchange forces.

FERRIMAGNETISM
Ferrimagnetism is outwardly very similar to ferromagnetism; indeed it is very difficult to distinguish between the two properties even using magnetic measuring techniques. Ferrimagnetic materials carry a remanent magnetisation below a critical temperature, termed the Curie or **Néel temperature**, and like ferromagnets they are paramagnetic above this temperature.

The magnetic behaviour of ferrites (ferrimagnets) depends on their particular crystal structure. Ferrites are commonly iron oxides with a **spinel** (close-packed face-centred cubic) structure containing two types of magnetic sites which have antiparallel **magnetic moments** of different magnitudes. Therefore the elementary magnetic moments of a ferrite are regularly ordered in an antiparallel sense, but the sum of the moments pointing in one direction exceeds that in the opposite direction (Fig. 2.1) leading to a net magnetisation. The imbalance in lattice moments in ferrites may be due to different ionic populations on the two types of sites or to crystallographic dissimilarities between the two types of magnetic sites.

Ferrites have low electrical conductivities and have many industrial applications. For example, Mn and

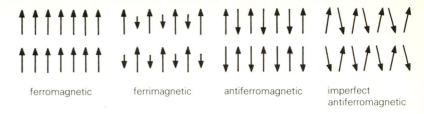

ferromagnetic   ferrimagnetic   antiferromagnetic   imperfect
                                                          antiferromagnetic

Figure 2.1 Arrangement of magnetic moments in ferromagnetic, ferrigmagnetic, antiferromagnetic and imperfect antiferromagnetic materials.

Net magnetisation

Zn ferrites are used in radiofrequency cores, while Mn mixture ferrites are used in computer memories. Magnetite is an example of a natural ferrite.

ANTIFERROMAGNETISM

In antiferromagnetic materials there are again two magnetic sublattices which are antiparallel, but their magnetic moments are identical, and so the material exhibits zero bulk spontaneous magnetisation (Fig. 2.1). Antiferromagnetic ordering is also destroyed by thermal agitation at the Néel temperature.

Modification of the basic antiferromagnetic arrangement can, however, lead to a net spontaneous magnetisation. Two such imperfect antiferromagnetic forms are parasitic ferromagnetism which may result from heterogeneities due to impurities or lattice defects, and by spin canting, which arises from a slight modification of the true antiferromagnetic anti-parallellism. Spin canting is illustrated in Figure 2.1. The mineral haematite is an example of a natural crystal with an imperfect antiferromagnetic structure caused by spin canting.

2.3 Hysteresis

The magnetic state of an iron bar depends on both the magnetic field to which it is subjected and the history of the bar. The field dependence of magnetisation can be described with the aid of Figure 2.2 which plots magnetisation on the vertical axis against magnetic field on the horizontal axis. Starting with an unmagnetised piece of iron it is found that its magnetisation increases slowly as a small field is applied and that if this field is removed the magnetisation returns to zero. However, on applying a stonger field, beyond a certain critical field, it is found that an

important change in magnetic behaviour takes place. The magnetisation is now no longer reversible in the straightforward way of the very low fields; instead, on removal of the field, a phenomenon referred to as **hysteresis** develops. In short, changes in magnetisation associated with the removal of the field now differ from those that occurred during the preceding increase of the field, in such a way that the magnetisa-

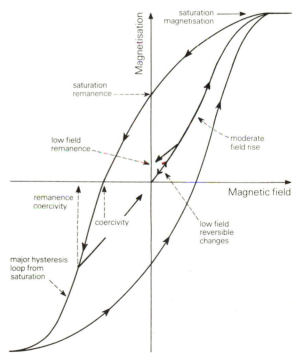

Figure 2.2 Magnetic hysteresis loop and initial magnetisation curve showing saturation magnetisation, saturation remanence, coercivity, remanence coercivity and low field magnetisation changes.

tion changes lag behind the field. Furthermore, it is found that on complete removal of the field, i.e. in zero field, the iron is no longer unmagnetised but has a remanent magnetisation. At moderate fields (Fig. 2.2) magnetisation rises sharply with increasing field and at still higher fields saturation of the magnetisation sets in and the magnetisation curve flattens out (Fig. 2.2). A complete hysteresis loop is obtained by cycling the magnetic field from an extreme applied field in one direction to an extreme in the opposite direction and back again.

Many of the simple magnetic properties used in later chapters to characterise materials can be classed as hysteresis parameters and the interrelationships between these properties can be best understood in terms of hysteresis loops (Fig. 2.2). Consider five of the most important hysteresis parameters. **Satura-**

tion **magnetisation**, M_S, is the magnetisation induced in the presence of a large (> 1 T) magnetic field. Upon removal of this field the magnetisation does not decrease completely to zero. The remaining magnetisation is called the **saturation remanent magnetisation**, M_{RS}. By the application of a field, in the opposite direction to that first used, the induced magnetisation can be reduced to zero. The reverse field which actually makes the magnetisation zero, when measurement is made in the presence of the field, is called the saturation **coercivity** $(B_0)_C$. An even larger reverse field is needed to leave no remanent magnetisation after its subsequent withdrawal. This reverse field is called the **coercivity of remanence**, $(B_0)_{CR}$. The gradient of the magnetisation curve at the origin of Figure 2.2 is the **initial susceptibility**, κ.

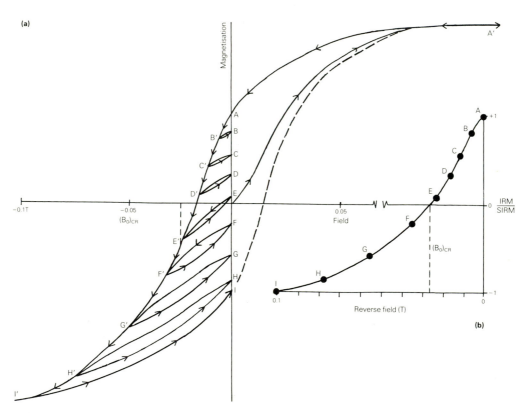

Figure 2.3 (a) Magnetic hysteresis changes performed by a specimen subjected to a forward saturating field (A') followed by a series of increasing back fields (B' to I'). The changes take the form of a set of minor hysteresis loops (B'B, C'C . . . H'H). The outer envelope (dashed) marks the major hysteresis loop of the material. (b) Measurements of the remanent magnetisations (A to I) left after each magnetisation step plotted against field as a 'coercivity' curve (or remanent hysteresis curve). The reverse field at which the 'coercivity' curve crosses the zero horizontal axis is the coercivity of remanence, $(B_0)_{CR}$. The remanent magnetisation (A) is the saturation remanence, M_{RS}.

HYSTERESIS OF REMANENCE MEASUREMENTS

Remanent magnetisations produced by exposure to fields of various strengths are widely used in the studies described in the latter half of this book as measures of magnetic mineral concentration and of magnetic stability. A typical sequence of magnetising fields and remanence measurements is illustrated in Figure 2.3 by way of a series of minor hystercsis loops (B'B) to (H'H). In our experiments reported in Chapters 8–12 and 16 an initially unmagnetised specimen is firstly subjected to a powerful forward **field** which magnetises it to saturation (A' in Fig. 2.3a). On removal of the specimen from the 'saturating' field its magnetisation decreases to point A in Figure 2.3a, i.e. to its saturation remanence value. This remanence can be easily measured (6.2). Figure 2.3a plots the further effects of placing the specimen in reverse fields of increasing field strengths (B' to H'). Each time the specimen is withdrawn from the field its new remanence (B to H) is measured. The normalised remanence measurements (B/A, C/A, ... H/A) are plotted against reverse field in order to construct a 'coercivity' curve changing from +1 to −1 (Fig. 2.3b). The reverse field, $(B_0)_{CR}$, required to reduce the saturation to zero remanence can be found from inspection of the 'coercivity' curve. This whole process of assessing hysteresis properties through remanence measurements can be performed quite quickly as each magnetisation step takes only a fraction of a second and each remanence measurement occupies less than a minute.

2.4 Effects of crystal size, shape and structure

2.4.1 Anisotropy

The magnetic properties of crystals are modified and controlled by magnetic anisotropy. In general, any specimen will be magnetically anisotropic, that is to say its magnetic properties will vary with direction. In fact many crystal properties such as elasticity and refractive index are anisotropic, varying with direction.

There are three forms of magnetic anisotropy. One form arises from the internal geometry of crystals, i.e. their lattice properties, and is referred to as magneto-crystalline anisotropy. The various axes of a crystal have different magnetic properties. For example in a single crystal of iron it is harder to magnetise the

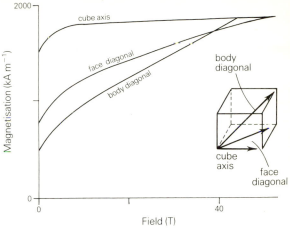

Figure 2.4 Magnetisation curves of a single iron crystal along its three principal crystallographic directions. Inset illustrates the geometry of the crystallographic axes.

crystal along certain axes (Fig. 2.4), although the saturation magnetisation is the same for all axes. Magnetocrystalline anisotropy is of considerable importance in one group of natural minerals, namely the imperfect antiferromagnets with their low spontaneous magnetisations. In these minerals, such as haematite and goethite, magnetocrystalline anisotropy leads to very high coercive forces. It also leads to characteristic low temperature magnetic behaviour as a result of extreme sensitivity of magnetocrystalline anisotropy to temperature. Magnetocrystalline anisotropy, along with spontaneous magnetisation and Curie temperature, is an important intrinsic magnetic property depending only on crystal structure and composition.

A second form of anisotropy is connected with the shape of magnetic bodies. A familiar example of this shape anisotropy is provided by the long, needle-like form of an ordinary magnetic compass. A compass needle is invariably magnetised parallel with its long axis as this is the easy direction of magnetisation. On magnetising a body (Fig. 2.5), magnetostatic forces produce magnetic charges on the end faces. (These concentrations of charge are what we usually regard as the poles of the magnet.) The field which would be produced by this apparent surface pole distribution acting in isolation is called the demagnetising field (Fig. 2.5c). This demagnetising field opposes the magnetisation and causes bodies to be magnetised most easily parallel to their long axes. An example of a natural mineral in which shape anisotropy is of great importance is the strongly magnetic ferrite, magnetite.

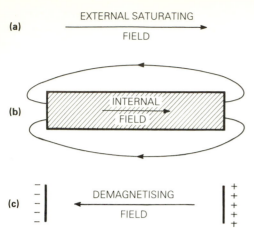

Figure 2.5 Schematic diagram illustrating the origin of the demagnetising field in a magnetised body. An external field (a) magnetises the body (b). The field acting inside the body (b), is modified by the demagnetising field (c), i.e. by the field which would be produced by the apparent surface pole distribution, caused by the magnetisation. The internal demagnetising field may be viewed as the internal field which would result from replacing the magnetised specimen by a surface distribution, or array of north and south monopoles which would produce an identical field pattern to that of the magnetised specimen.

A third form of magnetic anisotropy is referred to as strain anisotropy. It may be induced by mechanical stress through the phenomenon of magnetostriction. Magnetostriction results in an alteration of the size of a magnetic specimen when it is subjected to a magnetic field. Conversely, an inverse magnetostrictive effect (the magnetic equivalent of the piezo-electric effect) occurs when stress is applied to a magnetic specimen and its magnetic properties are altered. It arises from the distortion of the crystal lattice and may be viewed as a modification of crystalline anisotropy, although it is usually treated as a separate effect.

2.4.2 Domains and domain walls

The hysteresis properties of ferromagnets are largely related to the arrangements of magnetic **domains**. Consequently, magnetic domains play an important role, along with anisotropy, in controlling the magnetic properties of natural magnetic materials (Dunlop, 1981).

The concept of magnetic domains was proposed by Weiss (1907) in order to explain how a material with spontaneous magnetisation can exist in a **demagnet-**ised state. He suggested that the material could be split up into many domains, or regions, each spontaneously magnetised in one direction, and that the domains might be magnetised in different directions (as illustrated in Fig. 2.6a) so that the sum of the domain magnetisation could be zero. Bloch (1930) later suggested that the magnetic domains were separated by zones of finite thickness. These finite boundary regions, illustrated schematically in Figure 2.6b, have become known as domain or Bloch walls.

The reason why magnetic domains form is that they produce a state of lower total energy. They achieve this state by establishing a balance between various competing energies. The most advantageous balance is achieved by domains of about one micrometre in size.

The thickness of domain walls is also a compromise between opposing influences. In this case the balance of energies leads to a zone of about 100 atoms, some 10^{-8} to 10^{-7} m thick, within which the direction of the magnetic spins changes continuously from that of one domain to that of the adjacent domain (Fig. 2.6b).

(a)

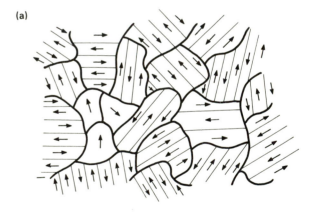

(b)

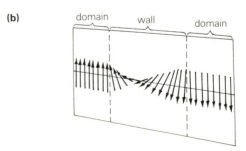

Figure 2.6 (a) Domain arrangement in a polycrystalline specimen. (b) Progressive rotation of spins through a 180° domain wall.

2.4.3 Multidomain behaviour

The magnetisation curves and hysteresis properties of large **multidomain** grains can be satisfactorily explained in terms of the movements of domain walls (Fig. 2.7). On the application of an external field to a large multidomain grain, domain wall translations will take place by favouring the growth of domains with a magnetisation component in the direction of the applied field. The induced magnetisation will change as the field is slowly increased and the boundary walls move from their minimum energy positions, e.g. position I in Figure 2.7. The initial magnetisation changes will be a reversible function of field. Eventually however, the boundary translations will reach positions such as (II) in Figure 2.7 where their equilibrium is unstable and the boundaries will move spontaneously, with discontinuous and irreversible changes in magnetisation to new equilibrium positions such as (IV). Further reversible and irreversible changes will continue with any further

field increases (Fig. 2.7). On removal of the external field the boundary walls may be trapped as they return towards their initial locations in local energy minima such as (IV).

It is in this way, by the trapping of domain walls, that a remanent magnetisation can arise from the application of a magnetic field to a multidomain grain. Jumps out of the local traps cannot occur until sufficiently strong fields are applied. Irreversible domain wall movements are known as Barkhausen jumps. In specimens containing many domains the boundary movements merge and the discontinuous nature of the magnetisation changes is blurred to produce smooth hysteresis curves. By applying sufficiently strong magnetic fields, magnetisation changes are also brought about by the twisting of domain magnetisations into the applied field direction. Such twisting corresponds to the approach to saturation in the hysteresis loop of Figure 2.2. A multidomain grain is finally saturated when all the domain magnetisations are aligned in the applied field direction.

2.4.4 Single-domain behaviour

The magnetic properties of single-domain grains are quite different from those of multidomain grains since domain wall motions do not play a rôle in their magnetisation cycle. The magnetic remanence of single-domain grain assemblages is much higher and more stable than that of multidomain grain assemblages and there are also clear differences in their hysteresis properties.

The absence of domain walls leads to characteristic, single-domain hysteresis loops. The shape of the loop of an individual single-domain grain depends on the orientation of the grain with respect to the applied field. For example, the hysteresis loop of a single-domain grain with its easy magnetisation direction (long axis) parallel to the applied field will have the rectangular shape of the heavy line in Figure 2.8. The magnetisation simply flips through 180° when a field exceeding a critical field, called the coercive force, is applied in the direction opposite to that of the magnetisation. In the case of a single-domain grain with its easy magnetisation direction perpendicular to the applied field the hysteresis loop has the completely reversible form of the dotted line in Figure 2.8. On application of a field the magnetisation turns towards the field direction, but it returns to its original 'easy' axis direction perpendicular to the field direction on

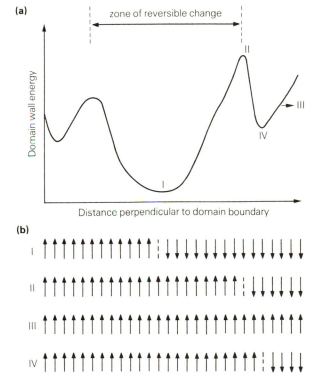

Figure 2.7 Domain wall energy as a function of position. (a) Domain wall energy depends on the interaction of the wall with local crystal defects and imperfections. (b) Schematic diagram of spin arrangements for four positions of the wall.

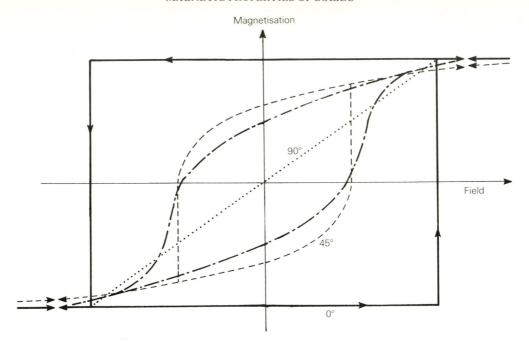

Figure 2.8 Calculated magnetisation curves for single-domain grains. The solid, dashed and dotted curves plot magnetisation changes for grains with their long axes orientated at 0°, 45° and 90° to that of the applied field. The dashed and dotted curve shows the calculated hysteresis loop of a collection of grains orientated at random (after Stoner & Wohlfarth 1948).

removal of the field. On application of a field in excess of the coercive force the magnetisation swings completely into the field direction. The dashed line of Figure 2.8 depicts the hysteresis changes of a grain with its easy axis directed at 45° to the field direction. In most natural samples containing single-domain grains we are dealing with assemblages with random orientations of their easy axes. The net hysteresis loop for such random assemblages takes on the modified form of the dashed and dotted curve in Figure 2.8.

2.4.5 Superparamagnetism

When ferro- or ferrimagnetic grains are extremely small, about 0.001–0.01 μm in diameter, they have thermal vibrations at room temperature which have energies of the same order of magnitude as their magnetic energy. As a consequence of this equivalence of energies these ultrafine-grained magnetic materials do not have a stable remanent magnetisation and do not exhibit hysteresis as their magnetisation is continually undergoing thermal reorientation. In the presence of an applied field they do, however, have an

overall magnetic alignment, i.e. an apparent magnetisation.

The type of behaviour these ultrafine grains exhibit is termed **superparamagnetic** (Néel 1955, Creer 1959, Vlasov *et al.* 1967). It is similar to, but much stronger than, paramagnetic behaviour. It is interesting that the susceptibility of superparamagnetic grains turns out to be much greater than that of an equivalent amount of mineralogically comparable stable single-domain or multidomain grains (Bean & Livingston 1959). This means that the presence of a small proportion of superparamagnetic grains in a natural sample can have an important effect on its susceptibility.

Superparamagnetic behaviour strongly depends on temperature. Indeed, if grains that behave superparamagnetically at room temperature are cooled sufficiently they will exhibit the usual ferro- or ferrimagnetic properties of **stable single-domain grains.**

2.4.6 Critical grain sizes

There are two important magnetic grain-size boundaries. These are (a) the division between ultrafine superparamagnetic grains and small stable single-domain grains and (b) the division between multidomain and stable single-domain grains.

Figure 2.9 illustrates the dependence of these critical grain sizes for magnetite on grain shape. The superparamagnetic/stable single-domain boundary occurs, at normal temperatures, in magnetite grains of around 0.03 μm diameter (Dunlop 1973a). The precise limit between multidomain and single-domain behaviour is complicated to estimate. A likely boundary and its variation with grain shape is drawn in Figure 2.9. Theoretical calculations show that magnetite grains with diameters in excess of 1 μm will certainly be multidomain (Kittel 1949). Spherical grains become multidomain at somewhat smaller sizes than elongated grains.

Haematite is found to have a much larger multidomain/single-domain transition size than magnetite, mainly because of its lower saturation magnetisation (Chevallier & Mathieu 1943). This large critical grain size of over 0.1 cm ensures that the great majority of haematite grains found in nature are single-domain.

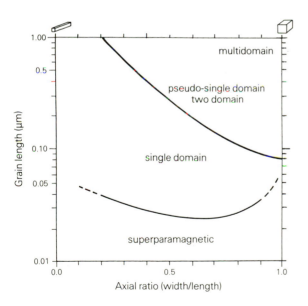

Figure 2.9 Multidomain, stable single-domain and superparamagnetic regions of magnetite grains as a function of grain length and axial ratio. A small change in grain length produces a large change in relaxation time at the superparamagnetic/single-domain boundary (data after Butler & Banerjee 1975).

2.5 Time dependence of magnetisation

The effects of time on magnetic phenomena are of great importance, especially when geological timescales are involved. Changes in magnetisation with time are known as viscous changes. They can arise through a variety of effects. The most important **viscosity** phenomenon from our point of view is that of thermal activation (Dunlop 1973b, Mullins & Tite 1973).

In large multidomain grains thermal activation causes magnetic viscosity by exciting domain wall movements and allowing the walls to cross otherwise impenetrable barriers. Thermal agitation can also cause the magnetic moments of single-domain grains to rotate from one minimum energy position to another across potential energy barriers which otherwise would be too high to allow any change in moment.

Magnetic viscosity can lead to either the growth or the loss of magnetisation. A viscous sample stored in a magnetic field will increase its magnetisation and may even acquire a **viscous remanent magnetisation**. Loss of viscous remanence can be brought about by leaving a viscous sample in a zero field environment. In many cases viscous magnetisation changes are found to be proportional to the logarithm of time.

2.6 Grain interactions

If magnetic grains lie close to each other then magnetostatic interaction can arise between them and modify their overall magnetic behaviour. Such magnetostatic grain interactions tend to lower bulk coercivity (Davis & Evans 1976). Indeed coercivity is found to decrease regularly with increased packing, where packing is defined as the fraction of the total volume occupied by magnetic particles. The magnetisation and demagnetisation curves of magnetic assemblages are also altered by interactions. Magnetism tends to be made more difficult by grain interactions whereas demagnetisation tends to become more easy. Another example of the effect of grain interactions is that the critical grain size of the multi-/single-domain boundary is increased by interactions because of the change in magnetostatic energy. Finally, the susceptibility of a magnetic assemblage may be lowered by magnetic interactions, particularly with interactions which involve superparamagnetic grains.

2.7   Summary

All magnetic effects can be explained by the flow of current. On the atomic scale small circulating currents are referred to as spin, electron spin being the most important cause of magnetic phenomena in 'natural materials. Atoms in which more electrons spin one way than the opposite way behave like small magnets. The electron spins of most materials are paired or continually disordered by thermal agitation so that they only give rise to the relatively weak magnetic effects of diamagnetism and paramagnetism. Atomic magnetic moments can, however, be aligned in some crystals by quantum mechanical exchange energy forces giving rise to strong spontaneous magnetisations. Such ferromagnetic or ferrimagnetic materials can carry a remanence, i.e. retain a magnetisation even in the absence of an applied field, and they can exhibit magnetic hysteresis properties.

Large crystals of these ferro- and ferrimagnetic materials usually split up into domains in such a way as to reduce their overall magnetisation. The crystals can easily be magnetised but they lose their induced magnetisation on removal of the magnetising field. Small grains of the same materials, too small to allow the formation of domain walls, always have strong magnetisations and make excellent permanent magnets. Even finer grains, however, are poor magnets as thermal agitations continually switch their magnetisations around.

Further reading

General books

Lorrain and Corson 1978. *Electromagnetism: principles and applications.*
Chikazumi 1964. *Physics of magnetism.*
Craik 1971. *Structure and properties of magnetic materials.*

Advanced books

Crangle 1977. *The magnetic properties of solids.*
Nagata 1953. *Rock magnetism.*
Bates 1961. *Modern magnetism.*

[3]
Natural magnetic minerals

Iron is the most abundant metal in the Universe.

Lepp 1975
Geochemistry of Iron

3.1 Iron and its abundance

Iron, the pre-eminent seat of magnetism in natural minerals (Section 2.2.4), is the fourth most abundant element in the Earth's crust. Consequently it is an important constituent of the majority of rocks found at the Earth's surface. Along with the commonest crustal metal, aluminium, it combines with the two most plentiful crustal elements, oxygen and silicon, to build up many of the common rock-forming minerals (Table 3.1).

The carriers of the magnetic properties of rocks are the more or less pure oxides of iron such as magnetite, titanomagnetite, haematite and maghaemite.. These iron oxides make up only a few percent of the volume of rocks and in most rocks they are well dispersed

amongst felsic minerals. Rock magnetic properties largely depend on these strongly magnetic iron oxides. However, if they happen to be unusually scarce then the iron sulphides or manganese oxides may become important. The weak paramagnetism of silicate or hydroxide minerals containing Fe^{2+}, Fe^{3+} or Mn^{2+} ions is generally swamped by the stronger magnetism of the less abundant iron oxides, but it can become noticeable in special situations.

3.2 Iron oxides

The iron oxides can be treated as ionic crystals that consist of an oxygen framework with cations in the

Table 3.1 Major minerals of the Earth's continental crust.

	Mineral	Igneous continental crust %	Chemical formulae
normally Fe-free	plagioclase	42	$NaAlSi_3O_8$–$CaAl_2Si_2O_8$
	K-feldspar	22	$KAlSi_3O_8$
	quartz	18	SiO_2
Fe-bearing	amphibole	5	$NaCa_2(Mg, Fe, Al)_5Si_8O_{22}(OH)_2$
	pyroxene	4	$Ca(Mg, Fe, Al)(Al, Si)_2O_6$
	biotite (chlorite)	4	$K(Mg,Fe)_3AlSi_3O_{10}(OH)_2$
	magnetite, ilmenite	2	Fe_3O_4, $FeTiO_3$
	olivine	1	$(Mg, Fe)_2SiO_4$

interstices. In studies of natural magnetic minerals two main groups are of interest. One group consists of minerals such as magnetite which crystallise with a spinel structure, while the other group is made up of minerals such as haematite which crystallise with a corundum structure.

3.2.1 The spinel group

A large number of oxide minerals crystallise with the spinel structure as it is an extraordinarily flexible structure in terms of the cations it can accept. The spinel unit cell is a face-centred cube. The oxide spinels contain 32 oxygen ions in the unit cell which form a nearly cubic close-packed framework, the cations occupying 16 octahedral sites and 8 tetrahedral sites within the oxygen framework. In spinels the cation sites are fixed whereas the positions of the 32 oxygen sites can vary. The entire oxygen framework can expand or contract in size in order to accommodate cations of various radii. This great flexibility of the oxygen framework allows a large number of elements to occur as important cations in natural oxide spinels or for vacancies to occur in the spinel lattice.

MAGNETITE (Fe_3O_4)

Magnetite is a very common magnetic mineral. It is found in the vast majority of igneous rocks and many metamorphic and sedimentary rocks and is one of the most abundant and ubiquitous of oxide minerals. It has the cubic inverse spinel structure (Table 3.2) and is ferrimagnetic (Néel 1948). The unit cell has eight tetrahedral sites filled with Fe^{3+} cations and sixteen octahedral sites, half of which are filled with Fe^{3+} cations and half with Fe^{2+} cations. The magnetite Curie temperature of 580 °C corresponds to a transition from ferrimagnetic ordering to disorder. The transition is accompanied by other physical changes such as a maximum in the coefficient of thermal expansion and the specific heat. At low temperatures, near −150 °C, magnetite undergoes another magnetic transition (Verwey & Haayman 1941) involving a decrease in crystallographic symmetry and an associated change in electrical conductivity.

ULVOSPINEL (Fe_2TiO_4) AND THE TITANOMAGNETITES

Ulvospinel has the same inverse spinel structure as magnetite but a different composition (Table 3.2). Ti^{4+}

Table 3.2 Spinel group iron oxides.

		a (Å)	
magnetite	$Fe^{3+} [Fe^{2+} Fe^{3+}] O_4$	8.396	I
magnesioferrite	$Fe^{3+} [Mg^{2+} Fe^{3+}] O_4$	8.383	I
jacobsite	$Fe^{3+} [Mn^{2+} Fe^{3+}] O_4$	8.51	I
chromite	$Fe^{2+} [Cr_2^{3+}\ \ ] O_4$	8.378	N
hercynite	$Fe^{2+} [Al_2^{3+}\ \ ] O_4$	8.135	N
ulvospinel	$Fe^{2+} [Fe^{2+} Ti^{4+}] O_4$	8.536	I

a = Cell edge.
N = Normal cation distribution $X[Y_2] O_4$ where [ ] indicates octahedral cations.
I = Inverse cation distribution $Y[XY] O_4$.

occupies half of the octahedral sites, which are filled by Fe^{3+} in magnetite. Charge balance is maintained in ulvospinel by FE^{2+} filling all the remaining cation sites. This arrangement means that ulvospinel has the chemical formula $FE_2 TiO_4$ and is antiferromagnetic, having eight Fe^{2+} cations on the octahedral sites, eight Fe^{2+} cations on the tetrahedral sites and a resulting net moment of zero. Solid solutions between magnetite and ulvospinel are commonly called titanomagnetites (Fig. 3.1). Variations in the unit cell size parameters 'a' and the Curie temperature along the titanomagnetite series are shown in Figure 3.2. Notice how the Curie temperature falls with increasing titanium content. The saturation magnetization similarly falls with increasing titanium content owing to the reduced exchange interactions, as does the susceptibility of single-domain titanomagnetite grains (Fig. 3.3). However, for multidomain titano-

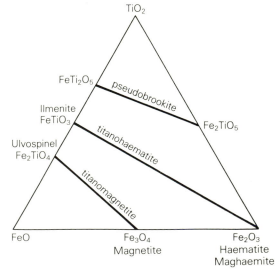

Figure 3.1 Ternary phase diagram of iron and titanium oxides showing the solid solution series.

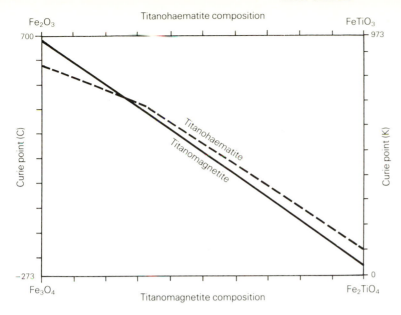

Fe₂O₃   Titanohaematite composition   FeTiO₃

Figure 3.2 Variation in Curie temperature with composition in the titanomagnetite (solid line) and titanohaematite (dashed line) solid solution series.

magnetites, which are more common in nature, the susceptibility changes little with composition (Fig. 3.3). The Curie temperature also falls with inclusion of Mg, Ca, Al, Cu, V and Si impurities in the spinel structure. Inclusion of 10% aluminium, for example, reduces the Curie temperature to 535 °C (Pouillard 1950).

MAGHAEMITE (Fe_2O_3)

Maghaemite is an extreme example of a cation-deficient spinel. It has the same chemical composition as haematite (see below) but the spinel structure of magnetite. Maghaemite's stability and structure are not well determined. A possible structural formula is $Fe^{3+}[Fe^{3+}_{5/3}\square_{1/3}]O_4$. Maghaemite can also contain some structural hydroxide; indeed it forms most readily in the presence of water. A characteristic property is that it inverts on heating above about 300 °C to haematite. On account of this instability its Curie temperature cannot be measured exactly. Impurities such as Na and Al stabilise the maghaemite structure and decrease its Curie temperature. In addition to being an important natural magnetic mineral maghaemite is widely used in the magnetic tape industry.

TITANOMAGHAEMITES

Titanomaghaemites have spinel structures occupying the region between the magnetite–ulvospinel and haematite–ilmenite join in the ternary diagram of Figure 3.1. Their chemical composition and hence their location on the ternary diagram can be conveniently expressed in terms of two parameters. One parameter, called x, expresses the Fe : Ti ratio; the other parameter called z, indicates the degree of oxidation. Natural titanomaghaemites can form above 1000 °C by solid solution and below 600 °C by metastable oxidation.

3.2.2 The corundum group

The corundum group unit cell is rhombohedral. Haematite and ilmenite form a solid solution series of rhombohedral minerals. In haematite all the cation layers are made up of Fe^{3+} ions whereas in ilmenite Fe^{2+} layers alternate with Ti^{4+} layers.

HAEMATITE (αFe_2O_3)

Haematite is a significant magnetic mineral in oxidised igneous rocks and sediments formed in oxidising conditions. When present as fine grains haematite has a distinctive blood-red colour.

The haematite (Fe_2O_3) structure can be thought of as being made up of tetrahedral Fe–O_3–Fe packages. This description of the haematite structure is most helpful for describing its magnetic structure. The pair of iron ions of each such unit are coupled with antiparallel spin moments, as a result of exchange (Section 2.2.3) interactions through the oxygen iron triplets. This spin arrangement gives haematite its

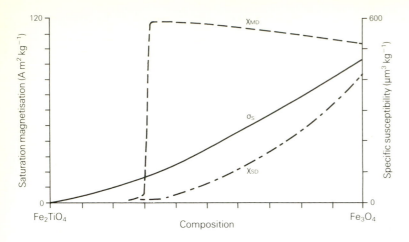

Figure 3.3 Dependence of saturation magnetisation (σ_s) and susceptibility of multidomain (χ_{MD}) and single-domain (χ_{SD}) grains on titanomagnetite composition. Substitution of titanium in the magnetite lattice reduces the saturation magnetisation but increases the hardness of the magnetisation of single-domain grains.

basic antiferromagnetic (Section 2.2.4) magnetic structure.

Adjacent iron layers in the haematite structure are coupled antiferromagnetically. The spins within each layer are parallel to each other, but the spins of adjacent planes are not exactly antiparallel and so a weak net magnetic moment results. Deviations of just 10^{-4} rad in the alignment of the spins (spin canting) can satisfactorily account for the net haematite magnetic moment (Dzyaloshinsky 1958). Although this imperfect antiferromagnetic structure results in a weak magnetisation the corundum structure ensures an extremely high stability. The intrinsic magnetic stability of haematite is of paramount importance in palaeomagnetic studies.

The Curie transition temperature of haematite is around 675 °C. Below room temperature another magnetic transition (Honda & Sone 1914) occurs when the spins move out of the plane of the iron atom layers. The transition occurs at −10 °C in pure synthetic crystals, but at lower temperatures in impure or imperfect crystals.

TITANOHAEMATITES

Minerals intermediate in composition between haematite (Fe_2O_3) and ilmenite ($FeTiO_3$) are commonly found in nature and are referred to as titanohaematites. In the ilmenite end-member, layers of Fe^{2+} alternate with layers of Ti^{4+} about the oxygen layers. Adjacent iron atom layers are magnetised antiparallel with each other. This alternation produces a structure which is essentially superparamagnetic (Section 2.4.5) at room temperature. Ilmenite has a Néel temperature of −218°C. In the titanohaematite

series the Curie temperature decreases fairly uniformly with increasing titanium content (Fig. 3.2). Intermediate composition titanohaematites are metastable and iron-enriched phases may form in them. These phases can lead to self-reversal properties.

3.3 Pyrrhotite (FeS) and the iron sulphides

The next most important magnetic minerals after the iron oxides are the iron sulphides. The most magnetic of these sulphides is pyrrhotite. It is ferrimagnetic with a monoclinic structure and an approximate composition of FeS. It always contains slightly less

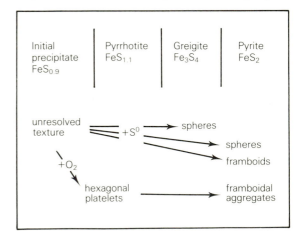

Figure 3.4 Forms of iron sulphides formed in sediments. Schematic sequence suggests textures developed during sulphurisation reactions (after Jones & Bowser 1978).

iron than suggested by this chemical formula, a common natural composition being Fe_7S_8. The majority of natural pyrrhotites have compositions within the range Fe_7S_8 to Fe_9S_{10}. The ordered vacancies in these pyrrhotite structures lead to inequalities in the magnetic ions on the antiferromagnetically coupled sublattices and hence to net resultant ferrimagnetic moments. Fe_7S_8 has a Curie temperature of 320 °C while Fe_9S_{10} has a slightly lower Curie temperature of 290 °C. Addition of impurities, such as nickel, into the pyrrhotite lattice causes the Curie temperature to be further lowered. Pyrrhotite has rarely been found to carry a useful palaeomagnetic record of the ancient geomagnetic field, but it has a high susceptibility (Clark, 1984). Its susceptibility can cause large local anomalies of the geomagnetic field which can be most helpful in mineral prospection.

Figure 3.4 illustrates the relationships between the iron sulphides and suggests a possible iron sulphide formation sequence to be found in sediments owing to sulphurisation. The formation of such authigenic iron sulphides is more common in saline than freshwater environments (Jones & Bowser 1978), although iron sulphides may be produced in sediments of productive freshwaters when oxygen deficiencies occur. Pyrite (FeS_2) is a common mineral and is paramagnetic (Table 3.4). Greigite (Fe_3S_4), intermediate in composition between pyrrhotite and pyrite, has rarely been found in natural samples, occurring only as an authigenic mineral in some freshwater carbonate sediments (Dell 1972).

3.4 Iron hydroxides and oxyhydroxides

GOETHITE (αFeOOH)
Goethite is yellowish brown to red in colour and has an orthorhombic structure. It is a very common mineral, typically formed as a weathering product, and is the stable iron oxide in soils of humid climates. Most goethite is antiferromagnetic, but owing to uncompensated spins produced by oxygen ion vacancies some goethite is weakly magnetic, with a Néel temperature of 120 °C. On cooling through its Néel temperature goethite can acquire a weak but stable **thermoremanent magnetisation**. On heating to higher temperatures of around 300 to 400 °C goethite dehydrates to haematite.

LEPIDOCROCITE (γFeOOH)
Lepidocrocite is brownish in colour and orthorhombic

Table 3.3 Magnetic properties of remanence-carrying natural minerals.

Mineral	Composition	Curie temperature (°C)	Room temperature (20°C) saturation magnetisation M_s (A m^2 kg^{-1})
magnetite	Fe_3O_4	585	93
ulvospinel	Fe_2TiO_4	−153	—
haematite	αFe_2O_3	675	0.5
ilmenite	$FeTiO_3$	−218	—
maghaemite	γFe_2O_3	~740	85
pyrrhotite	~Fe_7S_8	~300	~20
iron	αFe	780	200
goethite	$\alpha FeO.OH$	120	~1
lepidocrocite	$\gamma FeO.OH$	−196	—
magnesioferrite	$MgFe_2O_4$	440	21
jacobsite	$MnFe_2O_4$	310	77

in structure. It is less common than goethite. It has a Néel temperature of −196 °C and so cannot carry a magnetic remanence at normal temperatures. On heating it breaks down to form maghaemite at temperatures between 250 and 350 °C.

LIMONITE
Limonite is a geological field term for hydrated iron oxides of poorly crystalline, amorphous character. It consists mainly of cryptocrystalline goethite or lepidocrocite. Soil scientists prefer the term ferrihydrite.

3.5 Other magnetic minerals

A few other naturally occurring minerals are capable of carrying a magnetic remanence.

IRON
Iron, the archetypal ferromagnet, is found occasionally in natural samples, although it tends to be restricted to extraterrestrial samples such as meteorites and lunar samples returned by the Apollo and Luna missions (Fuller 1974). Iron being ferromagnetic has a high saturation magnetisation and a high Curie temperature (Table 3.3). It can carry a strong remanent magnetisation. The easy magnetisation axes of iron are the $<100>$ cube axes (Fig. 2.4).

FERROMANGANESE MINERALS
Some ferromanganese oxides and hydroxides can carry a remanent magnetisation. The jacobsite solid solution series with end members hausmannite (Mn_3O_4) and magnetite (Fe_3O_4) is known to be ferri-

magnetic (Table 3.3). The Curie temperature decreases continuously with increasing manganese content while the saturation magnetisation rises to a maximum value at the intermediate composition of $MnFe_2O_4$. Jacobsite is a relatively rare mineral.

Manganese oxyhydroxides such as todorokite and birnessite may be capable of carrying a remanence when formed with well developed crystal habits (Henshaw & Merril 1980). However, little experimental or theoretical work has been carried out on their magnetic properties and we have not found these oxyhydroxides to be remanence carriers. Authigenic ferromanganese oxides and hydroxides form as crusts or nodules in freshwater or marine environments (Murray 1876). Manganese-rich oxyhydroxides with crystalline habits tend to be more abundant in deep-sea environments. Much of the material in lacustrine deposits is poorly crystalline or amorphous and is paramagnetic not ferrimagnetic.

PARAMAGNETIC MINERALS

Iron- and manganese-bearing minerals not mentioned above tend to be incapable of carrying a magnetic remanence, at least at normal temperatures, and to be paramagnetic. Their failure to exhibit spontaneous magnetisation arises because their crystal structures are not conducive to positive exchange interactions, holding the iron or manganese ions apart at inappropriate distances or crystal geometries. Table 3.4 lists the paramagnetic susceptibilities of a number of common iron minerals. The susceptibility of these paramagnetic minerals simply depends on the number of free electrons, that is, it depends directly on the number of iron or manganese ions per gram of material.

3.6 Formation of natural magnetic minerals

3.6.1 Igneous rocks

The cooling and crystallisation of hot molten rock, called **magma**, leads to the formation of bodies of igneous rock. Magma that emerges at the Earth's surface gives rise to lava flows or, when violently ejected, to lava fountains and ash clouds. The rapid cooling of lava and ash produces fine grained volcanic rock. Magma that remains within the Earth cools slowly to form coarse grained plutonic rocks. Nearly all igneous rocks are made up of various combinations of the seven common silicate minerals of Table 3.1, with iron oxides as a small but ubiquitous accessory component. Rocks such as granites with a lot of silica are termed acidic. They consist mostly of quartz and feldspar. Basic rocks, poor in silica, are largely made up of feldspar, pyroxene and olivine crystals. Basic rocks tend to be darker in colour than acidic rocks and to contain higher concentrations of the iron oxides.

In many igneous rocks magnetite has the perfect shape and regular outline of early formed crystals. In others it is found embedded in the ground mass, having been one of the last minerals to crystallise out of the parent magma. Magnetite crystallising directly from cooling magma is generally titanium-rich. It can be produced by alteration of iron-bearing silicates and is then found as exsolved rods and as reaction rims. It often occurs with ilmenite as exsolution lamellae formed by subsolidus oxidation of titanomagnetite. Haematite is less common than magnetite and titanomagnetite in igneous rocks. When present it normally occurs as finely disseminated pigment giving the rock

Table 3.4 Specific susceptibilities of various minerals.

Remanence-carrying minerals (10^{-8} m^3 kg^{-1})		Other iron-bearing minerals (10^{-8} m^3 kg^{-1})		Other minerals and materials (10^{-8} m^3 kg^{-1})	
iron ($\propto$Fe)	2×10^7	olivines (Mg, Fe)$_2$SiO$_4$	$1 \rightarrow 130$	water (H$_2$O)	-0.9
magnetite (Fe$_3$O$_4$)	5×10^4	amphiboles (Mg, Fe, Al silicates)	$16 \rightarrow 100$	halite (NaCl)	-0.9
maghaemite (Fe$_2$O$_3$)	4×10^4	siderite (FeCO$_3$)	~ 100	quartz (SiO$_2$)	-0.6
pyrrhotite (Fe$_7$S$_8$)	$\sim 5 \times 10^3$	pyroxenes (Mg, Fe)$_2$Si$_2$O$_6$	$5 \rightarrow 100$	calcite (CaCO$_3$)	-0.5
ilmenite (FeTiO$_3$)*	~ 200	biotites (Mg, Fe, Al silicates)	$5 \rightarrow 95$	feldspar (Ca, Na, K, Al silicate)	-0.5
lepidocrocite (FeOOH)*	70	nontronite (Fe-rich clay)	~ 90	kaolinite (clay mineral)	-2
goethite ($\propto$FeOOH)	70	chamosite (Oxidised chlorite)	~ 90	montmorillonite (clay)	~ 5
haematite ($_2$O$_3$)	60	epidote (Ca, Fe, Al silicate)	~ 30	illite (clay mineral)	~ 15
		pyrite (FeS$_2$)	~ 30	plastic (e.g. perspex, PVC)	~ -0.5
		chalcopyrite (CuFeS$_2$)	~ 3		

* Only remanence carrying at temperatures well below room temperature.

a characteristic red colour. Pyrrhotite is uncommon in igneous rocks. Limonite and goethite occur as secondary minerals, produced by weathering of iron silicates and oxides.

BASALTS

The high magnetic concentrations and small grain sizes of these basic volcanic rocks make them ideal for palaeomagnetic remanence studies of the history of the **geomagnetic field** (Ch. 13). Consequently their magnetic mineralogy as well as their remanent magnetisation have been studied in great detail. Basalts typically contain between 2 and 6% iron oxide grains. Sea floor basalts have been rapidly quenched and so they generally contain homogeneous titanium-rich titanomagnetites. They are also often found to contain titanomaghaemites, formed by low temperature oxidation in the presence of water. Many continental basalts are found to have been subjected to high temperature oxidation, which can have completely altered the original mineral magnetic content they acquired during crystallisation. High temperature oxidation ($> 600\,°C$) progressively alters the primary titanomagnetites first to form magnetite and ilmenite, then titanohaematite and rutile and finally pseudobrookite. Oxidation at somewhat lower temperatures (400–600 °C) tends to produce titano-maghaemite. High temperature oxidation is related to volatile accumulation and is often pronounced in the interior of subaerially extruded basalt flows. On account of their palaeomagnetic importance such changes in magnetic mineralogy caused by high temperature oxidation have been investigated in great detail.

GABBROS

Some iron ore is present in all normal gabbros, generally in the form of ilmenite, titanomagnetite or magnetite. These coarse-grained basic rocks have undergone prolonged cooling so their iron oxide grains are generally very large. Such large iron oxide grains often exhibit distinctive granular or sandwich exsolution structures (Haggerty 1976). Gabbros make up a substantial proportion of the oceanic crust and their remanence may contribute to the magnetic anomalies found over the oceans. Titanohaematites may contribute to the magnetic anomalies found over the oceans. Titanohaematites are common in altered or weathered gabbros.

GRANITE

Slow cooling of granite plutons leads to the growth of large grains and the exsolution of ilmenite and titano-haematite. On the whole, granites are rather poor remanence carriers. Titanohaematite is generally the most important magnetic component of those granites which do carry a record of the ancient magnetic field. There is a tendency for the iron oxides in the more siliceous igneous rocks to be less titaniferous than those in basic or intermediate rocks.

3.6.2 Metamorphic rocks

In the low grade metamorphism of the green-schist facies, growth of chlorite and epidote consumes the primary iron oxides. Any natural remanence or magnetic susceptibility of the original rocks will, in general, be reduced by over an order of magnitude by the chemical changes associated with green-schist metamorphism. Iron sulphides have been found to be the major magnetic minerals in some slates. At higher grades of metamorphism magnetite may form as an equilibrium product. It can only accommodate an appreciable amount of titanium in its structure at the highest grades of metamorphism, i.e. in the granulite facies. So metamorphic magnetite is generally pure, free of inclusions and found as large crystals (Rumble 1976a). The lamellar intergrowths of titanomagnetite and ilmenite, characteristic of igneous rocks, are rarely found in metamorphic rocks. Magnetite and ilmenite assemblages have been intensively studied in many metamorphic rocks as they played an important early rôle in geothermometric studies. The grain size is always coarse so the minerals tend to be multi-domain and any palaeomagnetic remanence tends to be unstable. A further palaeomagnetic difficulty in metamorphic rocks is that their fabric can distort the direction of any magnetic remanence, although pronounced direction signals such as those associated with field reversals can still be recognised. Haematite–ilmenite solid solution series members and rutile can be found in all metamorphic grades while pseudobrookite is found in rocks formed by high temperature contact metamorphism.

3.6.3 Sediments

Red beds derive their distinctive colouring from haematite staining on their clastic particles and matrix. In addition to this red haematite pigment, large black iron oxide grains may be found. The black grains are usually haematite, often having formed by *in situ* oxidation of detrital magnetites. The haematite pigment is clearly diagenetic in origin. Possible sources of iron for the haematite pigment of red beds include release of ferric hydroxides from clay minerals, dehydration of goethite and lepidocrocite, and diagenetic breakdown of clastic ferromagnesian particles. Black and green shales contain iron sulphides rather than iron oxides. Many red, haematite-bearing sediments have turned out to be excellent recorders of the ancient field.

Limestones, although having very low concentrations of magnetic minerals, can carry a highly stable remanent magnetisation. Detrital magnetite is often the main carrier of this remanence but both geothite and haematite can also contribute to the magnetic properties of limestones.

Ocean sediments generally contain a complex magnetic mineralogy of both titanomagnetites and titanohaematites with a wide range of oxidation states and titanium contents. The magnetic grains of most ocean sediments are considered to be detrital and to have been derived from long distance atmospheric or ocean current transport. Post-depositional low temperature oxidation can lead to the formation of authigenic titanomaghaemites which carry unstable secondary remanences, particularly in slowly deposited ($<3\,\mu m\,a^{-1}$) sediments).

An excellent summary of the varied iron oxide and iron sulphide mineralogy to be found in lake sediment environments is given by Jones and Bowser (1978).

The oxygen content of a sediment at the time of its deposition is a very important factor in determining iron oxide mineralogy. For example sediments formed in oxygen-rich waters with little organic matter and with $Fe^{3+}:Fe^{2+}$ ratios greater than one are generally oxidised to form red bed sediments. In contrast the iron minerals of sediments formed in water with free oxygen but under reducing conditions are protected against the influence of the oxygen and they faithfully reflect their source materials. This situation is found in many lake sediments and in some marine turbidites on account of the rapid deposition of organic remains. A further example of the strong influence of redox conditions on iron mineralogy is provided by the dissolution of iron oxide and the formation of iron sulphides in reducing conditions. Such conditions arise for instance in suboxic hemipelagic muds and sapropel sediments which accumulate in regions of reducing environment caused by high primary production of organic matter coupled with slow bottom water circulation rates.

3.7 Summary

The commonest minerals of the Earth's crust, such as quartz and feldspar, are diamagnetic. The most widespread strongly magnetic minerals are the oxides of iron. Magnetite (and its close relations the titanomagnetites), maghaemites and titanomaghaemites are strong ferrimagnets and are found in a great variety of rock types. Haematite and its close relations the titanohaematites can hold very stable imperfect anti-ferromagnetic magnetisations and these iron oxide minerals are also to be found in many rocks and sediments. The next most important group of natural magnetic minerals is the iron sulphides. In particular pyrrhotite can be a strong ferrimagnet. The iron hydroxide goethite is also capable of carrying a remanence through an imperfect antiferromagnetism. Ferromanganese oxides and hydroxides can also be magnetic and can dominate the magnetic properties of natural materials from certain environments. other iron- and manganese-bearing minerals tend to be paramagnetic and while they are unable to carry a magnetic remanence they can contribute significantly to magnetisation properties such as susceptibility.

Further reading

General journal paper

Clark 1983. Comments on magnetic petrophysics.

Advanced journal papers

O'Reilly 1976. Magnetic minerals in the crust of the Earth.

Advanced article

Jones and Bowser 1978. The mineralogy and related chemistry of lake sediments.

Advanced books

Nagata 1953. *Rock magnetism.*
Rumble 1976. *Course notes on oxide minerals.*

[4]
Magnetic properties of natural materials

Walking around this little island (of Chaul) . . . a wonderful thing happened to me . . . placing the needle on top of a big boulder . . . the rose turned . . . it occurred to me that such a strange fact was due to the quality and nature of the rock.

Extract from log book of Joao de Castro 1538

4.1 Introduction

The main impetus for the study of the magnetic properties of rocks and sediments came with the realisation that many natural materials record valuable information about the Earth's ancient field by retaining a magnetic remanence acquired close to the time of their formation. The majority of magnetic studies of natural samples have thus been carried out by palaeomagnetists investigating natural remanent magnetisations. These studies have involved many rock types spanning all ages, and collected from all parts of the world. In association with these researchers into the Earth's fossil magnetism, studies of synthetic materials have also been carried out. These are often referred to as rock magnetic studies (Nagata 1953). They have concentrated on providing background information of use to the palaeomagnetist, and have dealt mainly with natural iron minerals commonly capable of carrying a stable remanence, such as haematite and the titanomagnetites (e.g. Day *et al.* 1977, Bailey & Dunlop 1983). A major endeavour of rock magnetists has been to elucidate the intricacies of thermoremanent magnetisation through laboratory studies of synthetic samples.

When dealing with natural samples we shall inevitably be involved with mixtures of magnetic materials. This chapter is largely concerned with setting a framework within which the magnetic properties of such mixtures can be interpreted by drawing on and summarising the results of a wide variety of rock magnetic studies.

In magnetic studies of an environmental nature, as described in Chapters 7–12 and 16, we have found mineral magnetic investigations to be invaluable in allowing unprecedentedly large quantities of material to be analysed and so we have concentrated on simple, flexible, rapid magnetic measurements which enhance this aspect of the work. Such reconnaissance mineral magnetic studies can then be followed up by more detailed and more diagnostic measurements on selected, or key, samples. The environmental mineral magnetic properties of greatest use have turned out to be magnetic susceptibility and isothermal remanence. Our interest has largely lain with magnetic properties at room temperature, despite the intrinsic connection between temperature and magnetic properties long known to the physicist. This is because in many of our environmental samples chemical changes readily take place at elevated temperatures.

21

Table 4.1 Units in magnetism and their relationships.

Quantity	SI	CGS (emu)	Relationship
induction in free space (field)	B_0 tesla (T)	B gauss (G)	$1\,T = 10^4 G$
magnetic force (field)	$H\,A\,m^{-1}$	H oersted (Oe)	$1\,A\,m^{-1} = 4\pi \times 10^{-3}\,Oe$
permeability of a vacuum	$\mu_0 = 4\pi \times 10^{-7}\,H\,m^{-1}$	$\mu_0 = 1$	$1\,H\,m^{-1}$ equivalent to $10^7/4\pi\,G\,Oe^{-1}$
induction in free space (field)	$B_0 = \mu_0 H$	$B = H$	$1\,T$ equivalent to $10^4\,Oe$
induction in medium	$B = B_0 + \mu_0 M$	$B = H + 4\pi I$	$1\,T = 10^4\,G$
magnetisation per unit volume	$M\,A\,m^{-1}$	$I\,G$	$1\,A\,m^{-1} = 10^{-3}\,G$
magnetisation per unit mass	$\sigma = M/\rho\,A\,m^2\,kg^{-1}$ ρ = density	$\sigma = I/\rho\,G\,cm^3\,g^{-1}$	$1\,A\,m^2\,kg^{-1} = 1\,G\,cm^3\,g^{-1}$
susceptibility per unit volume	$\kappa = M/H$	$\kappa = I/H$	1 (SI unit) $= 4\pi\,G\,Oe^{-1}$
susceptibility per unit mass	$\chi = \kappa/\rho\,m^3\,kg^{-1}$	$\chi = \kappa/\rho\,G\,Oe^{-1}\,cm^3\,g^{-1}$	$1\,m^3\,kg^{-1} = 4\pi \times 10^{-3}\,G\,Oe^{-1}\,cm^3\,g^{-1}$

4.2   Units

A diversity of magnetic unit systems are currently in use. The system we are following is an SI system based on Crangle (1975). Table 4.1 summarises the SI units in this system and their relationship with the older CGS units. Table 4.2 is a practical table illustrating the differences between the two systems. It tabulates the magnetic properties of single- and multidomain magnetite and haematite using both SI and CGS units.

4.3   Magnetic remanence

4.3.1   Natural magnetic remanences

Rocks, sediments and soils can acquire remanent magnetisations by natural processes. Such natural remanences are not as intense as those which can be artificially imparted in strong laboratory fields but they can be just as stable. Somewhat surprisingly very stable remanences can be acquired in weak magnetic fields such as the Earth's field. This can happen when a suitable natural event occurs to change the coercivity of the magnetic minerals from very low to very high values. For example, in igneous rocks, the event is simply the cooling of the magnetic minerals through their Curie and **blocking temperatures**; in red sediments it is the chemical growth of secondary magnetic minerals through their critical **blocking volumes**; and in deep-sea sediments the event is consolidation, which locks the tiny detrital, micronsized magnetic minerals firmly into the bulk of the sediment.

THERMOREMANENT MAGNETISATION
A thermoremanent magnetisation is acquired simply by a magnetic mineral cooling from above its Curie

temperature in a magnetic field. If the sample is isotropic the magnetisation is parallel to the applied field and, for small fields, the **intensity** of the remanence is proportional to the field. Furthermore, the remanence is remarkably stable and this means that thermoremanences can survive with little change through geological time. Thermoremanent magnetisations can thus accurately record the direction and intensity of weak magnetic fields at remote times in the past as described in Chapter 13.

The intensity of the thermoremanence of multidomain grains is much lower than that of singledomain grains (Fig. 4.1). Many rocks have magnetic grains, which, although they can be seen under the microscope to be of multidomain size ($> 20\ \mu m$), behave magnetically as though they contain a remanence carried by single-domain grains. There are a number of possible explanations for this behaviour, including the microscopic subdivision of large grains to single-domain size, inclusions of submicroscopic magnetic particles, and pseudo single-domain behaviour. The remanence of mixtures of multidomain, single-domain and superparamagnetic grains is overwhelmingly influenced by the single-domain fraction. Consequently, the thermoremanence of stable single-domain grains plays an important rôle in many palaeomagnetic studies.

CHEMICAL REMANENT MAGNETISATION
When a magnetic mineral is produced by chemical changes at temperatures below its Curie temperature, it can acquire a remanence in the direction of the ambient field. A **chemical remanent magnetisation** is locked into a magnetic grain when it grows larger than a critical size called its blocking volume. If chemical growth were to continue well above the blocking volume, a grain could eventually become multidomain, passing through a pseudo single-domain stage.

Table 4.2 Magnetic properties of magnetite and haematite in SI and CGS units.

Quantity		Single-domain magnetite		multidomain magnetite		Haematite	
saturation magnetisation	M_S	480 kAm^{-1}	(480 G)	480 kAm^{-1}	(480 G)	2.5 kAm^{-1}	(2.5 G)
saturation magnetisation per unit mass	σ_S	92 A m^2 kg^{-1}	(92 G cm^3 g^{-1})	92 A m^2 kg^{-1}	(92 G cm^3 g^{-1})	0.5 A m^2 kg^{-1}	(0.5 G cm^3 g^{-1})
saturation remanence	M_{RS}	50 kA m^{-1}	(50 G)	5 kA m^{-1}	(5 G)	1 kA m^{-1}	(1 G)
saturation remanence per unit mass	σ_{RS}	10 A m^2 kg^{-1}	(10 G cm^3 g^{-1})	1 A m^2 kg^{-1}	(1 G cm^3 g^{-1})	0.2 A m^2 kg^{-1}	(0.2 G cm^3 g^{-1})
susceptibility of ARM per unit mass	χ_{ARM}	800 µm^3 kg^{-1}	(0.07 cm^3 g^{-1})	100 µm^3 kg^{-1}	(0.01 cm^3 g^{-1})	10 µm^3 kg^{-1}	(0.001 cm^3 g^{-1})
susceptibility	κ	2.4	(0.19)	2.8	(0.22)	125×10^{-6}	(10×10^{-6})
susceptibility per unit mass	χ	450 µm^3 kg^{-1}	(0.04 cm^3 g^{-1})	530 µm^3 kg^{-1}	(0.04 cm^3 g^{-1})	6 µm^3 kg^{-1}	(50 µcm^3 g^{-1})
coercive force	$(B_0)_C$	10 mT	(100 Oe)	2 mT	(20 Oe)	0.4 T	(4000 Oe)
coercivity of remanence	$(B_0)_{CR}$	33 mT	(33 Oe)	15 mT	(150 Oe)	0.7 T	(7000 Oe)
density	ρ	5250 kg m^{-3}	(5.25 g cm^{-3})	5250 kg cm^{-3}	(5.25 g cm^{-3})	5000 kg cm^{-3}	(5 g cm^{-3})

Chemical remanence is of a lower magnitude than thermoremanence because the saturation magnetisation and anisotropy energy are lower at temperatures well below the Curie temperature. The stabilities of the two remanence types are however similar. Much less laboratory work has been carried out on the magnetic properties of chemical remanences although they are common in nature and have been widely used for investigating ancient field directions, particularly in sedimentary rocks.

DETRITAL REMANENT MAGNETISATION

Detrital magnetic particles (with a previously acquired thermal or chemical remanence) can align themselves with an applied magnetic field while falling through water, and lead to a **detrital remanence**. However, it is more likely that the detrital remanence of most sediments is formed after the particles have come to rest on the substrate, by the rotation of the magnetic grains in water-filled interstices. This so-called post-depositional remanent magnetisation is locked into the sediment by consolidation due to either compaction or to the growth of authigenic minerals or organic gels. In some slowly deposited sediments it may be many tens of thousands of years after deposition before the detrital magnetic particles become consolidated or buried and record the ambient field. In other sediments the post-depositional remanence can be locked into the material in only a few days or weeks after the time of deposition. The origin of natural remanent magnetisation in sediments is described further in Chapter 13 along with related laboratory redeposition experiments.

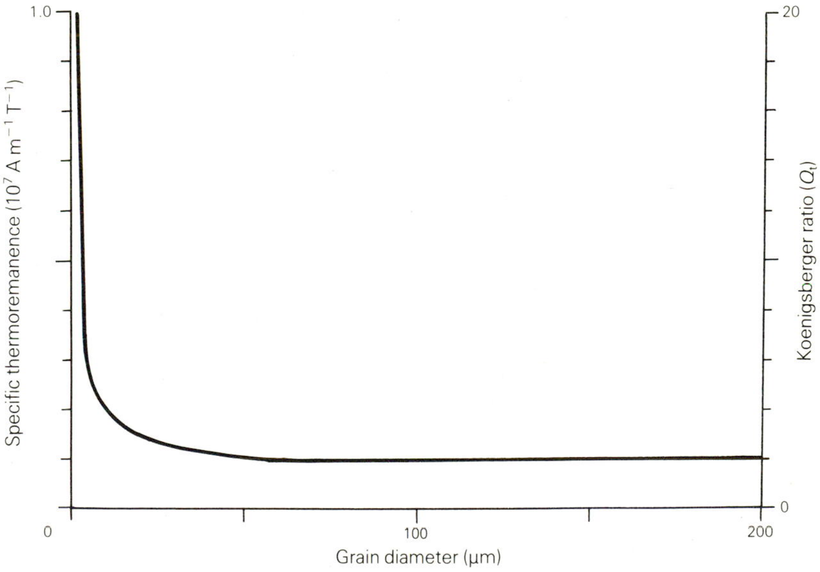

Figure 4.1 Dependency of low field thermoremanence intensity, per tesla, with magnetic grain size. The Koenigsberger ratio, Q_t, expresses the strength of the thermoremanence. It is defined as the ratio of the thermoremanence acquired in a low field to the magnetisation induced by the same low field at room temperature.

23

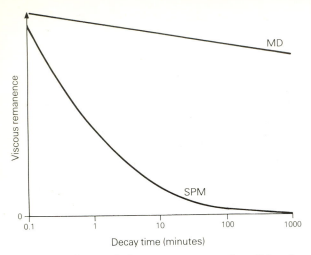

Figure 4.2 Decay of viscous remanence of multidomain (MD) and superparamagnetic (SPM) grain assemblages in zero field.

4.3.2 Laboratory-imparted remanences

VISCOUS REMANENT MAGNETISATION

A viscous remanence can be acquired when a sample is exposed to a new magnetic field. The remanence is, by definition, time dependent (2.5). In the nineteenth century Ewing (1900) observed that viscous remanence had a logarithmic time dependence.

The logarithmic time dependence can be predicted theoretically for both multidomain and interacting assemblages of single-domain grains (Dunlop 1973b). The stable single-domain/superparamagnetic boundary, however, truncates the logarithmic relationship and leads to characteristic time–remanence curves (Fig. 4.2) which contrast with those of multidomain materials. Magnetic grains near the stable single-domain/superparamagnetic boundary with low relaxation times are probably the most important contributors to viscous remanences in natural materials (Mullins & Tite 1973, Dunlop 1983b).

Decay of natural remanence with time during zero field storage indicates the presence of 'soft' components of viscous remanence. Such components can often be removed by waiting a few days or weeks for the natural remanence to stabilise. Such soft remanences are commonly found in recent sediments which have been held in a core store for some time and especially in those which have partially dried. These storage remanences are generally due to a small portion of the magnetic particles gradually realigning themselves in the new field direction of the core store,

rather than to superparamagnetic or domain wall thermal activation effects.

ISOTHERMAL REMANENT MAGNETISATION

The remanent magnetisation acquired by deliberate exposure of a material to a steady field at a given temperature (most commonly room temperature) is called an **isothermal remanence**. The magnitude of the remanence depends on the strength of the steady field applied. This dependence can be readily demonstrated in the laboratory by placing a specimen in stronger and stronger fields and measuring the remanence after each field exposure. Figure 4.3 illustrates the increase in isothermal induced *magnetisation* with field, while Figure 2.3 illustrates changes in isothermal *remanence* with field. In later chapters when we are describing experiments which involve subjecting specimens to fields of different strengths almost all our measurements will be of the isothermal remanent magnetisation remaining in the specimens after they have been withdrawn from the influence of the applied fields, rather than their magnetisations while in the fields.

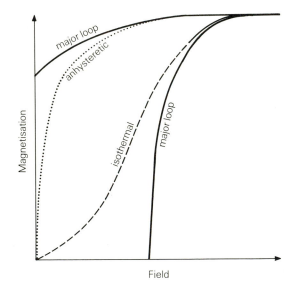

Figure 4.3 Schematic diagram illustrating acquisition curves for anhysteretic (dotted) and isothermal (dashed) magnetisations. The isothermal magnetisation acquisition curve increases slowly in low fields, then more rapidly before saturating. The anhysteretic magnetisation curve follows a simpler path, which closely parallels the right-hand side of the major hysteresis loop, before saturating. On the anhysteretic, or ideal magnetism curve, the domain walls achieve positions of true equilibrium. So the anhysteretic curve always lies above the initial magnetisation curve.

The maximum remanence which can be produced is called the saturation isothermal remanent magnetisation (expressed by the alternative symbols M_{RS}, σ_{RS}, $\mathcal{J}_{RS}$ and SIRM). The field at which saturation is reached depends on composition and microstructure (e.g. grain size). The highest isothermal remanence which it is practical to produce using laboratory equipment may not actually be the true saturation remanence but may instead only be an isothermal remanence produced on the approach to saturation. This circumstance arises, for example, when haematite is a constituent mineral. We denote our laboratory measurements by the symbol SIRM rather than σ_{RS} (the physicists' symbol for saturation remanence) in order to indicate that our measurements are of the highest isothermal remanence we can produce with our equipment but recognising that our SIRM will, in certain circumstances, fall somewhat short of the true saturation remanence.

Isothermal remanence is found naturally in materials which have been struck by lightning, as large magnetic fields associated with lightning strikes can induce strong, although generally low stability, isothermal remanences.

ANHYSTERETIC REMANENT MAGNETISATION
Anhysteretic remanent magnetisation is generally imparted by subjecting a sample to a strong alternating field which is smoothly decreased to zero in the presence of a small steady field. Anhysteretic (free from hysteresis) remanence is sometimes referred to as an ideal remanence.

Anhysteretic remanence increases in strength with the application of either a stronger steady field or a stronger alternating field until saturation is reached (Fig. 4.3). A linear change in remanence with field strength is found for steady fields with magnitudes similar to the Earth's field (Patton & Fitch 1962). This linear rate of change is referred to as the susceptibility of anhysteretic remanence. In practice anhysteretic remanences can be easily produced in palaeomagnetic laboratories by using the same equipment as that used for alternating field demagnetisation (6.5.1). The sample can be magnetised rather than demagnetised simply be ensuring that a steady direct field acts on the sample throughout the experiment.

4.4 Magnetic susceptibility

Susceptibility is a measure of the ease with which a material can be magnetised. Volume susceptibility is defined by the relation $\kappa = M/H$, where M is the volume magnetisation induced in a material of susceptibility, κ, by an applied field, H. By using this definition we have volume susceptibility as a dimensionless quantity. SI values of volume susceptibility are 4π times larger than CGS values (Table 2.1). **Specific susceptibility,** χ, is defined as volume susceptibility divided by density $\chi = \kappa/\rho$ (Table 4.1) and has units of $m^3\ kg^{-1}$. Susceptibility is generally measured in small fields, of strength less than 1 mT. At these low fields it is found that susceptibility is reasonably independent of applied field intensity.

Magnetic susceptibility measured by the usual methods (6.3) is an apparent value because of the self-demagnetising effect described in Section 2.4 in connection with anisotropy. When a substance is magnetised its internal magnetic field is less than the externally applied field. κ_i, the intrinsic susceptibility, relates the induced magnetisation to the internal magnetic field, whereas κ_e, the extrinsic susceptibility which we actually observe, relates the induced magnetisation to the externally applied field. The relationship between the two susceptibilities can be easily derived and shown to be

$$\kappa_e = \kappa_i/(1 + N\kappa_i)$$

where N is the demagnetisation factor.

For a strongly magnetic mineral such as pure magnetite, $N\kappa_i > 1$. Hence we have $\kappa_e \approx 1/N$ and, if N is known, then there is a very simple relationship between the susceptibility we measure and the concentration of ferrimagnetic grains in a sample. This is the situation for natural samples where the concentration of ferrimagnetic grains is generally small, being a few percent or less. In terms of a volume fraction of ferrimagnetic grains, f, the measured susceptibility, κ, is given by $\kappa = f\kappa_e$ and so to a good approximation $\kappa = f/N$. In practice for natural samples it is found that N is reasonably constant with a value close to $\frac{1}{3}$, i.e. that expected for a sphere. So if the magnetic grain shapes are roughly spherical and the dominant magnetic mineral is magnetite, as can be expected in most natural samples, the volume fraction ($f \ll 1$) can be estimated directly by dividing the volume susceptibility by three. This simple relationship between susceptibility and magnetite concentration is extensively used in subsequent chapters.

A list of susceptibilities of common natural ferrimagnetic and antiferromagnetic minerals and of some other common minerals is given in Table 3.4. A

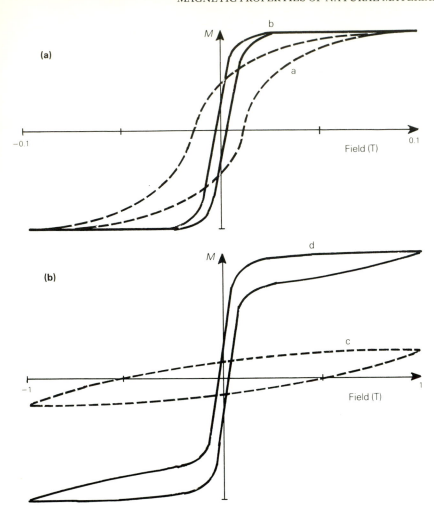

Figure 4.4 Magnetic hysteresis loops. (a) Loop a: assemblage of single-domain magnetite grains (dashed); loop b: multidomain magnetite (solid). The heights of the single-domain and multidomain magnetite loops are the same because saturation magnetisation does not vary with grain size. (b) Loop c: haematite (dashed); loop d: mixture of hard and soft minerals (solid).

number of studies have established that the magnetic susceptibility of natural materials mainly depends on their magnetite content (Mooney & Bleifuss 1953, Puranen 1977, Currie & Bornhold 1983). The inset of Figure 4.9 illustrates this simple linear relationship between susceptibility and magnetite content. When ferrimagnetic minerals such as magnetite are scarce and the susceptibility is consequently very weak, paramagnetic and even diamagnetic minerals can make substantial contributions to it (Prasad & Ghildyal 1975). **Frequency dependent susceptibility** can help identify samples with grains spanning the superparamagnetic/stable single-domain boundary (Stephenson 1971, Mullins & Tite 1973) and is discussed in Section 6.3 along with susceptibility instrumentation.

4.5 Anisotropy of susceptibility

The relative ease with which a material can be magnetised in various directions is expressed by its **anisotropy of susceptibility**. Such anisotropy (Section 2.4) in natural samples can arise through external stresses, such as natural tectonic stresses, or be intrinsic due to the shape or crystalline structure of the ferromagnetic minerals in the samples. The first type of anisotropy is of practical interest because of its contribution to changes in the local geomagnetic field as precursors of volcanic eruptions (Johnston & Stacey 1969) or earthquakes (Undzendor & Shapiro 1967, Johnston *et al.* 1976). The second type has application in fabric analyses (Hamilton & Rees 1970).

Table 4.3 General hysteresis properties.

Rock type	χ ($\mu m^3 kg^{-1}$)	SIRM ($mAm^2 kg^{-1}$)	SIRM/χ (kAm^{-1})	$(B_0)_{CR}$ (mT)	$\dfrac{IRM_{-100\ mT}}{SIRM}$	NRM ($\mu Am^2 kg^{-1}$)
ultrabasic	0.2	0.4	2	15	0.9	200
gabbro	1.0	2.0	2	15	0.9	500
basalt	1.8	70	40	35	0.8	5000
diorite	0.5	1	2	20	0.7	10
granite	0.2	0.3	1.5	35	0.6	1
limestone	<0.001	0.005	>10	70	0.4	0.5
sandstone	0.1	0.5	7	30	0.8	5
siltstone	0.2	1.4	8	30	0.9	10
clay	0.1	1	10	40	0.9	50
gneiss	0.05	7.5	150	50	0.7	5
schist	0.01	0.01	1	20	0.9	1
slate	0.01	0.007	0.5	50	0.6	1

For natural samples to be anisotropic the shape or crystalline axes of the assemblage of magnetic grains must be aligned to form a fabric. The effect of grain shape alignment will dominate when the magnetic anisotropy is due to ferrimagnetic grains such as magnetite. The shape effect (Section 2.4) of ferrimagnetic grains results from the demagnetising field being less along the long axes of magnetic grains, so that the susceptibility is higher parallel to the long or major axes.

Crystalline anisotropy is more important than shape anisotropy in rhombohedral and hexagonal iron oxide (Section 3.2.2) and sulphide (Section 3.3) minerals as their spontaneous magnetisation is low. In these minerals the direction of minimum susceptibility is perpendicular to the basal crystallographic plane. In some metamorphic rocks the observed magnetic anisotropy results from alignment of the basal planes of pyrrhotite grains (Fuller 1963).

4.6 Magnetic hysteresis

As briefly described in Section 2.3 the magnetisation of a ferromagnetic material lags behind the external inducing magnetic field. This important magnetic phenomenon is known as hysteresis. It can be conveniently displayed for a specimen by plotting a graph of the magnetisation produced by a cyclic field. The graph takes the form of a symmetric loop (Figs 2.2 & 4.4). The area inside the loop is a measure of the hysteresis.

The hysteretic properties of natural materials of different composition vary widely (e.g. Cisowski 1980). For example magnetite produces tall, thin hysteresis loops (e.g. loop b in Fig. 4.4a), whereas haematite produces flat, fat loops (e.g. loop c in Fig. 4.4b). Variation in grain size also plays an equally important rôle in determining the shape of a crystal's hysteresis loop. Multidomain and superparamagnetic grains produce much thinner loops than single-domain grains. Paramagnetic and diamagnetic minerals do not exhibit hysteresis loops; instead they simply show a reversible linear relationship between magnetisation and applied field.

Cycling the magnetic field to values well below saturation produces quite different shaped hysteresis loops from those of saturation. Such minor hysteresis loops characteristically exhibit elliptical shapes. Laboratory hysteresis measurements of rhombohedral ferromagnets (e.g. haematite) tend to be limited to minor loops (loop c in Fig. 4.4b) as haematite saturation fields cannot be produced with conventional equipment. Most natural ferrimagnets saturate in lower fields, well within the capabilities of laboratory magnets so that in practice we observe their saturation or major hysteresis loops.

The single-domain loop of Figure 4.4a has been drawn for an assemblage of grains. An individual single-domain grain has a quite differently shaped hysteresis loop from that of the average loop of Figure 4.4a. As shown in Figure 2.8 the hysteresis loop of an individual grain depends on the orientation of the grain with respect to the applied field. Calculations of compound magnetisation curves for collections of prolate ellipsoids orientated at random (Stoner & Wohlfarth 1948) readily explain the observed hysteresis loops of natural assemblages of single-domain ferrimagnetic grains. Superparamagnetic ferrimagnetic grains do not display hysteresis but show a steep magnetisation versus field curve which saturates at low fields. The saturation magnetisation of superparamagnetic grains is the same as that of multidomain or single-domain grains.

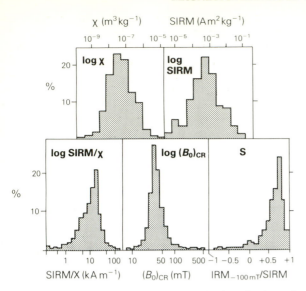

Figure 4.5 Histograms showing the distribution of values of five mineral magnetic parameters in 1000 natural samples. Susceptibility (χ) and saturation remanence (SIRM) are closely log normally distributed.

Investigations of natural materials nearly always have to take into account mineral mixing effects. A number of experimental (Wasilewski 1973) and theoretical studies (Kneller & Luborsky 1963) have been carried out on the systematics of hysteresis properties of mixtures. The most striking effect of extreme mixing is the production of constricted or wasp waisted loops (loop d in Fig. 4.4b). A related effect is that a small addition of a high coercivity material can produce a large increase in coercivity and in coercivity of remanence (Kneller & Luborsky 1963).

4.7 General magnetic properties of natural materials

The main magnetic parameters that are used in later sections for quickly characterising the magnetic mineralogy and granulometry of natural samples are susceptibility, saturation remanence, moderate field (~ 0.1 T) isothermal or anhysteretic remanences, remanence coercivity and saturation magnetisation. The main characteristics of these magnetic properties are described below and the typical values and ranges of five of those room temperature magnetic parameters are illustrated by measurements on a set of 1000 natural samples selected as being representative

of the wide assortment of rocks, soils, minerals and recent sediments encountered during our environmental studies (Table 4.3).

SUSCEPTIBILITY

As described in Section 4.4, the susceptibility of natural materials is mainly a consequence of their magnetite content so that susceptibility can often be used as a rapid, surrogate measure of magnetite concentration. Susceptibility values of rocks are commonly found to be log normally distributed (Irving *et al.* 1966, Yanak & Uman 1967). Such a distribution can be seen for both susceptibility and saturation remanence (Fig. 4.5) in our 1000 sample collection. The form of the distribution reflects the log normal heavy mineral content distribution of natural materials. Susceptibility ranges from weak negative values (~ -1 nm³ kg⁻¹) for diamagnetic rocks such as quartzites to high values ($\sim 10\,000$ nm³ kg⁻¹) in iron

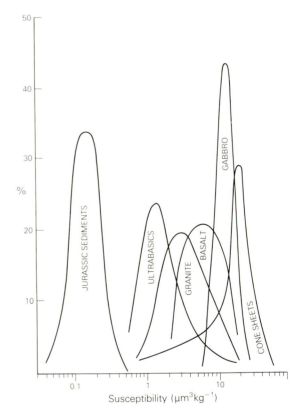

Figure 4.6 Variability of susceptibility for various rock types illustrated using 288 *in situ* measurements made on the Isle of Skye, Scotland.

Table 4.4 Average hysteresis properties of 1000 natural samples.

Parameter	Median value	Range	
		minimum	maximum
χ	43 nm^{3}kg^{-1}	-1	10000
SIRM	0.58 mAm^{2}kg^{-1}	0.002	100
SIRM/χ	13 kAm^{-1}	0.11	320
$(B_0)_{CR}$	40 mT	8	570
IRM$_{-100\,mT}$/SIRM	0.7	-0.92	1.00

ores (Table 4.4). An idea of the variation of susceptibility within various rock types is given in Figure 4.6 by the results of a survey of *in situ* susceptibility measurements on six rock types on the Isle of Skye.

ARTIFICIALLY IMPARTED REMANENCES
The simplest artificial remanence to grow is iso-

thermal, room temperature, remanent magnetisation. Some workers prefer to use anhysteretic rather than isothermal remanences (Banerjee *et al.* 1981), although the range of magnetising fields available tends to be more restricted. In detail, anhysteretic remanences are more difficult to interpret than isothermal remanences (Dunlop 1983a) as they display somewhat more complicated variations with grain size and they are reduced more by grain interactions (Jaep, 1971). The most useful single artificially imparted remanence is probably the saturation isothermal remanence (Cisowski 1980), which although mainly a measure of magnetite content, is also dependent on grain size and can be strongly influenced by other magnetic minerals such as haematite.

Figure 4.7a illustrates the variation of saturation remanence with grain size for magnetite as deduced

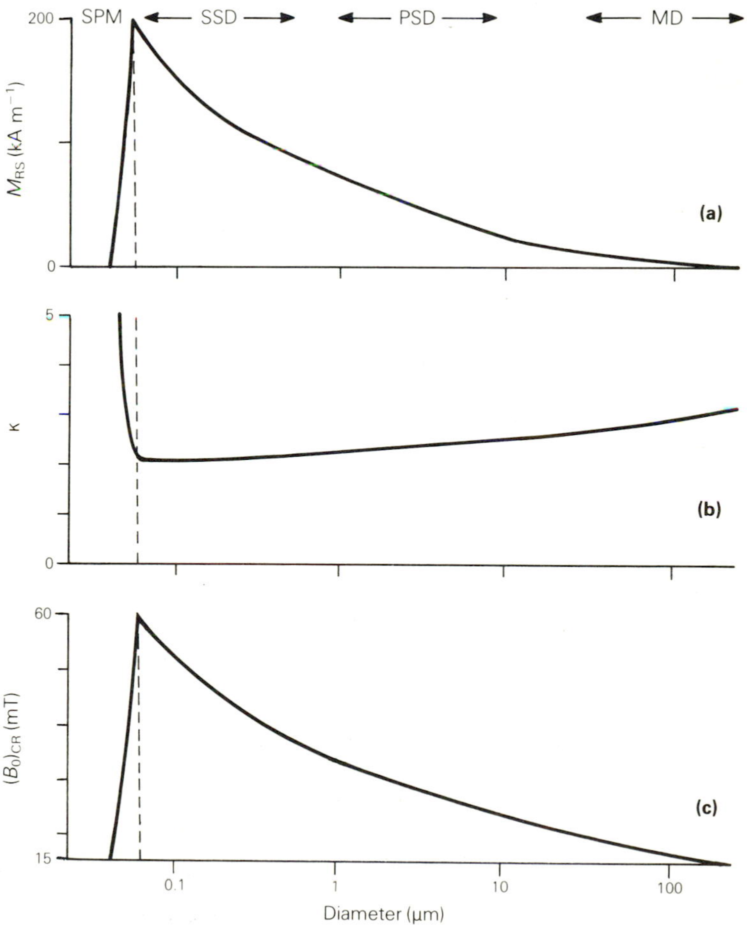

Figure 4.7 Summary diagram of the variation of (a) saturation remanence, (b) susceptibility and (c) coercivity of remanence with magnetite grain size. The dashed lines highlight the changes in trend of mineral magnetic properties at the superparamagnetic/stable single-domain boundary.

from experiments on dispersed magnetite powders (Parry 1965). Notice how saturation remanence is much higher in single-domain grains of around 0.1 μm diameter than in large multidomain grains and how the range of variation through the multidomain and stable single-domain regions is much greater for saturation remanence than for susceptibility (Figs 4.7a & b). Saturation remanence values range from virtually zero, in materials devoid of iron or manganese minerals, to over 100 mA m² kg⁻¹ in finely disseminated iron oxide ores or heavy mineral bands (Table 4.2). The histogram of saturation remanence values of our 1000 samples (Fig. 4.5) has a median value of 0.58 mA m² kg⁻¹ and the usual log normal distribution of a magnetic parameter that mainly depends on concentration.

SATURATION REMANENCE TO SUSCEPTIBILITY RATIO

The dependence of the two parameters saturation remanence and susceptibility upon each other is displayed in the scattergram of Figure 4.8. The clear

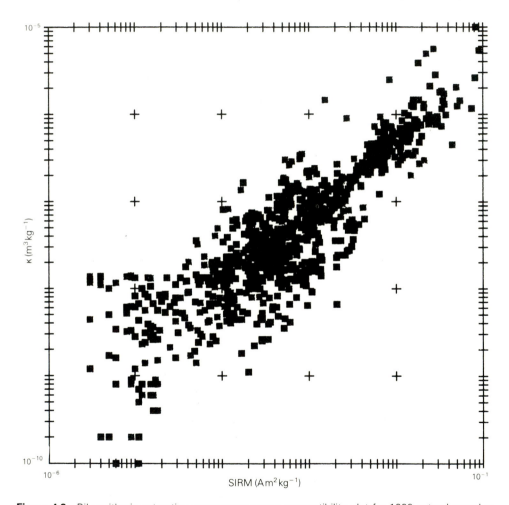

Figure 4.8 Bilogarithmic saturation remanence versus susceptibility plot for 1000 natural samples. Squares denote results from individual specimens. High magnetite concentrations plot towards the top right and low concentrations towards the bottom left corners of the diagram. Specimens lying below and to the right of the main group tend to have high haematite to magnetite ratios. Low SIRM, but moderate κ specimens, plotting at the far left have relatively high susceptibilities on account of paramagnetic contributions. Multidomain magnetites and assemblages with high concentrations of superparamagnetic magnetite plot on the upper side of the main group.

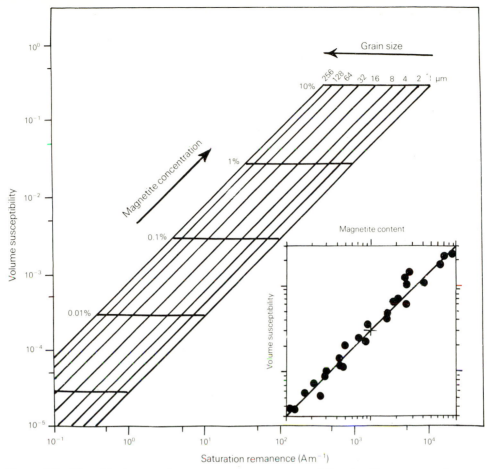

Figure 4.9 Bilogarithmic saturation remanence versus susceptibility plot. The concentration/grain size grid is for pure magnetite. Inset illustrates the linear relationship commonly observed, in sets of natural samples, between susceptibility and magnetite concentration.

grouping along the diagonal corresponding to a SIRM/κ ratio of 10 kAm^{-1} in Figure 4.8 results from magnetite having an average effective grain size of 5 μm in natural samples, and from both susceptibility and saturation remanence dominantly reflecting magnetite concentration. The Pearson correlation coefficient of this correlation between the logarithms of susceptibility and saturation remanence for the 1000 natural samples is 0.86. Samples with unusually high haematite to magnetite ratios plot to the right of the general trends as they have SIRM/κ ratios of around 1000 kAm^{-1}.

The ratio of saturation remanence to susceptibility can be used as a rough estimate of magnetic grain sizes in crystals larger than a few hundred angstroms. The grid, based on the data of Parry (1965) and

Dankers (1978) and unpublished results of the authors, on the bilogarithmic plot of Figure 4.9 illustrates how the concentration and grain size of magnetite crystals in a rock can be estimated from measurements of susceptibility and saturation remanence. The magnitude of the magnetic measurements gives the concentration and the ratio gives the grain size.

An important factor in many soils and soil-derived sediments is the presence of ultrafine iron oxides in the superparamagnetic size range (Mullins 1977 and Ch. 8). Superparamagnetic particles have distinctively high susceptibilities and low SIRM/κ ratios. Mixtures of superparamagnetic and single-domain grains thus have similar SIRM/κ ratios to those of multidomain grains and plot in the larger grain size region of Figure

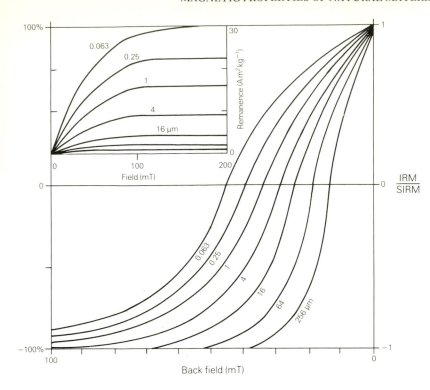

Figure 4.10 Coercivity plots of isothermal remanence versus back applied field for magnetite grains ranging in diameter from 0.063 to 256 μm. The back coercivity curves are normalised with respect to saturation remanence SIRM in order to display the variation in stability with grain size. Inset shows un-normalised acquisition of isothermal remanence curves for the same range of magnetite grain sizes, emphasising the dependence of saturation remanence on grain size.

4.9. Further hysteresis measurements, as described below, are needed to distinguish between large multi-domain magnetite grains and samples containing mixtures which include very small superparamagnetic grains. The increase in scatter in the lower left-hand part of Figure 4.8 reflects the importance of super-paramagnetic and paramagnetic effects in samples with very low magnetite concentrations.

REMANENCE COERCIVITY
Coercivity of remanence $(B_0)_{CR}$, is a relatively straightforward, very useful hysteresis parameter which can be used in determining magnetic mineralogy and grain size and in helping to characterise magnetic mixtures. It is the field which reduces the saturation isothermal remanence to zero (Fig. 2.2). For crystals of one mineralogy coercivity of remanence is a sensitive indicator of grain size, varying for magnetite from less than 10 mT for multi-domain grains to almost 100 mT for small elongated grains (Figs 4.7c & 4.10).

Examples of remanence hysteresis curves for basalts in various states of oxidation are shown in Figure 4.11. The basalts range from low oxidation states (type I) containing large homogeneous titano-magnetites with remanence coercivity of around 10 mT, through moderately oxidised lavas bearing exsolved spinels containing small magnetite crystals (type III) with higher coercivities of about 40 mT, to highly oxidised lavas in which the main magnetic crystals are haematites (type VI) characterised by very high remanence coercivities of over 300 mT.

REMANENCE RATIOS
The ratios of remanences produced in different laboratory fields can also be profitably used for characterising samples. One ratio that we have found particularly helpful is the ratio of a moderate (100 mT) back field isothermal remanence to the saturation remanence. This ratio (often referred to as S) has been found to be expedient for recognising samples with unusual haematite to magnetite ratios (Stober & Thompson 1979). The basic idea behind the S ratio is that in practice the magnetisation of most ferri-magnets will have saturated in fields below 0.1 T and that the major differences in high field remanence (i.e. $SIRM - IRM_{100mT}$) will be due to the imperfect anti-ferromagnets such as haematite and goethite.

Some instruments (e.g. most portable discharge induction coils (Section 6.6.2)) do not produce fields

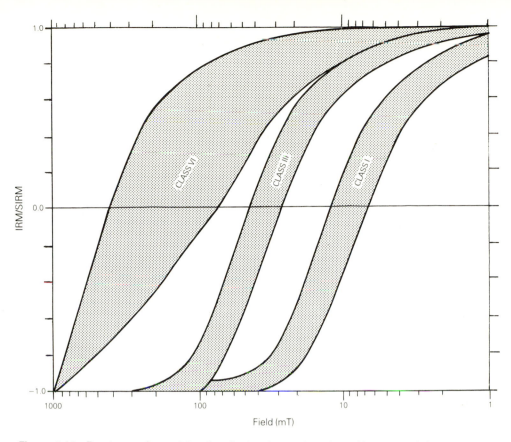

Figure 4.11 Envelopes of coercivity plots for basalt samples of low (I), medium (III) and high (VI) oxidation states.

as high as 1 T and so isothermal remanence ratios similar to S but employing fields other than 1 T and 0.1 T have been used in later chapters.

The ratio of saturation remanence to anhysteretic remanence, SIRM/ARM, has been found to be particularly useful for detecting grains slightly larger than superparamagnetic in size on account of their low SIRM/ARM ratios. Gillingham and Stacey (1971) and Kneller (1980) have discussed theoretical relationships between SIRM and ARM for multi-domain and single-domain grains respectively.

SATURATION MAGNETISATION
Saturation magnetisation is unaffected by grain size, pure magnetite and haematite having saturation magnetisations of 480 kA m^{-1} and 2.5 kA m^{-1} respectively (Table 2.2). Saturation magnetisation is thus theoretically a most attractive mineral magnetic parameter. Unfortunately absolute measurements of

the saturation magnetisation of the low magnetic mineral concentrations with which we have largely been concerned have not proved straightforward and so examples of absolute measures of saturation magnetisation have tended to be restricted to strongly magnetic igneous rocks (e.g. Nagata 1953) or synthetic minerals (e.g. Parry 1965, Dunlop 1973a). Saturation magnetisation can however be usefully used in ratio with saturation remanence or susceptibility (Fuller 1974).

MAGNETISATION AND COERCIVITY RATIOS
Two ratios that are frequently employed to characterise magnetic materials are the magnetisation ratio and the coercivity ratio. Both can be obtained from instruments that plot out hysteresis loops (Sections 6.4.3 & 6.4.4).

The magnetisation ratio of saturation remanence to saturation magnetisation, M_{RS}/M_S, is a sensitive indicator of magnetisation state. Theoretical computations show that ideally M_{RS}/M_S is exactly 0.5 for an assemblage of randomly orientated non-interacting uniaxial magnetic monodomain grains (Stoner & Wohlfarth 1948). In contrast the magnetisation ratio M_{RS}/M_S is less than 0.1 for multidomain grains and even lower for superparamagnetic grains (Bean & Livingston 1959).

The coercivity ratio $(B_0)_{CR}/(B_0)_C$ relates the coercivity of remanence to the saturation coercivity. The lower limit of this coercivity ratio is 1.0. Stoner and Wohlfarth's random assemblage model predicts a coercivity ratio of 1.09 for uniaxial single-domain grains no matter what value their demagnetising coefficients take. Multidomain grains are expected to have coercivity ratios in excess of 4.0, while superparamagnetic grains will have ratios in excess of 10.0 (Wasilewski 1973). Rhombohedral ferromagnetic components can have ratios of 3.0 or more.

Magnetic materials can be neatly classified by determining their magnetisation and coercivity ratios (Day et al. 1977). Figure 4.12 illustrates how single-domain, pseudo-single-domain and multidomain grains can be recognised through their magnetisation and coercivity ratios. Indeed such hysteresis ratios

and parameters have been widely used in the magnetic tape, permanent magnet and transformer core manufacturing industries for product grading.

A further magnetic ratio, that of SIRM/κ to $(B_0)_{CR}$, can be helpful in distinguishing mixtures containing superparamagnetic grains (Bradshaw & Thompson 1985). The reason for this is that superparamagnetic grains contribute to susceptibility, but not to saturation remanence and furthermore they do not affect coercivity of remanence. Consequently, samples with a large superparamagnetic component have unusually high susceptibilities which lead to low SIRM/κ to $(B_0)_{CR}$ ratios. Figure 4.13 sketches the main types of magnetic mixtures found for a wide range of SIRM/κ and $(B_0)_{CR}$ values.

It will be apparent that hysteresis parameters can be combined in a large number of ways and that specific situations or special investigations may justify particular combinations. The hysteresis parameters and ratios used in classifying natural samples will to some extent have to depend on the availability of local equipment. In later chapters we have endeavoured to use ratios that can be measured on simple, readily available instruments and that appear to differentiate assemblages of soils and sediment samples very sensitively although they may permit no more than rather general qualitative estimates of the basis of the mineral magnetic variation. In the final prospects chapter, an approach to quantifying the interpretation of general mineral magnetics is discussed.

4.8 Temperature dependence of magnetic properties

Competition between magnetic ordering energy and thermal randomising energy leads to the basic thermomagnetic properties observed in ferromagnets.

SATURATION MAGNETISATION
The thermal change of spontaneous magnetisation is theoretically a fundamental magnetic property since it depends only on a ferromagnet's composition and crystal structure. It is independent of variables such as crystal size, shape and internal stress. Consequently M_S–T curves have been widely used for mineral identification in rock magnetic studies. Spontaneous magnetisation, which at most temperatures can be distinguished from saturation magnetisation only with difficulty, has its largest value at absolute zero

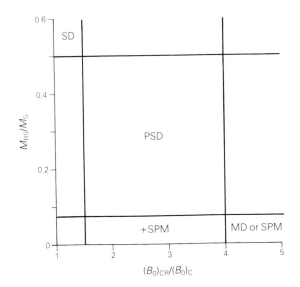

Figure 4.12 Classification of magnetic minerals in terms of magnetisation and coercivity ratios. Single-domain (SD) grains plot in the upper left corner, multidomain (MD) in the lower right while pseudo-single-domain (PSD) lie in between. Superparamagnetic grains (SPM) plot in the lower part of the diagram (after Day et al. 1977).

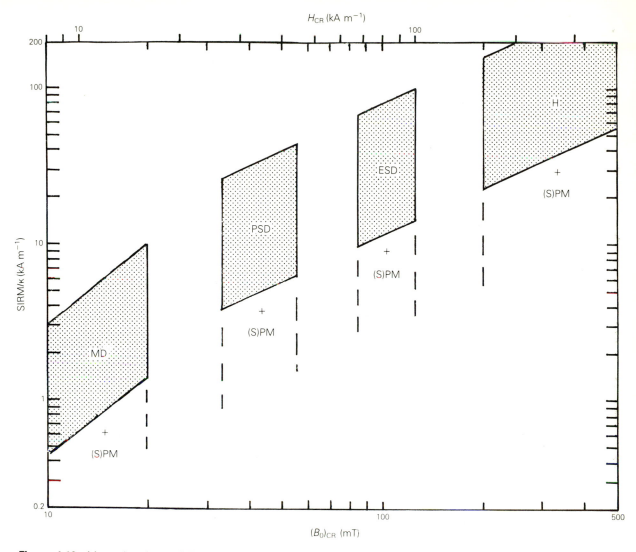

Figure 4.13 Magnetic mixture SIRM/χ versus $(B_0)_{CR}$ diagram. The grid schematically divides the diagram into magnetic mineralogies and magnetisation states. Multidomain (MD), pseudo single-domain (PSD) and elongated single-domain (ESD) magnetites fall in the upper left to centre of the diagram. Haematite (H) lies in the upper right corner; mixtures containing (super)paramagnetic grains lie further towards the lower right.

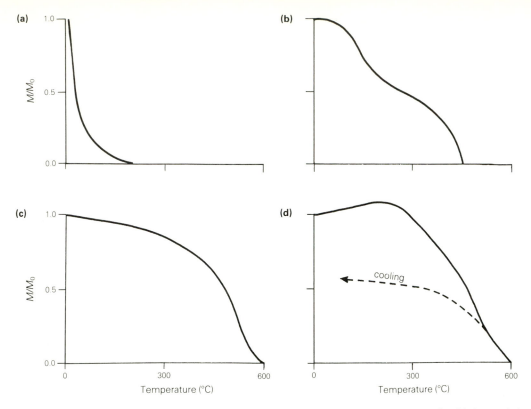

Figure 4.14 Strong field thermomagnetic curves for basalts with spinel minerals in various states of oxidation and alteration. (a) Fresh homogeneous low Curie temperature titanomagnetites; (b) exsolving titanomagnetites with two magnetic minerals; (c) magnetites produced by deuteric oxidation; (d) maghaemite in deep-sea basalt.

and falls at an increasing rate with increasing temperature to become zero at the Curie temperature.

Examples of the variation of saturation magnetisation with temperature of basalts containing different magnetic minerals are given in Figure 4.14. Abrahamsen *et al.* (1984) describe several of these types of variation of saturation magnetisation with temperature from one deep drill core through Tertiary basalts on the Faeroe Islands. Equipment for producing such thermomagnetic curves is described in Section 6.4.

SUSCEPTIBILITY

Susceptibility, κ, varies with temperature in a more complicated manner than saturation magnetisation. The reason for this difference in behaviour is that susceptibility is sensitive to various parameters, such as internal stress and crystalline anisotropy, which change with temperature.

Figure 4.15 plots typical changes of susceptibility with temperature for 'magnetite' crystals of various grain sizes and various compositions. All the curves have been normalised so that the value of susceptibility at 0 °C is plotted as unity. The susceptibility of single-domain magnetite tends to vary with temperature in a manner similar to the dashed curve of Figure 4.15 with only small susceptibility changes, except at the Curie temperature. For multidomain magnetites (solid curve) more pronounced susceptibility changes are observed especially on cooling below room temperature. Multidomain titanium-rich magnetites (dotted curve) have characteristic κ–T curves with low Curie temperatures and with a steady fall in susceptibility on cooling. The susceptibility of paramagnetic minerals follows the Curie–Weiss law of $\kappa = C/T$, i.e. that paramagnetic susceptibility is inversely proportional to absolute temperature, C being a constant. Superparamagnetic crystals follow a similar law at temperatures above their blocking temperature when their susceptibility falls at a rate slightly greater than $1/T$. This κ–T behaviour is illustrated by the curve of

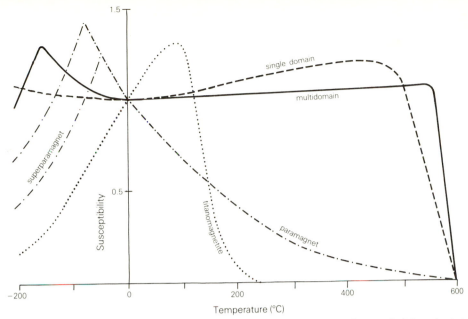

Figure 4.15 Low field susceptibility versus temperature curves for various 'magnetite' domain states and compositions. Schematic diagram with curves arbitrarily normalised to equivalent susceptibilities at 0°C. Titanomagnetite (dotted), single-domain magnetite (dashed), multidomain grains (solid), superparamagnetic grains (dashes and dots). (The temperature of the peak susceptibility of superparamagnetic grains depends on their volume.)

dots and dashes in Figure 4.15. On cooling below the blocking temperature the susceptibility of superparamagnets drops sharply as a stable magnetisation is blocked in and the coercivity increases. This blocking effect is diagrammatically illustrated in Figure 4.15 by the additional curves of dots and dashes falling away at low temperatures.

Measurements of susceptibility at low temperature are easy to make (Radhakrishnamurty *et al.* 1978) and can provide valuable additional magnetic information, particularly about the likely relative importance of the contribution of paramagnetic or superparamagnetic crystals to the bulk susceptibility.

REMANENCE

In palaeomagnetic studies of natural remanent magnetisation, blocking temperatures can be of more interest than Curie temperatures as they control long-term magnetic stability at elevated temperature.

Blocking temperatures are determined by heating the remanence in a zero field and observing the temperature at which the remanence is destroyed. A remanence *v.* temperature plot derived from such an experiment is shown in Figure 4.16. The plot compares the thermal demagnetisation results of a natural remanence and a laboratory thermoremanence. We can see that the natural remanence has lost part of its primary magnetisation. The missing remanence (shaded region of Fig. 4.16) is that of the low stability, low blocking temperature grains. The high stability, high blocking temperature magnetisation is seen to be the same in both the natural remanence and the artificial laboratory-produced thermoremanence.

Somewhat surprisingly, magnetic remanence can be destroyed by cooling as well as by heating. The magnetisation of pure haematite crystals undergoes a transition at −10 °C (Section 3.2.2), while magnetite shows a thermomagnetic effect at −150 °C. For both of these minerals these low temperature effects are caused by a change in magnetocrystalline anisotropy (Section 2.4.1). At the transition temperature the anisotropy disappears as it changes sign and there is an associated loss of remanence which can be diagnostic of mineral type.

Irreversible thermomagnetic remanence effects can even be employed in the recognition of magnetic minerals (Dunlop 1972). Maghaemite, for example, is

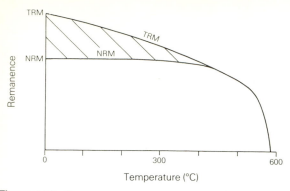

Figure 4.16 Remanence versus temperature demagnetisation plot for a laboratory thermoremanence (TRM) and a natural remanence (NRM) in the same specimen. Low blocking temperature magnetisations have been lost from the NRM.

metastable (Section 3.2.1) and on heating to 300 °C inverts to haematite. Its saturation magnetisation, susceptibility and remanence all show decreases around 300 °C which are not recovered on subsequent cooling. This irreversible behaviour is often used to infer the occurrence of maghaemite in natural samples (Fig. 4.14d). Pyrrhotite (Section 3.3) characteristically shows a distinctive rise in magnetisation on heating to 200 °C, due to a reordering of lattice vacancies, followed by a fall at the Curie temperature of 320 °C.

4.9 Summary

The magnetic properties of natural materials can be used as a petrological tool and as a method of characterising samples. Artificially imparted remanences and magnetisations can be used in order to distinguish between various types of magnetic minerals and their grain sizes. Low field magnetisation, or susceptibility, is a particularly easy property to measure and is a quick first guide to magnetite concentration. Hysteresis and thermomagnetic properties can provide further information about the types of magnetic minerals and magnetic mixtures in natural materials.

Geological materials can acquire natural remanence magnetisations through natural processes. Some such naturally grown remanences are stable and can record information about the ancient magnetic field, with little alteration, for millions of years.

Further reading

General journal paper

Clark 1983. Comments on magnetic petrophysics.

Advanced journal papers

Stacey 1963. The physical theory of rock magnetism.
O'Reilly 1976. Magnetic minerals in the crust of the Earth.

Advanced books

Nagata 1953. *Rock magnetism.*
O'Reilly 1984. *Rock and mineral magnetism.*

[5]

The Earth's magnetic field

Magnus magnes ipse est globus terrestris.

W. Gilbert 1600

5.1 Geomagnetism

Scientists have been measuring the Earth's magnetic field for several centuries. They have built up an increasingly more detailed worldwide picture of the field and its variation with time. Field measurements are now made from ships, aeroplanes and satellites as well as on the ground. For older records of the geomagnetic field and its changes with time we must turn away from historical documents to the study of the fossil magnetism in the Earth's rocks and sediments. The palaeomagnetic signature of the geomagnetic field, recorded by the remanent magnetisation of rocks and sediments, naturally produces a sparser and less detailed picture than the one we can build up for the present field, but the strength of this record is that it extends the time of observation of the field by seven orders of magnitude to cover 3000 million years. As the field originates in the Earth's molten core, the palaeomagnetic record is crucial to the understanding of both the origin and evolution of the geomagnetic field and the Earth's deep interior.

5.1.1 Description of the geomagnetic field

In a rough way the field outside the Earth resembles that outside a uniformly magnetised sphere as illustrated in Figure 5.1. This approximation of the geomagnetic field to that of a simple dipole was first recognised by William Gilbert, physician to Queen Elizabeth I, and published in his treatise *De Magnete* in 1600. Gellibrand in 1635 was the first to

demonstrate that the geomagnetic field varied not only regionally but also with time.

The direction and strength of the geomagnetic field can be most simply determined using a magnetic needle. Freely pivoted it will swing into a position parallel to the local lines of force of the geomagnetic field, pointing approximately to the north and, if free to tilt, taking up an inclined position. In the northern hemisphere the magnetised needle, in its equilibrium position will point downwards towards the north, whereas in the southern hemisphere it will point upwards towards the north. The field intensity can be

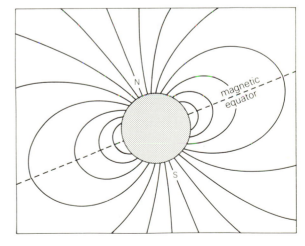

Figure 5.1 The magnetic lines of force of a geocentric dipole field. The Earth's magnetic axis is presently tilted about 10° from its spin axis.

39

determined by turning the needle through an angle of 90° from its rest position. The torque required to restrain the needle from returning to its equilibrium position is then a measure of the field intensity.

To describe completely the magnetic field, which is a vector, we need three numbers. These numbers may be chosen in different ways (Fig. 5.2). It is generally convenient to take the intensity of the field (F), the angle of dip of the field below the horizontal plane, i.e. the **inclination** (I), and the angle between the horizontal component of the field and the true, geographical north, i.e. the **declination** (D). The total intensity (F), declination (D) and inclination (I) then completely define the magnetic field at any point. Another method of specifying the field is to use the field components along the northward (x), eastward (y) and downward (z) directions (Fig. 5.2). These

three components also completely define the field. These various field components along with the horizontal field intensity (H) are related by the following equations:

$$H = F \cos I \qquad z = F \sin I \qquad \tan I = z/H$$

$$x = H \cos D \qquad y = H \sin D \qquad \tan D = y/x$$

$$F^2 = H^2 + z^2 = x^2 + y^2 + z^2$$

Variations of the declination and inclination of the geo-magnetic field over the surface of the Earth are illustrated in Figure 5.3 for the year 1980. The line along which the inclination is zero in Figure 5.3b is called the magnetic equator. The points where the inclination is +90° and −90° are the north and south magnetic poles respectively. The field intensity varies from roughly 7×10^{-5} T at the magnetic poles to 2.5×10^{-5} T at the magnetic equator.

The dipole field that best fits the actual field of the Earth today has its poles $11\frac{1}{2}°$ from the geographic poles. It is only an approximation accounting for 90% of the Earth's field. In some places the difference between the actual field and that of the best fitting dipolar field reaches 20% of the real field. The intersections of the best fitting dipole axis with the Earth's surface are called the geomagnetic poles. The north and south geomagnetic poles lie at $78\frac{1}{2}°$N, 70°W and $78\frac{1}{2}°$S, 110°E respectively and should not be confused with the magnetic poles in Figure 5.3a. Subtraction of the best fitting **dipole field** from the actual field leaves the non-dipole field. We find that the present magnetic field differs substantially from the dipole field over half a dozen areas a few thousand kilometres across. These anomalous regions show no obvious relationships with geographical or geological features. Indeed the features of the non-dipole field have changed markedly over the last 400 years (Yukutake & Tachinaka 1968). In contrast the dipole has hardly changed its orientation during this period (Barraclough 1974) although its strength has decreased, on average, at a rate of 0.05% per year (Yukutake 1979).

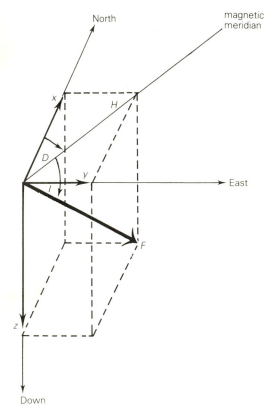

Figure 5.2 Relationship of the magnetic elements. The magnetic field at any point on the Earth's surface can be fully described by three parameters. Geomagnetists often use the orthogonal north (x), east (y) and down (z) components. Palaeomagnetists, involved mainly with angular data, tend to use declination (D), inclination (I) and intensity (F).

5.1.2 Secular variation

The importance of changes in direction of the geo-magnetic field over the time of a few hundred years is illustrated in Figure 5.4 which shows the changes in field direction at London. The first measurement of

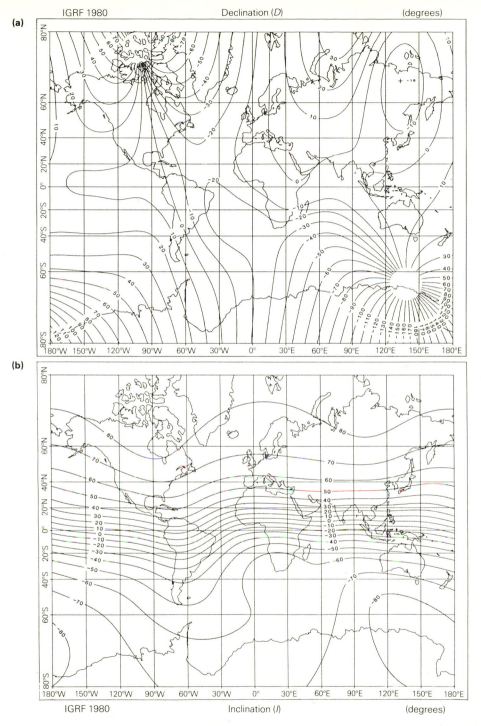

Figure 5.3 The Earth's magnetic field (1980). (a) The declination (variation) from the geographical meridian (10° intervals). (b) The inclination (dip) from the horizontal (10° intervals). (From Fabiano *et al.* 1983)

1580
11° east of north

1660
Due north

1820
24° west of north

1985
5° west of north

Figure 5.4 The variation in direction of a compass needle at London since 1580 AD due to slow secular changes of the Earth's magnetic field (after Press & Siever 1974).

the declination of a compass needle at London was made around 1570 AD when it was found that the needle pointed 11° to the east, in 1660 AD it pointed due north and by 1820 AD it had swung round to 24° to the west (Fig. 5.4 and Bauer 1896). Since then the declination has steadily decreased and now, at London, is 5°W and is decreasing at the rate of 9 minutes per year. The inclination of a dip needle at London has also varied with time: in 1700 AD it reached a maximum of over 74° and is now close to 66°. These secular or gradual variations of field direction at London are plotted in Figure 5.5 along with the historical variations measured at Rome and at Boston. Notice how the secular variations at London and Rome, some 1500 km apart, have been relatively similar whereas the variations at Boston, some 5000 km from Europe, have been very different. Geomagnetic field intensities as well as directions have been found to have changed appreciably over the past few hundred years. For example, at Cape Town the horizontal component of the field decreased by 45% in

140 years. The worldwide average (root mean square) secular intensity change amounts to some 50 nT per year.

Sudden worldwide changes in the rate of secular variation, termed jerks or impulses (Courtillot *et al.* 1978, Malin & Hodder 1982) occur at intervals. During this century globally synchrous jerks occurred in 1912–13 and 1969–70, each had a duration of around one year.

World **secular variation** maps of the rate of change of components of the geomagnetic field show a general resemblance in character to maps of the non-dipole field and it is generally felt that the secular variation reflects changes which are taking place in the non-dipole field. Halley in 1692 noticed that a large proportion of the features of the secular variation could be explained by a westward movement. This general westward drift averages about 0.2° per year, for non-dipole features, and in places has been as high as 0.6° per year. Both the non-dipole foci and the secular variation foci tend to move westwards. In addition these two kinds of foci form, deform and decay, rather like eddies in a stream of water, implying that the non-dipole field is temporary, always changing its features over lifetimes of a few hundred years (Fig. 5.6). Secular variation then is a complex regional phenomenon with some local centres drifting rapidly, some apparently remaining stationary, some being short lived and others being more persistent (Thompson 1984).

5.1.3. Origin of the geomagnetic field

Gauss (1839) first developed the mathematical method of analysing the geomagnetic field in terms of a potential and representing it as an infinite series of spherical harmonic functions. He used the method with observations of the geomagnetic field made on the surface of the Earth and was able to demonstrate that the field was predominantly of internal origin. We now know that there is a very small net external contribution, of about 0.1%, to the geomagnetic field at the Earth's surface. The external contributions which rise somewhat during magnetic storms are due to currents circulating in the upper atmosphere. They tend to vary rapidly with periods ranging from a few seconds to years. Our main concern in this and later chapters about the geomagnetic field is with the Earth's internally generated field.

The geomagnetic field pattern bears little relation

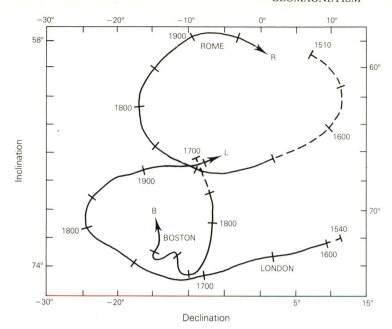

Figure 5.5 Secular changes in magnetic declination and inclination as observed in London, Rome and Boston (solid lines with tick marks every 50 years). The dashed curves begin at the time of the first declination measurements at each locality. The early inclination changes are based on archaeomagnetic data. Secular direction changes largely result from fluctuations in the Earth's non-dipole field which vary from locality to locality.

to geological features and varies much more rapidly than geological processes of uplift or continental drift and so we must look beneath the Earth's crust and mantle to motions in the fluid core of the Earth for the origin of the geomagnetic field. The most promising theory for the origin of the Earth's field is the dynamo theory which was developed by Elsasser (1946) and Bullard (1948) in the 1940s and 1950s.

The basic idea of the dynamo theory is that the Earth's magnetic field arises in the Earth's dense metallic liquid core, owing to the circulation of electric currents (Hide & Roberts 1956). A system of currents once generated in the Earth's core could be expected to run for some time because of the effects of self-induction. The electric energy would be gradually dissipated through Joule heating and after about 10^5 years any current system would have virtually disappeared. We know from palaeomagnetic studies that the Earth has had a magnetic field comparable in form and strength to that of the present field for at least 3×10^9 years. As this length of time is much longer than the 10^5 year free decay time there must be a mechanism controlling and maintaining the electric currents in the Earth's core. The most plausible mechanism is the interaction of magnetic fields with the flow of electric currents arising from the fluid motion of the Earth's core, i.e. a self-exciting dynamo. Theoretical calculations suggest that in the Earth's

core, fluid motions across an existing magnetic field will produce their own magnetic fields and induced electric currents and can be very important hydro-dynamically. Fluid motions could thus reinforce and maintain the geomagnetic field through a self-exciting dynamo mechanism (Elsasser 1946, Bullard & Gellman 1954). Although the basic mathematical formulation of such motions and interactions (which form the subject of magnetohydrodynamics) involves no more than the laws of classical mechanics, thermo-dynamics and electromagnetism, the subject is extremely complex because the mathematical problems are mostly non-linear and there are many physical unknowns and few observational constraints. So only a few solutions have been obtained to the mathematical problems and these relate to particularly simple solutions, but the general consensus is that any sufficiently complicated and vigorous motion in the Earth's core will operate as a dynamo. The primary energy source of the geomagnetic dynamo is thought to be either the radioactive decay of elements in the Earth's core (e.g. Verhoogen 1973) or the gravitational energy released by the sinking of heavy material in the outer core (Braginsky 1963). The resulting thermal or compositional instabilities lead to the formation of convection currents which in turn, through their magneto-hydrodynamic actions, drive the dynamo (Lowes 1984).

The fluid motions of the Earth's core may also

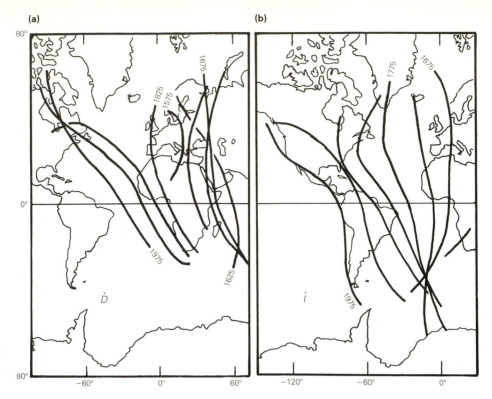

Figure 5.6. Examples of geomagnetic change between 1575 and 1975 AD illustrated by the locations of the (a) $\dot{D} = 0$ and (b) $\dot{I} = 0$ isopores (i.e. the locations where declination and inclination are passing through maxima or minima). The zero secular change isopores, in common with many features of the geomagnetic field, change position with time. For example, in central Europe in 1575 AD declination is reaching an easterly maximum (Fig. 5.5). The position of the easterly maximum moves slowly eastwards with time. By 1675 AD inclination in central Europe passes through a maximum (Fig. 5.5), the position of the maximum drifting steadily westwards to cross the Atlantic by 1825 AD. Around 1800 AD a westerly declination maximum grows up and crosses central Europe and Africa as marked by the successive positions of the Zero declination change isopore.

account for the temporal fluctuations in the Earth's magnetic field in addition to the overall origin of the field. The maintenance of the geomagnetic field involves a range of electromagnetic interactions, feedback systems and transfers of kinetic, magnetic and electrical energies. So eddying or turbulent motion would result in complicated and varying disturbances of the magnetic field and furthermore local irregularities in fluid flow or current systems near the surface of the core could be expected to produce magnetic disturbances on the scale of the non-dipole field. The magnetic field of a shallow electric current loop seated in the top layer of the core would resemble that of a radial magnetic dipole buried in the core. The non-dipole field has often been modelled in terms of radial

dipoles rather than in terms of spherical harmonic components. The number, location and magnitude of the dipoles used in the models vary considerably as no unique model can be derived from the observational data on the Earth's surface. Models with as few as eight radial dipoles can be constructed which produce a magnetic pattern at the Earth's surface which closely resembles that of the geomagnetic field. A further problem in constructing dipole models is that the amount of magnetic shielding or screening in the conducting fluid of the Earth's core is difficult to estimate. The modern geomagnetic approach to the question of secular magnetic variations is in terms of the dynamics of the redistribution of flux lines crossing the core/mantle boundary.

5.2 Palaeomagnetism

The history of the geomagnetic field can be extended into the geological past through the study of fossil magnetisation since natural materials can acquire a remanence (fossil magnetisation) in the Earth's magnetic field and so record its ancient direction and intensity (Section 4.3.1.). Many natural materials are able to retain such a signature of the geomagnetic field through later field changes or geological events. If we can find such materials, reconstruct their palaeo-orientation, measure their remanent magnetisation and date the time of origin of the magnetisation, we have a measure of the geomagnetic field in the past. Palaeomagnetists have assembled a remarkable picture of the ancient geomagnetic field using such fossil remanent magnetisations from many parts of the world and for ages ranging back to those of the very oldest rocks found on Earth.

The youngest materials studied, such as archaeological pottery shards or lake sediments, produce a palaeomagnetic picture which ties in well with the historically documented observations of the geomagnetic field. Palaeomagnetic results from rocks and sediments a few million years old show that the ancient field, although of similar form and intensity to the present field, was often completely reversed. This surprising phenomenon has been particularly well documented. It can be shown that the field has **reversed polarity** many hundreds of times and that the field has been in a 'reversed' state, when a compass needle anywhere on the Earth's surface would have pointed south, as often as it has been in the present, 'normal' state. Even older rocks with ages of tens and hundreds of million years reveal a further palaeomagnetic feature called polar wander. The palaeomagnetic pole positions of such rocks, in which the effects of secular variation have been averaged out, do not coincide with the present geographic pole, but are found to lie on a path which gradually leads away from the present pole as older and older rocks are considered. Such paths are called apparent polar wander paths as the movement of the palaeomagnetic **pole position** reflects the drift of continental blocks rather than the movement of the geomagnetic pole.

Palaeomagnetic phenomena, in addition to extending our knowledge of the geomagnetic field, can have important stratigraphic, tectonic and climatic implications. Interpretation of palaeomagnetic records can however be problematical and subjective although the collection and measurement of palaeo-

magnetic samples is generally straightforward. Many old rocks have undergone complex geological histories so that their original magnetic records may have been modified by later magnetisations and geological events. In consequence, a number of geological and laboratory tests must be applied in order to extract useful geomagnetic information. The equipment used in such palaeomagnetic measurements and testing is described in Chapter 6. The palaeomagnetic method is examined in more detail in Chapters 13 and 14, along with some applications and implications of palaeomagnetic results. The remainder of this chapter summarises the main geomagnetic information which has been won from studies of fossil magnetism, starting with results from recent materials and then proceeding deeper into the geological past.

5.2.1 Archaeomagnetism

Archaeomagnetism is the study of the magnetic remanence of archaeological artifacts and structures. Most archaeomagnetic results have been derived from baked materials such as pottery or tiles, which acquired a thermoremanent magnetisation on cooling. Such baked materials may also be used to find the

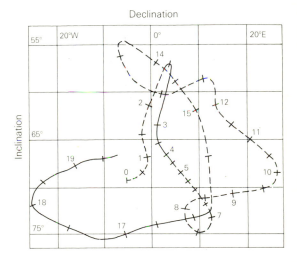

Figure 5.7 Archaeomagnetic variation in declination and inclination at London since 0 AD. The dashed line shows the secular direction changes deduced from the thermoremanence of bricks, tiles and pottery. The solid curve shows the direct historically documented variation of Fig. 5.5. Although the motion of the magnetic field vector is complex, showing both clockwise and anticlockwise looping, the average direction is close to that which would be produced by a geocentric axial dipole (data after Aitken 1974, Clark 1980).

ancient intensity of the geomagnetic field. Figure 5.7 illustrates the way in which archaeomagnetic records from France (Thellier 1981) and Britain (Aitken 1974, 1978) have extended the historical pattern of field direction changes back to the times of the Roman empire. Historical records of declination and inclination generally performed a clockwise loop during the period of direct observations (Fig. 5.5), a motion that is attributed to westward drift of the non-dipole field (Runcorn 1955). The clockwise sense of motion is also seen in the European archaeomagnetic record from 1300 AD to the present day (Fig. 5.7), but at earlier times an anticlockwise motion is found suggesting that in the past the non-dipole field has at times drifted eastwards (Aitken 1974, Kovacheva 1982).

5.2.2 Palaeosecular variation

Variations of the geomagnetic field, with characteristic timescales of 100 or 1000 years, are found in the palaeomagnetic records of rapidly deposited sediments. Organic-rich muds at the bottom of lakes have been found to carry continuous records of magnetic field changes spanning tens of thousands of years (Mackereth 1971). Figure 5.8 shows the last 5000 years of the palaeomagnetic record recovered from the bed of Loch Lomond. The familiar declination and inclination features of the historical and archaeomagnetic records can be seen in the uppermost sediments along with magnetic variations of similar amplitude and character in the lower sediments. The clockwise looping of the historical declination and inclination vector and the eastward looping of the archaeomagnetic vector are preserved in the uppermost Lomond record. The deeper sediments show that before AD 0 looping was once again mainly clockwise. Figure 5.8 also demonstrates that the secular variation of the geomagnetic field has been following a complex rather than a regular pattern. No simple periodicity is revealed although the characteristic timescale of the variations has a length of between 2000 and 3000 years (the spectral energy peaks at a period of 2700 years). Lake sediment records from other parts of the world show variations with similar characteristic timescales but different declination and inclination patterns. The potential of these limnomagnetic records as a magnetostratigraphic tool is discussed in Chapter 14.

Palaeointensity investigations of lava and very old archaeological samples have been used to estimate the changes in geomagnetic dipole moment over the last 11 000 years (Fig. 5.9). Variations from different parts of the world and rapid local fluctuations have been averaged out to reveal pronounced changes in dipole intensity with durations of order 10^4 years. The dipole field intensity, for instance, varied by a factor of two between a minimum at 6500 years BP and a maximum at 2800 years BP (Fig. 5.9).

5.2.3 Polarity timescale

Palaeomagnetic results show that the ancient geomagnetic field has reversed itself at frequent, but highly irregular, intervals in the geological past. Detailed radiometric dating studies have also shown that the polarity reversals take place simultaneously all around the world (Cox et al. 1963, Cox 1969). Averaged over a long sequence of polarity reversals the total amount of time spent by the geomagnetic field in the reversed state is equal to that spent in the normal state. Furthermore, the average field intensity of the two states is also similar. The energy levels of the geomagnetic dynamo in its normal and reverse states thus appears to be very similar.

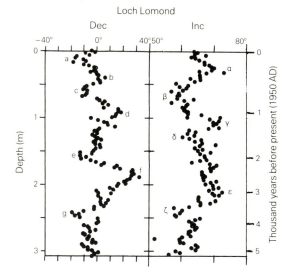

Figure 5.8 Palaeomagnetic directions recorded in the top 3 m of sediment of Loch Lomond. The declination measurements are centred on zero. ^{14}C age determinations provide the timescale at the right. Dots mark the natural remanence directions of individual 10 ml volume palaeomagnetic subsamples. Major palaeomagnetic turning points labelled following the convention of Mackereth (1971) (data from Turner & Thompson 1979).

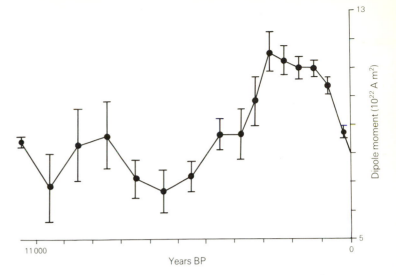

Figure 5.9 Variation in dipole moment of the Earth's magnetic field during the past 11 000 years as deduced from archaeomagnetic palaeointensity studies. The archaeomagnetic results, with their standard deviations shown by vertical bars, are world averages over 1000 year intervals (500 year intervals after 3000 years BP) (data after McElhinny & Senanayake 1982).

Polarity reversals have been found preserved in the thermoremanence of lava flows, in the detrital remanence of deep-sea sediments and the magnetic anomalies charted over the mid-ocean ridges. All these diverse recordings can be combined into one consistent polarity timescale. Marine anomaly profiles provide a particularly detailed and nearly continuous record of polarity reversals over the past 170 million years as changes of polarity of the geomagnetic field are preserved in the ocean crust, as it spreads away from the mid-oceanic ridge crests as part of the plate tectonic process (Vine & Matthews 1963, Larson & Pitman 1972).

Over the past fifty million years the geomagnetic reversal process appears to have been almost completely random. Characteristic periods associated with individual reversals have been about two hundred thousand years while the time to complete a transition between polarity states is estimated to be between one and ten thousand years. Over 50 million years ago the dynamo process was less symmetrical with respect to polarity and long periods of single polarity are found, the longest being a period of reversed polarity which persisted for 85 million years. Changes in the average frequency of reversals have been found to occur at intervals of 50 million years or longer. Such geomagnetic periods of tens and hundreds of millions of years have a geological ring about them which suggests that their physical origin is to be sought in the Earth's solid mantle rather than the fluid core. Mantle convection currents could alter the physical properties of the core/mantle interface by producing hot or cold spots or even bumps over timescales of hundreds of millions of years. These slow temperature or topographic changes could affect the operating characteristics of the dynamo and govern the long period timetable of geomagnetic polarity changes, such as the changes in average frequency of geomagnetic reversals.

5.2.4 Apparent polar wandering

When averaged over some hundred thousand years the geomagnetic dipole axis has been found to coincide with the Earth's spin axis. Indeed palaeomagnetic evidence suggests that throughout the Earth's history the geomagnetic field has been predominantly dipolar and axial. This simple relationship between the geomagnetic axis and the spin axis has allowed palaeomagnetic measurements to be used to make continental reconstructions and to chart the drift of continents over the surface of the Earth (Runcorn 1956, Irving 1977).

The directions of fossil magnetisation in rocks from different parts of the world can be most easily compared by plotting the magnetic results in terms of palaeomagnetic pole positions. Figure 5.10 shows the change in palaeomagnetic pole positions of rocks from Europe and North America over the past 500 million years. The shapes of the two polar wander paths can be explained very well by the drift of the continents associated with formation of the Atlantic ocean.

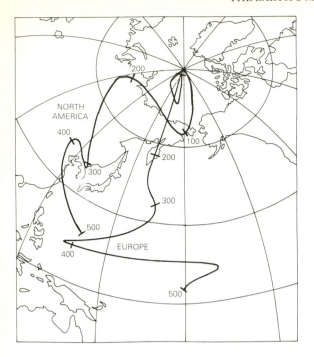

Figure 5.10 Apparent polar wander paths for North America and Europe (tick marks every 100 million years). The general trend of the paths from equatorial regions to the north pole reflects the northward drift of both continents during the past few hundred million years. The longitudinal difference between the two paths is a measure of the separation of the two continents which has taken place with the formation of the North Atlantic ocean.

During the past 100 million years the Atlantic ocean has been gradually opening as Europe and America have moved apart. The continents and the floor of the Atlantic ocean have also been slowly drifting northwards. These plate motions combined with a time-averaged geocentric axial dipole field produce the polar wander paths of Figure 5.10. It is possible to use palaeomagnetic data to make detailed, quantitative reconstructions of the past positions of the continents. The palaeolatitude, λ, is given by the palaeomagnetic inclination, I, through the dipole field formula $\tan I = 2 \tan \lambda$. The palaeomeridian is given directly by the palaeomagnetic declination. The palaeolongitude cannot be found from palaeomagnetic data, although relative palaeolongitudes can be derived by comparing sections of apparent polar wander paths.

5.3 Summary

Stars and large planets generate their own magnetic fields through dynamo actions deep in their interiors. The Earth has been generating a field for at least the last three billion years. On account of Coriolis forces planetary magnetic fields average out to be symmetrical about their rotation axis. Throughout the Earth's history its field has closely corresponded to that of a dipole with its axis aligned with the spin axis.

The geomagnetic field can be generated with either its north magnetic pole or its south magnetic pole in the northern hemisphere and it has frequently switched between these two stable polarity states. The present polarity of the Earth's field is referred to as normal. Saturn currently has a reversed field while Jupiter's field is also reversed. In addition to the major reorganisations of the Earth's field lines which take place when the whole field switches polarity, minor field reorganisations are continuously taking place and are referred to as secular changes. These long-lasting fluctuations affect both the intensity and the direction of the field and have been documented over the past few hundred years by observations of changes in the declination of magnetic compasses and the inclination of dip needles.

Palaeomagnetic studies of the fossil magnetism of rocks and sediments extend our knowledge of the geomagnetic field into the ancient past and can be used for investigating geomagnetic dynamo behaviour, for geological dating purposes and for continental drift and microplate tectonic studies.

Further reading

General books

McElhinny 1973. *Palaeomagnetism and plate tectonics.*
Strangway 1970. *History of the Earth's magnetic field.*
Irving 1964. *Palaeomagnetism and its application to geological and geophysical problems.*
Tarling 1983. *Palaeomagnetism.*
Merrill and McElhinny 1983. *The Earth's magnetic field.*
Aitken 1974. *Physics and archaeology.*

Advanced book

Jacobs 1975. *The Earth's core.*

[6]
Techniques of magnetic measurements

It is a capital mistake to theorize before one has data.

Sir Arthur Conan Doyle,
Scandal in Bohemia

6.1   Introduction

Following the electronics revolution, particularly the production of integrated circuits and microcomputers, instrumentation for magnetic measurements has improved dramatically. Advances have included increases in sensitivity, speed of measurement, portability, availability and simplicity of operation. Magnetic susceptibility equipment is now commercially available at reasonable cost largely owing to the technology developed for the amateur metal-detecting market. Such commercial instruments are now more sensitive, reliable and accurate than many which were purpose built for use in palaeomagnetic research laboratories. Being battery operated the commercial instruments are not tied to the model conditions of the laboratory bench and are ideal for field surveys. Furthermore, their digital displays, electronic calibration and automatic push button zeroing enable reliable susceptibility measurements to be made at a rate of several per minute by inexperienced users. Apparatus for magnetic remanence measurements has similarly been greatly simplified in terms of convenience and speed of operation. These various electronic advances in instrumentation have been fundamental to the recent increase in scope of application of magnetic investigations to environmental projects and especially to the use of magnetic

measurements as a preliminary reconnaissance tool at the start of site investigations. Magnetic instrumentation has thus become attractive to many earth scientists and environmentalists and is no longer solely used by the palaeomagnetic specialists.

Not all magnetic instrumentation has been streamlined by commercial attention. So on entering a magnetic laboratory one may still be faced by a bewildering array of large coils, magnets, cables, furnaces, dewars and spinning shafts and be tempted to reflect that an apt synonym for palaeomagnetism is indeed palaeomagic. Nevertheless the bulk of the environmental magnetic measurements discussed in Chapters 8–16 have been made on just two types of instrument. These are the susceptibility bridge of Section 6.3.1 and the fluxgate magnetometer of Section 6.2.3. Both instruments are very simple, extremely quick to use and are to be found in most palaeomagnetic research laboratories. A third important piece of equipment is either a pulse discharge magnetiser (Section 6.6.2) or an electromagnet (Section 6.6.2). Electromagnets or pulse discharge units are needed for the generation of high magnetic fields such as those used in the study of isothermal remanence. Again such apparatus is relatively simple to use, and exceptionally fast in operation; hundreds of magnetically saturated samples can be produced in an hour.

Table 6.1 Uses of instrumentation.

Subject of investigation	Magnetic property	Section	Example of application	Chapter	Instrumentation	Section
concentration	initial susceptibility	4.4	DSDP core 514	12	susceptibility	6.3
concentration	saturation magnetisation	4.3.2	N. Atlantic sediment cores	12	induced magnetisation	6.4
mineralogy	magnetite, haematite	3.2	sources of suspended sediments	9	remanence/field generation	6.2/6.6
domain state	hysteresis, coercivity	4.6	Plynlimon bedload tracing experiment	9	induced magnetisation	6.4
viscosity	superparamagnetism	4.3.2	enhanced soils on archaeological sites	8	pulsed induction meter	6.7.3
temperature dependence	Curie temperature	4.8	atmospheric fallout in the English midlands	11	Curie balance	6.4.1
natural remanence	thermoremanence	4.3.1	Iceland lava flows	13	remanence	6.2
multicomponent remanence	demagnetisation	2.5	Gass lake drying remanence	14	magnetic cleaning	6.5

There are a wide variety of experimental techniques available to the investigator of the magnetic properties of minerals. A comprehensive and detailed review of them all is beyond the scope of this chapter. The aim of the chapter is directed more towards indicating the range of methods, to outlining the most important physical principles upon which the instrumentation is based, discussing some experimental problems and limitations, and describing the main instruments featured in the later application (Chs 8–16). Table 6.1 outlines for a range of magnetic properties (a) the main magnetic instrumentation sections of the book connected with their measurements, (b) the associated introductory theory and (c) an example of their use.

The first group of instruments described are those which measure magnetic remanence (Section 6.2). The second group described have those in which an applied magnetic field is used to produce a magnetic signal. In low applied fields we have magnetic susceptibility (Section 6.3) and at high applied fields we have induced magnetisation (Section 6.4). A third group of instruments are used for magnetic cleaning and these are described in Section 6.5. Ancillary equipment for the production and reduction of magnetic fields is described in Section 6.6. The final Section (6.7) covers the major types of portable equipment which are used for field surveys.

6.2 Measurement of remanent magnetisation

The classic instrument capable of detecting weak remanent magnetisations is the astatic magnetometer (Section 6.2.2). Considering its simplicity, it is a remarkably sensitive piece of apparatus (Blackett 1952). The fluxgate magnetometer (Section 6.2.3) is in general more sensitive and faster to use than the astatic magnetometer. Fluxgate magnetometers can also tolerate magnetically and vibrationally noisy working environments. They are available commercially, complete with an on-line microcomputer (Molyneux 1971) programmed to perform statistical calculations and spherical trigonometric calculations. Thus fluxgate magnetometers can be operated very easily and successfully, even in the field, without need for any specialised palaeomagnetic training. We have made extensive use of fluxgate magnetometers in our environmental investigations.

In principle, calibration of instruments for measuring remanent magnetisation is straightforward. A coil, of the same dimensions as the samples under investigation, carrying a d.c. current can be used to provide a known magnetic moment (measured in Am^2) (Collinson 1983). In practise spinner magnetometers are difficult to calibrate and great care must be paid to sample shape and position.

6.2.1 Generator magnetometer

The remanent magnetisation of a sample can be measured by spinning it at a high rate in a coil (Johnson & McNish 1938). As the sample spins the field lines of its dipole moment cut the windings of the coil and induce an alternating electromotive force (e.m.f.) in accordance with Faraday's Law, just as occurs in the operation of a small dynamo. The amplitude of the induced electrical signal is proportional to the sample magnetic moment and its

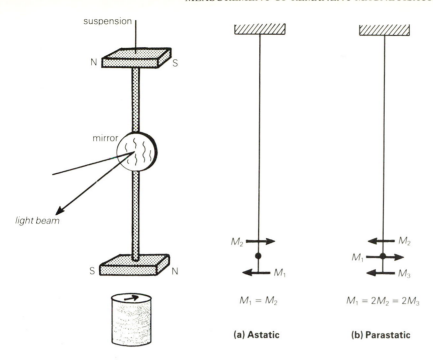

Figure 6.1 (a) Astatic suspension, two magnets with identical moments fixed antiparallel to each other on a rigid stem. An optical lever system is used to detect small rotations of the system caused by a sample placed beneath the lower magnet. (b) Parastatic suspension, central magnet has twice the moment of the upper and lower magnets.

$M_1 = M_2$

(a) Astatic

$M_1 = 2M_2 = 2M_3$

(b) Parastatic

phase is determined by the direction of the sample moment. Since only the component of sample magnetisation perpendicular to the pick-up coil axis contributes to the rotating magnetic field and to the alternating electric signal, it is necessary to spin a sample in more than one orientation in order to determine its remanence fully. In normal generator magnetometer operation a predetermined sequence of six different orientations is used for each sample. The main limiting factor in the generator method of remanence measurement is the accumulation of electrostatic charges on the sample holder. In addition there is the practical drawback of the break up of moderately fragile samples at the high rotation speeds.

6.2.2   Astatic and parastatic magnetometers

The astatic magnetometer is a very simple and reliable instrument. As it can be operated either in zero field or in a low magnetic field it can be used to measure low field magnetisation, initial susceptibility and anisotropy of susceptibility, as well as magnetic remanence. Despite the low cost involved in constructing an astatic magnetometer, its requirement of a magnetically quiet and vibration-free environment

has led to its replacement by other instruments in palaeomagnetic laboratories around the world.

The astatic magnetometer is based on the principle that the torque on a suspended magnet depends on the applied magnetic field. So the field associated with a magnetised sample can be detected by bringing the sample close to a suspended magnet and watching the magnet twist. The key to measuring the weak magnetisation of natural samples lies in making the magnetic suspension sensitive to magnetic field gradients, but at the same time insensitive to changes in magnetic field. This situation is achieved in the astatic magnetometer by using two magnets, each with the same magnetic moment, but aligned antiparallel to each other, as shown in Figure 6.1a.

Another sensitive arrangement of the magnet suspension (Fig. 6.1b) is the parastatic system (Thellier 1933). In this system (Fig. 6.1b) three magnets are used: the central magnet of moment $2M$ is balanced by two antiparallel magnets of moment M fixed above and below it. With both astatic and parastatic magnetometers a sequence of measurements is needed in order to find the total magnetic remanence of a sample. In some experimental arrangements as many as 32 measurements are needed per sample, each involving a different sample position. However, recent advances which include spinning the sample near the magnet suspension, electronic monitoring of the optical system, and on-line data processing combined with the parastatic configuration, enable the remanence of fairly weak samples to be measured in a town environment at a rate of tens of samples per hour.

6.2.3 Fluxgate magnetometer

The sensitivity, flexibility and ease of operation of fluxgate magnetometers have made them the work horses of modern palaeomagnetism. A fluxgate probe is normally about 50 mm long and consists of a high permeability core, such as a strip of mu-metal, on which a primary and secondary coil have been wound to make a transformer (Fig. 6.2a). In operation an alternating current in the primary coil drives the core around its hysteresis loop in and out of saturation. Any direct magnetic field along the axis of the probe offsets the hysteresis loop (Fig. 6.2b). This distortion due to the applied field introduces asymmetry or even harmonics into the waveform (Fig. 6.2c) and can be detected in the output of the secondary coil. In routine operation a fluxgate probe can detect fields of 1 nT (1 γ) and in practice the noise level is so low that when measuring the remanence of natural samples the practical limit is generally set by the purity of the sample holder.

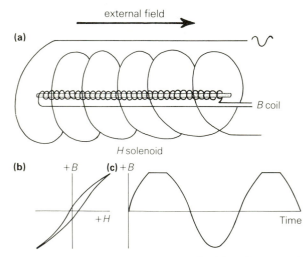

(a)

external field

B coil

H solenoid

(b) +B **(c)** +B

+H

Time

Figure 6.2 Schematic diagrams of: (a) primary and secondary windings of a fluxgate probe; (b) offset B–H loop; (c) asymmetrical output and saturation caused by an external magnetic field.

Instrument sensitivity can be increased and the effect of any sample inhomogeneity significantly reduced by using a double-probe gradiometer arrangement. The double-probe layout is essentially the same as that of the astatic magnetometer, antiparallel fluxgates being used rather than anti-parallel magnets. Further sensitivity is produced by using a ring-shaped gradiometer probe, by spinning the sample and by shielding the fluxgate probe from the Earth's magnetic field. All these features are incorporated in the whole core magnetometer (Molyneux *et al.* 1972) of Figure 6.3. Spinning the sample provides a convenient method of taking numerous readings by triggering rapid computer sampling of the fluxgate output using a photocell device (Fig. 6.3). For routine measurements of conventional 10 ml volume specimens the sample is spun six times, i.e. about three mutually perpendicular axes in both upright and inverted positions. Using this procedure remanent magnetisations as low as 10^{-4} A m^{-1}(10^{-7} G) can be measured in about five minutes. Fluxgate based equipment has been used for the great majority of lake sediment remanence measurements described in Chapter 14. It has also been used for many of the saturation remanence and coercivity measurements reported in Chapters 8–12. These strong laboratory remanences, however, can be measured in a few tens of seconds as one short spin is quite sufficient for each sample. Portable, battery-operated fluxgate magnetometers are now manufactured commercially by L. Molyneux with sensitivities, and speed and convenience of operation, equalling those of the older laboratory-based versions.

6.2.4 Superconducting magnetometer

The main advantages of cryogenic superconducting magnetometers (Goree & Fuller 1976) are their high sensitivities and fast response times. Figure 6.4b shows the usual layout of a superconducting magnetometer with a vertical access hole. The pick-up coils (Fig. 6.4a), flux transformer and SQUID (superconducting quantum interference device) sensor (Fig. 6.4c) are enclosed by a superconducting shield. The sample remains at room temperature and is lowered in to the magnetometer's sense region on a long plastic rod. Insertion of a sample into the superconducting magnetometer pick-up coil system initiates a direct current in its superconducting circuitry. This current

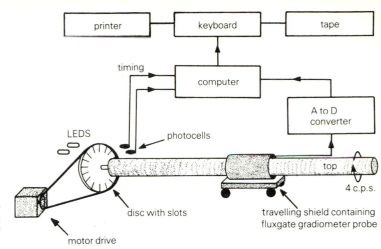

Figure 6.3 Schematic diagram of a computerised spinner magnetometer arranged for whole core declination and horizontal intensity measurements.

is fed via a flux transformer to another coil where it produces an amplified field which is detected by a SQUID. Measurement is independent of the rate at which the sample is inserted and, by using three mutually perpendicular pick-up coils, total remanent magnetisation can be found from one sample insertion.

SQUID sensors have been used to detect magnetic field fluctuations as low as 50 fT (0.000 05 γ) (Fig. 15.3 and Section 15.5) and to measure remanent magnetisations of 5×10^{-9} A m^2 kg^{-1} (5×10^{-9} G cm^3 g^{-1}) (Collinson 1983) with a one second averaging time. They can also be used for several other types of measurement. For example, by trapping a low magnetic field in the same region, measurements of viscous magnetisation and measurements of initial susceptibility and its anisotropy have been made. The superconducting magnetometer has exceptional potential in the magnetic investigations of natural samples, but it is likely to remain a tool operated by the specialist owing to the high costs of its helium consumption.

6.3 Measurement of initial susceptibility

Magnetic susceptibility equipment is the simplest of all magnetic instrumentation to use. Measurement, using the a.c. method (Section 6.3.1), involves sliding a sample into the instrument and reading a meter or dial or else pressing a button for a teletype print out. A hundred samples can easily be processed on such equipment in less than an hour. Portable susceptibility bridges are made with sensing heads of a variety of shapes and sizes, so that it is possible to measure the magnetic susceptibility of whole cores, soil profile faces and *in situ* bedrock as well as the more usual samples of 25 mm diameter rock cores or 10 ml volume plastic boxes. Several of the magnetometers, generally used for measuring magnetic remanence (Section 6.2), can be adapted to measure initial magnetic susceptibility (Section 6.3.2).

The change of susceptibility with temperature can be investigated with many of the instruments used for normal room temperature measurements, the only modification being to place a small non-magnetic furnace or dewar inside the sensing coils. The sample size which can be accommodated is, of course, considerably reduced and the instrument sensitivity lowered by about an order of magnitude.

6.3.1 The a.c. method

The most common method of measuring initial susceptibility involves the use of a balanced a.c. bridge circuit (Mooney 1952, Mooney & Bleifuss 1953, Aksenov & Lapin 1967, Molyneux & Thompson 1973). Bridge methods are in general very accurate and they are widely used for the measurement of small changes in inductance, capacitance or resistance. In susceptibility bridges the magnetising field is produced by a current-carrying solenoid, flat coil or Helmholtz coil pair. A balanced coaxial pick-up coil is used to detect the induced magnetisation. Insertion of a sample into the coil system alters its inductive balance and produces an out-of-balance signal, in the

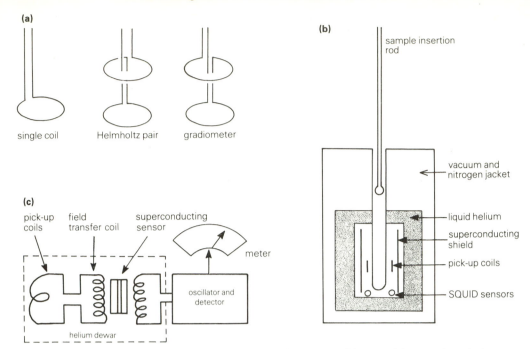

Figure 6.4 Superconducting magnetometer: (a) pick-up coil configurations; (b) general layout of vertical access magnetometer; (c) the heart of the magnetometer lies in its SQUID sensors which detect the d.c. current produced by the insertion of a sample into the pick-up coils.

pick-up coil, which is proportional to the total susceptibility of the sample. Depending on the type of a.c. bridge and coil arrangement used, the out-of-balance signal may be amplified and measured in millivolts, rectified and measured in microamperes or nulled using a low resistance high linearity potentiometer, for both the in-phase and quadrature (Section 6.3.4) components. The ultimate sensitivity of a.c. induction bridges is probably limited by the mechanical and thermal stability of the sensing coils.

An alternative approach is to use a balanced transformer circuit. Bruckshaw and Robertson (1948) used a double coaxial pick-up and Helmholtz pair. Likhite *et al.* (1965) and Radhakrishnamurty *et al.* (1968) employed a similar arrangement in constructing equipment which worked at different frequencies and could measure samples of different sizes. A further approach employed by Bartington (pers. comm.) is to detect the frequency change which is caused in a sharply tuned 'metal detector' oscillator circuit by the introduction of a sample (Lancaster 1966). Smit and Wijn (1954) in addition to describing four a.c. bridges for measuring the frequency dependence of both in-phase and quadrature susceptibility summarize

resonator and wave methods which can operate at substantially higher frequencies.

Modern a.c. instruments using peak magnetic field strengths of about 0.1 mT (1 Oe) at a frequency between 1 and 10 kHz have noise levels below 1×10^{-6} SI units (about 10^{-7} G Oe^{-1}). Natural susceptibilities vary from the weak negative susceptibility (diamagnetism) of unpolluted peat or of carbonate- and quartz-rich materials to the comparatively high susceptibility of 10^{-3} SI units (10^{-4} G Oe^{-1}) of basic igneous rocks. Susceptibility bridges are most easily calibrated by reference to paramagnetic salts such as copper sulphate ($CuSO_4.5H_2O$) or ferrous sulphate ($FeSO_4.7H_2O$) with susceptibilities of 7.4×10^{-8} m^3 kg^{-1} (5.9×10^{-6} G Oe^{-1} cm^3 g^{-1}) and 1.4×10^{-6} m^3 kg^{-1} (115×10^{-6} G Oe^{-1} cm^3 g^{-1}) respectively. Calibration samples should be of the same size and shape as the specimens under investigation.

6.3.2 The direct method

Astatic (Section 6.2.2), ballistic (Section 6.4.2) and superconducting (Section 6.2.4) magnetometers can be used to measure low field susceptibility.

Magnetisation measurements are made in the presence of a low direct field, rather than in the zero field condition of magnetometers used for remanence measurements. So both a remanent and an induced moment contribute to the magnetisation. The relative importance of these two moments can be found by reversing the orientation of the sample. The remanence rotates with the sample while the induced moment remains in the applied field direction. By calculating the sums and differences of the magnetisation for the two sample orientations the separate remanent and induced moments are found. The initial susceptibility is then given by the ratio of the induced moment to the applied field.

6.3.3 Anisotropy of initial susceptibility

It is possible to detect anisotropy of susceptibility using several types of instruments. However, as it is often necessary to measure 1% variations in susceptibility of samples with mean values around 10^{-8} SI units (10^{-7} G Oe^{-1}), instruments are needed which can measure susceptibility differences of 10^{-10} SI units (10^{-9} G Oe^{-1}). The most common methods of measuring anisotropy use a low field torque meter (Ising 1943, King & Rees 1962) or an alternating current bridge. Figure 6.5 illustrates the orthogonal coil

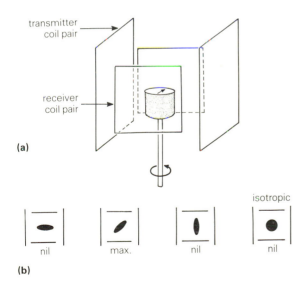

(a)

(b)

Figure 6.5 (a) Layout of orthogonal transmitter and receiver coils in anisotropy of susceptibility bridge. (b) The plan view diagrams schematically illustrate the coupling between the coils for three orientations of an anisotropic sample and for an isotropic sample.

arrangements and instrument responses of the alternating current method.

The size and shape of sample can have significant effects on anisotropy measurements. It is not simple to calculate the most appropriate sample dimensions as they depend in a complex way on the geometry of the magnetic sensor. In general however, the sample shape should approximate that of a sphere. So cylinders should be used with length-to-diameter ratios of about 0.9. Cuboid, irregularly shaped samples and samples departing from the recommended dimensions by more than a few percent will produce erroneous results which will reflect the sample's shape rather than its magnetic fabric.

6.3.4 Quadrature and frequency-dependent susceptibility

The time delay between the application of a field and the full magnetisation response can be investigated through the measurement of quadrature susceptibility. A.C. susceptibility can be divided into 'in-phase' and quadrature ('out-of-phase') components. The more pronounced the lag in the magnetisation response the more important the quadrature susceptibility. The conventional bridge arrangements of Section 6.3.1 can be used to measure quadrature susceptibility (or permeability) if suitable phase detection circuitry is incorporated in the out-of-balance electrical monitoring system.

Another approach used in investigating magnetic relaxation phenomena is that of measuring susceptibility at different frequencies. The variation of susceptibility with frequency is known as the susceptibility spectrum. At low frequencies, magnetisation remains in phase with field, so the in-phase susceptibility has a value close to that of the direct, static susceptibility while the out-of-phase susceptibility is effectively zero (Galt 1952, Smit & Wijn 1959). However as the frequency is increased, relaxation effects become more important and the in-phase component, after a small rise (Snoek 1948), decreases steadily, while the out-of-phase component rises, peaks and then also falls back to zero. The peak out-of-phase susceptibility and the most rapid decline in in-phase susceptibility theoretically occur at the same frequency. The overall trend is for susceptibility to fall with increasing frequency of measurement.

Bhathal and Stacey (1969) noted that the initial susceptibility of multidomain magnetites only exhibited very small susceptibility changes of 0.3% per decade of frequency at measurement frequencies of around 1kHz. The much larger falls of susceptibility with increased measurement frequency and the high quadrature (out-of-phase-components) that are observed in certain natural samples are thus unlikely to result from the movement of domain walls in large multidomain grains. The dominant cause of high quadrature readings or of pronounced changes in susceptibility with frequency will instead be the viscous effects of ferri- and ferromagnetic grains lying close to the superparamagnetic/stable single domain boundary with relaxation times of around 10^{-4} seconds. The overall fall of susceptibility with increasing measurement frequency observed in natural samples can be accounted for by the magnetisation of grains becoming 'blocked in' as the superparamagnetic/stable single domain boundary shifts to smaller volumes as the frequency of measurement is raised.

By judicious selection of frequency it is possible to investigate usefully the susceptibility spectrum by making just two susceptibility measurements. The instrument of Bartington (Section 6.3.1.), which is the main bridge we have used in making the frequency dependent susceptibility measurements of Chapters 8–11 and 16, uses two frequencies of 1 and 10 kHz and a peak alternating strength of 3×10^{-4} T(3 Oe) in the discrimination oscillator circuit. Introduction of a sample into the detection coil creates a small frequency shift. The difference in shift at 1 and 10kHz is taken as a measure of frequency dependent susceptibility and is given the symbol χfd in later chapters. Using this equipment with natural samples, the range of values we have encountered for frequency dependent susceptibility expressed as a percentage of total susceptibility (χfd/χ) is between 0 and 24%. The maximum change in frequency dependent susceptibility for coarse multi-domain magnetite was less than 0.26%, in excellent agreement with the results of Bhathal and Stacey (1969); the highest frequency dependent differences were found for dusts from deflated soils (see Section 11.8 and Fig. 11.16).

6.4   Measurement of induced magnetisation

A very great number of different experimental arrangements have been devised for the measurement of the induced magnetisation of gases, liquids and solids. Measurements of induced magnetisation in medium to high strengths are rather more difficult to carry out than measurements in low field strengths and quite sensitive equipment is needed in order to investigate the magnetisation properties of most natural materials. The need for sensitivity arises because the induced magnetisation of natural materials, although dominated by strongly magnetic ferrimagnetic crystals, is actually very weak on account of the low concentrations of ferrimagnetic crystals. Indeed the magnetisation of natural samples is often rather similar in strength to the weak magnetisation of paramagnetic substances.

Methods of measuring induced magnetisation may be divided into three main groups. These are:
(a)   measurement of the dipole field of a magnetised sample,
(b)   measurement of the force on a magnetised sample in a non-uniform magnetic field,
(c)   measurement of induction by use of coils.
The force method forms the basis of the Curie balance (Section 6.4.1), an instrument which is widely used in palaeomagnetic laboratories. The induction method forms the basis of the ballistic magnetometer (Section 6.4.2) and the vibrating sample magnetometer

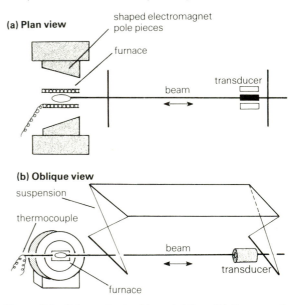

Figure 6.6   Schematic plan (a); and oblique (b) diagrams of a Weiss–Curie force balance. The sample is enclosed in a small evacuated silica capsule and is suspended between the shaped pole pieces of an electromagnet. It is restrained to move horizontally in a tube which can be heated or cooled by a small brass furnace/dewar. The magnetic force on the sample is detected by a transducer. Magnetic balances are the most common instruments that have been used for measuring the Curie temperatures of natural materials.

(Section 6.4.3). These two induction-based instruments are used for detailed investigation of the hysteresis properties of natural material.

6.4.1 The force method

Magnetic balances measure the force exerted on a sample placed in an inhomogeneous magnetic field. Many experimental arrangements, such as the Gouy (1889) method, use long, thin samples. Another arrangement, the Faraday–Curie method (Fig. 6.6), uses small, roughly spherical samples (Curie 1895, Foëx & Forrer 1926) and has been found very suitable for measuring natural materials in the form of small chips or powder pellets.

Magnetic balances can be used to determine acquisition of magnetisation curves, the saturation magnetisation and the hysteresis properties of ferromagnetic minerals and natural samples over a wide range of temperatures and atmospheres. They are routinely used in the estimation of Curie temperatures by recording the change of saturation magnetisation with temperature. Examples of saturation magnetisation versus temperature records of natural samples containing magnetite are shown in Figure 4.14. Construction of complete hysteresis curves using the

force method presents some experimental problems, also the work is somewhat time consuming so that other approaches to drawing complete hysteresis loops, such as using a vibrating sample magnetometer (6.4.3), are usually preferred.

6.4.2 Ballistic magnetometer

The basic instrument has great flexibility (West & Dunlop 1971, Nagata 1976). It has been used to determine the susceptibility of natural samples, as well as their minor hysteresis loops and their saturation magnetisation at both low and high temperatures. It is, however, somewhat limited in sensitivity and so has mainly been used in investigation of igneous rocks.

6.4.3 Vibrating sample magnetometer

The vibrating sample magnetometer (Foner 1959, Kobayashi & Fuller 1967) is the instrument most widely used in measuring the hysteresis properties of natural samples. It has a high sensitivity and the capacity for hysteresis measurements to be made over a wide range of temperatures. It operates on the

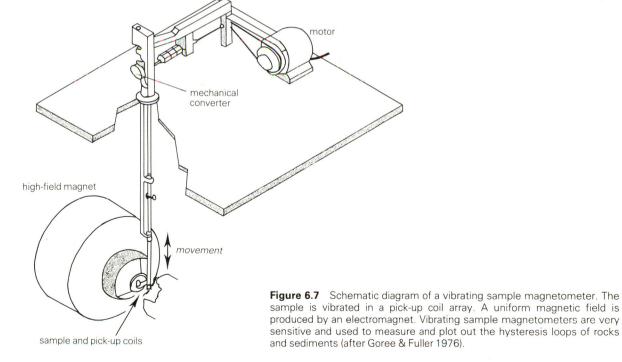

Figure 6.7 Schematic diagram of a vibrating sample magnetometer. The sample is vibrated in a pick-up coil array. A uniform magnetic field is produced by an electromagnet. Vibrating sample magnetometers are very sensitive and used to measure and plot out the hysteresis loops of rocks and sediments (after Goree & Fuller 1976).

induction principle, sample magnetisation being detected as an a.c. voltage in the pick-up coils (Fig. 6.7). The voltage, generated by vibrating a sample at a fixed frequency in the coil system, results from the flux change across the pick-up coils as the sample changes position. In the normal vibrating magnetometer configuration, depicted in Figure 6.7, the sample is vibrated perpendicular to the magnetic field direction.

Direct output can easily be arranged in the form of graphs of hysteresis loops. The magnetising field of the electromagnet is monitored by a Hall effect probe (Section 6.6.1) and the output fed to one channel of an x-y graph plotter. The magnetisation measured by the rectified output of the magnetometer pick-up coils is fed to the other channel. Hysteresis loops are traced out by varying the magnetic field. It takes some minutes to draw a complete hysteresis loop by smoothly cycling the magnetic field from its peak (saturating) value through zero to the opposite polarity peak and then back to the original value. So only two or three samples can be processed in an hour. Furthermore, extracting information from the graphical output can be rather time consuming, particularly when magnetic mixtures are being investigated.

Samples with saturation magnetisations greater than 10 mA m^2 kg^{-1} (10 mG cm^3 g^{-1}) such as igneous rocks or heat-enhanced stream bedload samples, cf Chapter 9, can be measured without difficulty on a vibrating sample magnetometer. The hysteresis properties of many sediments can be measured if care is taken with sample holder correction and long integration times are used.

Figures 4.12 and 16.5 illustrate the main magnetic parameters derived from vibrating sample magnetometer measurements and their uses in describing magnetic crystal assemblages. Such hysteresis diagrams and magnetic parameters provide the most comprehensive magnetic characterisation of natural materials which can be achieved without extravagant specialisation or excessive effort.

6.4.4 Alternating field method

A very neat and rapid method of obtaining hysteresis characteristics is the alternating magnetic field induction method (Bruckshaw & Rao 1950) in which hysteresis loops can be displayed directly on an oscilloscope screen (Likhite et al. 1965). The sample magnetisation is detected using the double-search coil

method of a.c. susceptibility bridges (Section 6.3.1). Magnetising fields of up to 0.4T (4000 Oe) have been obtained by using an electromagnet with a high permeability core. The rather low sensitivity of the a.c. method has so far restricted its possible applications with natural materials to the examination of basic igneous rocks. The instrument is very appealing, however, on account of the speed with which it can be used. Another useful feature of the alternating field method is the ease with which low temperature investigations can be performed. The change of hysteresis properties with temperature can be monitored by simply dipping a sample in liquid nitrogen, placing it in the detection coil and then allowing it to warm up to room temperature. Large collections of basalts have been characterised magnetically using this technique. A practical application of such magnetic characterisation has been the preselection of basalts for more time-consuming studies such as palaeointensity determinations.

6.5 Magnetic cleaning techniques

As rocks or sediments have had a complex magnetic history it is necessary to apply demagnetisation techniques to separate out their various components of remanent magnetisation. Two methods routinely used for this task are alternating field demagnetisation (As 1967, Creer 1959) and thermal demagnetisation (Thellier 1938, 1966). The former is used for samples in which magnetite carries the remanent magnetisation, the latter for haematite-bearing samples and samples which are chemically stable at elevated temperatures. The principle of magnetic cleaning or partial demagnetisation is that the less stable components of the remanent magnetisation are selectively removed to leave the more stable components.

As detailed demagnetisation studies can be very time consuming, a common practice when dealing with collections of hundreds of samples is to investigate the demagnetisation properties of up to about 10% of the samples in detail. These pilot samples are chosen to be representative of the main collection and their demagnetisation results are used to decide on the most efficient method of isolating the different components of magnetisation in the collection.

6.5.1 Alternating field demagnetisation

Demagnetisation is accomplished by subjecting a sample to an alternating field which is gradually reduced to zero. The alternating field is produced by a coil in a tuned a.c. circuit, which is generally driven at mains frequency. The field is smoothly reduced from its peak value (Fig. 6.8) by a liquid rheostat or motor-driven voltage regulator. The effect on the sample is to remove a part of the remanence by magnetic grains with coercive forces lower than the strength of the peak applied magnetic field. The part of the remanence carried by grains with higher coercive forces remains unaltered.

Pilot samples are generally demagnetised at steps of 5 or 10 mT (50 or 100 Oe) up to about 100 mT (1000 Oe). Demagnetisation at each step is carried out along three mutually perpendicular axes or alternatively a tumbling device, which presents all axes of the sample to the alternating field, can be used. In practice it is essential to ensure that demagnetisation is carried out in zero field and also that the alternating field is free from both transients and asymmetry. Without these safeguards an anhysteretic remanence (ARM) may be grown which can mask the natural remanence and invalidate the cleaning process. As tumbling devices speed up the demagnetisation process, and randomise unwanted anhysteretic remanences they are expedient and widely used.

6.5.2 Thermal demagnetisation

There are two methods of carrying out thermal demagnetisations. One, called continuous thermal demagnetisation, involves monitoring the change in remanence of a sample as it is heated (Wilson 1962). The other method, called progressive thermal demagnetisation, consists of a number of partial demagnetisation steps using successively higher temperatures (Thellier 1938). At each step the

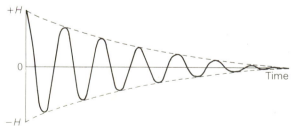

Figure 6.8 Decay with time of the waveform of the alternating field in a demagnetisation cycle.

sample's remanence is measured after having been heated and cooled in a zero magnetic field. With both thermal demagnetisation methods the components of remanence with low blocking temperatures are removed first thus revealing the higher stability portion of the remanence.

In the first method, a furnace is mounted within a magnetometer and the sample's remanence is measured while it is hot. In the second method, large separate furnaces, which can accommodate up to 50 samples, can be used. The progressive method is much faster than the continuous method when dealing with large collections. Temperature steps of 50 or 100 °C are commonly used for the treatment of pilot samples in the progressive method. Particular importance is attached to changes in remanence at the highest blocking temperatures. Samples may acquire a spurious TRM in the progressive method, when cooling down, unless the ambient field is strictly maintained at a low value < 10 nT ($< 10\ \gamma$).

6.6 Magnetic fields

All experimental techniques in palaeomagnetic or mineral magnetic investigations involve the production, cancellation or detection of magnetic fields. The field strengths with which we are routinely interested vary over eleven orders of magnitude from 0.01 nT($0.01\ \gamma$), for the field associated with a weakly magnetised sample, to 2 T (20 000 Oe) for the field in the air gap of an electromagnet. So a variety of apparatus is needed to measure or produce fields of different strengths.

6.6.1 Measurement

The easiest method of measuring low magnetic fields, i.e. in the range 1 nT ($1\ \gamma$) to 0.2 mT (2 Oe), is with the fluxgate probe (Section 6.2.3). The total intensity of the Earth's magnetic field can be measured very conveniently by the proton magnetometer. Stronger magnetic fields are measured using a Hall probe. Alternating magnetic fields can be most easily measured with a search coil method.

SEARCH COIL
The simplest way to measure a steady or alternating field in air is to use a search coil and an instrument for measuring a current pulse. A search coil consists of a number, N, of turns of wire on a small former of cross-

sectional area A. When placed in a magnetic field, B_0, the flux through it is NAB_0. If the flux is changed by switching off the field, removing the coil or turning it so that it encloses no flux then an e.m.f. is produced across the coil which can be measured by a fluxmeter or a ballistic galvanometer. The flux change is NAB_0; so from a measure of the flux and the dimensions of the coil we have a measure of the field. If the coil is rotated in the field or if the coil is held stationary and the field is alternated then an alternating e.m.f. is induced in the coil. The amplitude of the alternating e.m.f. is a measure of the field intensity and is equal to $NAB_0\omega$, where ω is the frequency of alternation.

HALL EFFECT PROBE

Hall probes can be used to detect magnetic fields which vary in strength from 50 μT to 3 T (0.5 Oe to 30 000 Oe). Metals and semiconductors when carrying an electric current perpendicular to a magnetic field produce an e.m.f. in the third perpendicular direction proportional to the magnetic field. This behaviour is a Hall effect . Hall effect probes are very simple and very convenient to use.

PROTON PRECESSION

There are two methods of obtaining a measurement of the Earth's magnetic field from atomic behaviour. One is to expose a sample such as an alkali vapour to radiation and determine its resonance precession frequency with optically pumped sensors and the other is to determine the rate of free precession of a particle such as a proton. The free precession method, although not as sensitive as the resonance method, is widely used in surveying and observatory instruments as it is reliable, stable and absolute. It measures the total field intensity rather than just one component of a field. The proton precession detector head consists of a bottle of about 300 ml volume containing a liquid with a high concentration of protons (e.g. alcohol) surrounded by a coil. Before each measurement the bottle is subjected to a strong polarising field. This tends to align the protons which then precess when the field is removed. The frequency of precession, which is measured, depends only on the product of the gyromagnetic ratio of the proton (a well determined quantity) and the field intensity. The gyration of the protons dies away in a few seconds allowing field measurements to be rapidly repeated. Proton magnetometers can measure magnetic fields in the range 20–80 μT (20 000–80 000 γ) with an accuracy of 1 nT (1 γ). However, they cannot operate in field gradients above 500 nT m^{-1} (5 γ cm^{-1}). Portable, lightweight battery-operated proton magnetometers are the most common instruments used in magnetic geophysical prospection.

6.6.2 Generation

All magnetic fields are generated in some way by the magnetic effects of electric currents. In the laboratory magnetic fields up to a few mT (10 Oe) in strength are most conveniently generated by current-carrying coils. Moderate laboratory fields of up to 2.5 T (25 kOe) can be supplied by iron-cored electromagnets and modest d.c. supplies of a few kilowatts power. Higher fields have rarely been used in the investigation of natural materials but if needed, superconducting magnets (Wilson 1983) can produce persistent d.c. fields up to 15 T (150 kOe), explosive pulse methods can be used to generate 50 T (500 k Oe) fields for the order of a microsecond (Rubin and Wolff 1984), while the electromagnetic flux-compression method can yield 280T (2.8 MOe) fields for magneto-optical or cyclotron resonance studies (Miura *et al.* 1979). Portable coil systems are produced commercially which can supply very uniform, repeatable short-duration fields of strengths up to 0.8 T (8000 Oe) using a pulsing discharge system similar to a flash light and are ideal for isothermal remanence studies. Mains-driven pulse discharge units can produce fields of up to 5 T (50 kOe) or even 10 T (100 kOe) in coils cooled to liquid nitrogen temperature (McCaig 1977).

COILS

Uniform direct fields of up to 1 mT (10 Oe) can be produced without any difficulty from circular coil windings in the form of a solenoid and a d.c. power supply or battery. The field inside a solenoid is equal to $\mu_0 Ni$ where N is the number of turns per metre and i the current in amps. The field is constant anywhere inside the solenoid but falls off sharply at its ends.

Helmholtz coils are also widely used in magnetic laboratories. These consist of a pair of circular coils arranged so that the spacing between them is equal to their radius, a. This Helmholtz configuration produces a remarkably uniform field. The field has an intensity of $0.9\mu_0 Ni/a$ at the midpoint and is uniform to 10% within a sphere of radius $0.1a$. Square coils can also be used to produce a similarly uniform field. They are arranged with a spacing equal to 0.544 times the length of one of their sides.

ELECTROMAGNETS

The principal parts of an electromagnet (Fig. 6.9c) are an iron yoke, iron-cobalt alloy pole pieces and two coils carrying direct current to build up and maintain a magnetic flux in the yoke. In most electromagnets the energising coils are placed close to the pole pieces as this produces a slightly larger field. The strength of the magnetic field which can be produced between the pole pieces greatly depends on the width of the gap between them. Consequently the pole piece gap is normally kept as small as possible. Flat-faced pole pieces with a truncated cone shape (Fig. 6.9a) are commonly used to concentrate the flux and produce strong uniform fields. Specially shaped pole pieces such as those shown in Figure 6.9b can be used to produce non-uniform fields. These shaped pole pieces are used with magnetic force balances where the rather unusual condition of a uniform product of the field and the field gradient is desirable.

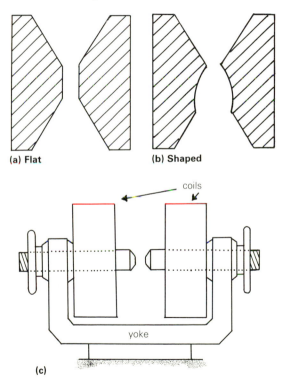

(a) Flat (b) Shaped

(c)

Figure 6.9 (a) Conical flat-faced pole tips of an electromagnet used to produce high uniform magnetic fields, for example in the vibrating sample magnetometer of Fig. 6.7. (b) Shaped pole pieces used to produce magnetic field gradients, e.g. in the magnetic force balance of Fig. 6.6. (c) Overall layout of an electromagnetic showing Y-shaped yoke and support, adjustable pole pieces and large exciting coil pair.

6.6.3 Shielding

The development of materials which can deflect magnetic fields has greatly simplified investigation of weak magnetic fields in town laboratories (Cohen 1967, 1970, Thomas 1968). Magnetic shields can screen the sensitive equipment needed for weak field measurement from spurious transient magnetic fields such as those produced by lifts and also from steady magnetic field gradients associated with stationary ferrous objects such as steel girders and pipes. Magnetic shields are made from high permeability materials for example the iron-nickel alloys of mu-metal and permalloy. The most efficient shielding is achieved by using a series of thin sheets rather than a single thick block. Triple-sheeted shields can be expected to reduce magnetic fields by a factor of 5000. The extremely high permeability of the shields can only be attained by annealing the alloy after it has been fabricated into its final shape. Straining of the alloys, for instance by a dent or a scratch, can significantly reduce their shielding ability. For this reason it is often prudent for the outer shield to be made of a somewhat lower permeability but higher strength material.

6.7 Portable instruments

A wide range of portable, battery-operated equipment has been developed for the geophysical prospector and the archaeological surveyor. One group of these portable instruments is classed as passive. Instruments in this class measure the magnitude of the Earth's magnetic field or one of its components. The proton magnetometer is the commonest instrument of this group. The fluxgate magnetometer is another common passive instrument which is often used to measure the gradient of the vertical magnetic field. The other main class of magnetic surveying instruments is made up of active equipment. Active instruments produce a magnetic field and then measure the response of the ground to the applied field. They are used to investigate near-surface features. The two main types of active instruments are those that operate on pulsed induction and on induction balance principles.

6.7.1 Magnetometers

Local anomalies in the Earth's magnetic field arise from changes in the remanent magnetisation and susceptibility of subsurface materials. The composition, shape, depth and orientation of the anomalous material all control the distortion of the Earth's magnetic field. Interpretation of total field magnetic anomalies is thus not a simple task, and often involves quite complicated mathematical modelling procedures. However, the width of a magnetic anomaly gives information about the depth of the source of the anomaly, and the magnitude gives an indication of the nature of the source. On archaeological sites large anomalies of up to 200 nT (200 γ) are associated with burnt materials which possess a remanent magnetisation, such as kilns (Aitken 1974). Smaller, but well defined anomalies, are found connected with soil features, such as silted-up pits and ditches. These are generally caused by the higher magnetic susceptibility of topsoil compared with its subsoil or bedrock and are typically between 5 and 100 nT (5 and 100 γ) in strength (see Section 8.10 and Fig. 8.7). Anomalies associated with igneous intrusions may be over 1000 nT (1000 γ) in amplitude.

When a magnetic survey is conducted, the spacing between observations and the height of the detector, above the ground surface, are adjusted to local conditions. A fairly standard arrangement, however, when looking for fairly shallow objects on an archaeological site would be a detector height of 0.3 m with a 0.5 or 1 m grid spacing.

During the time in which a survey is conducted the Earth's magnetic field will change in strength because of its diurnal variation. As these background diurnal changes are of the same order of magnitude as the strength of many anomalies, it is necessary either to operate a separate base station or to make repeated measurements at one locality. The survey readings can then be corrected for the background changes.

6.7.2   Gradiometers

The usual method of measuring gradients of the Earth's magnetic field is to measure the intensity of the field at two localities simultaneously. The gradient is given by the field difference divided by the distance between the measuring localities. Two fluxgate detectors mounted about 1 m apart on a staff form a lightweight instrument which is suitable for measuring the gradient of the vertical component of the Earth's magnetic field. A continuous digital output makes the equipment particularly suitable for rapid scanning.

Gradiometers react strongly to near-surface

features, as the lower detector responds much more vigorously to buried objects than the upper detector, while deep geological features affect both detectors equally. Consequently, the main application of portable gradiometers has been in the surveying of archaeological sites and the detection of the magnetic polarity of surface exposures of igneous rocks.

6.7.3   Pulsed induction

Portable pulsed induction instruments have been developed as very sensitive metal detectors (Colani 1966). They can detect magnetic materials as well as responding to metallic objects. Their extreme sensitivity enables them, for example, to detect differences in the thickness and magnetic content of topsoil. They operate by first transmitting a magnetic pulse into the ground for about 1 ms (Fig. 6.10). They then monitor the response of the ground to the pulse over the next few microseconds. A response is received from metallic objects because the primary magnetic

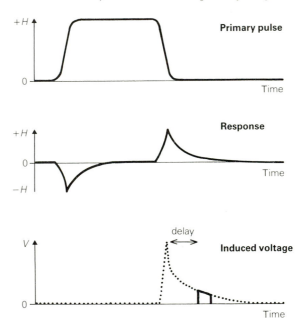

Figure 6.10 Operation of pulsed induction instrument. A primary magnetic pulse emitted by a transmitter loop induces eddy currents in metallic objects or magnetisation changes in viscous materials. These responses to the primary pulse changes increase rapidly but decay more slowly. The response initiated by the trailing edge of the primary produces an induced voltage in a receiver loop which is sampled and displayed a short time after the end of the main pulse (modified from Colani 1966).

pulse creates eddy currents in the metal. A changing secondary magnetic field is produced by the decaying eddy currents and this is picked up and integrated by the pulsed induction equipment. Superparamagnetic and viscous magnetic materials produce a similar response of a slowly decaying secondary magnetic field. Pulsed induction instruments thus respond strongly to metallic objects, but in their absence, measure magnetic viscosity.

Lightweight pulsed induction equipment consisting of a ground coil, handle and small electronics box is ideal for rapid scanning or surveying. Output is displayed on a meter mounted on the electronics box. On some soils, magnetic signals may be so strong that the coil has to be held a few centimetres above the ground level rather than directly on it. With pulsed induction equipment, as there is both a transmitted signal and a pick-up signal, sensitivity falls off as a sixth power of distance and so it is only possible to detect shallow objects no matter how strong the signal. The results of a pulsed induction survey of the magnetic content of topsoil of a Welsh lake catchment are described in Chapter 10.

6.7.4 Induction balance

The magnetic susceptibility of *in situ* natural materials can be measured using lightweight portable equipment based on the a.c. bridge method (6.3.1). This type of equipment is extensively used by amateurs interested in metal detecting (Lancaster 1966). A most useful facility of induction balance equipment is its ability to discriminate between magnetic materials and metallic objects through the phase of the out-of-balance signal. The principle of operation of the portable induction balance is the same as that of the a.c. susceptibility bridge (Section 6.3.1). The main difference between these pieces of equipment lies in the geometry of their transmitter and receiver coils. In the laboratory equipment, solenoids are used and the material under investigation is placed inside the coils. In the portable surveying equipment the material under investigation lies outside the coils. Elaborate arrangements of loops or flat coils have been designed to give both good sensitivity and precise source location in the surveying equipment.

As with all active instruments the sensing depth of the induction balance is fairly shallow. In the sandwich coil layout of 'metal detectors' the signal is dominated by material within a few centimetres of the coil system. A rather different coil design, which has been commercially developed for pipe finding, can be used to sense objects down to 1 m in depth. In this type of induction balance equipment the transmitter and receiver coils are separated by about 1 m and are mounted at right angles to each other on the ends of a shaft.

Continuous digital readout combined with the facility for metal discrimination makes the induction balance a very convenient surveying tool. Sensitive, battery-operated kits have been recently designed to serve as both laboratory and surveying equipment and to accept a remarkably wide range of detector coils and probes.

6.8 A basic environmental magnetic kit

Magnetic instrumentation is constantly being improved. Over the past few years new developments have created the situation in which portable equipment, easy to operate and maintain, has become available commercially at realistic prices. Sensitive palaeomagnetic and rock magnetic equipment can now be reliably operated by inexperienced investigators with little or no background in physics or electronics.

Chapter 16 presents a case study of the application of magnetic measurements to one catchment area. The flow diagram of Figure 16.2 illustrates the sequence in which mineral magnetic and palaeomagnetic measurements can most efficiently be applied to the bedrock, soils, river sediments and estuarine cores of such a catchment. The complete Rhode River/Chesapeake Bay study of Chapter 16 has drawn on a wide range of magnetic instrumentation, but the major part of the study has been performed with the aid of just three magnetic instruments. These are (a) an a.c. magnetic susceptibility unit (Section 6.3.1), (b) a fluxgate magnetometer (Section 6.2.3) and (c) a pulse discharge magnetiser (Section 6.6.2). These three pieces of equipment can be purchased for less than the cost of the average family car and form the basis of an 'environmental magnetic kit'.

Battery-operated, portable susceptibility equipment, in addition to its routine laboratory rôle, can also be used with suitable accessory, plug-in coils for field prospection, for whole core scanning and for frequency-dependent or quadrature susceptibility measurements. A susceptibility bridge is probably the most useful single piece of equipment for 'environmental' magnetic studies.

The fluxgate magnetometer has been refined into a highly sensitive, portable, battery-operated instrument which can be used for both palaeomagnetic natural remanence studies and for artificial remanence measurements in mineral magnetic studies. A fluxgate magnetometer costs around four times as much as a susceptibility bridge.

Portable pulse discharge magnetisers have only recently become available commercially. They routinely produce more repeatable, more accurate magnetisations than electromagnets and are simple to use and calibrate.

All three instruments in this basic kit can be easily transported. For example they can be taken as hand luggage on an aeroplane.

Looking to the future, one piece of equipment which it would be extremely useful to be able to add to this kit would be an instrument for measuring induced magnetisations. The equipment should, of course, be fully portable, very robust, totally reliable, foolproof in use and sensitive enough to determine the hysteresis characteristics of pre-nineteenth century peat samples (Section 11.5). If instrumental advances continue at the rate of the last ten years, and if applications for environmental magnetic studies continue to be found, such a dream instrument could soon become reality.

6.9 Summary

Advances in many branches of science can be linked with instrumental developments. Palaeomagnetism is an example of such a subject since the development of the astatic magnetometer system at the end of the nineteenth century permitted the earliest investigations of the natural remanence of rocks. A major step forwards took place in the 1950s when very sensitive astatic and parastatic systems were constructed around newly developed high coercivity, high remanence, permanent magnets. Computer control, especially of the fluxgate system, in the early 1970s greatly speeded the palaeomagnetic measuring process. Further increases in sensitivity and speed of measurement have recently been achieved through the production of super conducting SQUID magnetometers. Accompanying each of these instrumental developments have been new possibilities of investigating additional rock types and consequent progress in both geomagnetic and geological aspects of palaeomagnetic research.

Other instrumentation for mineral magnetic research has not shown the spectacular advances of the palaeomagnetist's magnetometer, and so measurements of mineral magnetic parameters such as coercive force, Curie temperature and saturation magnetisation remain rather specialised. Magnetic susceptibility measurements have however become much easier and more sensitive through recent electronic developments. Mineral magnetic experiments in 'environmental studies' have thus tended to centre around susceptibility bridges and the palaeomagnetist's magnetometers, and the measurements of initial susceptibility, isothermal remanence, anhysteretic remanence and remanence coercivity.

Further reading

General book

Aitken 1974. *Physics and archaeology.*

Advanced books

Collinson 1983. *Methods in rock magnetism and palaeomagnetism.*
Collinson, Creer and Runcorn 1967. *Methods in palaeomagnetism.*
Nagata 1953. *Rock magnetism.*

General journal papers

Collinson 1975. Instruments and techniques in palaeomagnetism and rock magnetism.
Banerjee 1981. Experimental methods of rock magnetism and palaeomagnetism.

[7]
Magnetic minerals and environmental systems

> Interdisciplinary research can result from two sorts of inquiry, one relating to common structures or mechanisms and the other to common methods.
>
> Jean Piaget
> *Main Trends in Inter-Disciplinary Research*

The major primary energy sources in the sun and in the Earth's interior ultimately drive all the systems of energy transformation and material flux which are the concern of environmental scientists. Whether we are studying the lithosphere, hydrosphere or atmosphere, or the cycles which involve the transfer of material between them, the movement of solid particles in some form or other is of major importance. The processes involved include entirely natural ones such as the major ocean circulation systems more or less unaffected by human activity, many others such as soil development and river erosion strongly modified by man's activity, and some, for example the particulate output associated with fossil-fuel combustion, which are entirely the result of modern technology.

7.1 Surface processes and magnetic minerals

The extent to which magnetic properties and the measurements used to characterise them are conservative within environmental systems depends on the nature of the processes to which the magnetic minerals have been subjected. Some of the most important effects are outlined below.

(a) *Chemical transformations.* Of primary interest here are those which take place as a result of weathering, soil formation and sediment diagenesis. They may lead to the conversion of paramagnetic iron to ferri- or antiferromagnetic forms. They may equally bring about transformations between magnetic mineral types or lead to the conversion of ferri- or antiferromagnetic oxide compounds to paramagnetic forms. In relation to the present parameters they may therefore be either 'constructive', 'transformative' or 'destructive'.

(b) *Physical comminution.* Physical weathering, erosion and transport by water or ice will often involve the comminution of material. Where this gives rise to reduced size, with or without changed shape in the magnetic minerals present, the magnetic parameters may be altered. Preliminary results from measurements of glacial drifts derived from homogeneous source areas suggest that they can be compared with the bedrock from which they were derived despite intervening comminution which has changed some grains from multidomain to stable and pseudo-stable single domain (Ch. 2).

(c) *Transport and deposition.* Where material is

moved with neither chemical transformation nor comminution, the main processes mediating between sediment source and the point of deposition are often sorting mechanisms which will affect both the magnetic and non-magnetic fractions in the material. In consequence, the magnetic parameters of sediments may differ in important ways from those of the source material from which they have been derived without any chemical or physical change affecting the magnetic domains themselves. Bjorck *et al.* (1982) give examples of the relationships between particle size and magnetic susceptibility in several till and sediment samples and illustrate the way in which selective depletion or enrichment of given size ranges during transport and sedimentation will affect the measurements. The distinction between *particle size* and *magnetic grain size* is crucial. Although the two can obviously never be completely independent of each other, the relationship will rarely be simple and direct. For example fine-grained haematite abounds as an important constituent of the cement coating sand grains. Equally, coarse shale particles may include stable single-domain and superparamagnetic magnetite.

(d) *Concentration and dilution.* Several processes will lead to either concentration or dilution of magnetic minerals without affecting their nature. For example primary magnetic minerals in the bedrock may be more persistent than other soil minerals during weathering and thus become concentrated in the upper part of the regolith. Conversely, the growth and accumulation of organic matter in soils, and peat and its deposition in lake sediment will dilute magnetic mineral concentrations, as will the deposition or precipitation of material such as calcium carbonate and diatom silica. Although where such processes of dilution and concentration are significant, they will give rise to variations in susceptibility and remanence values, they will not affect interparametric ratios.

7.2 Primary and secondary magnetic minerals

In practice, it is convenient to distinguish between primary and secondary magnetic minerals within the context of the regolith, the weathering zone at the Earth's surface. *Primary* magnetic minerals we consider to be those present in the parent material upon which weathering and soil formation may be taking place, irrespective of whether the parent material is igneous, metamorphic or sedimentary bedrock, or drift. *Secondary* magnetic minerals are those formed from 'primary' iron by chemical processes or biogenic effects. In most situations the distinction is reasonably clear in theory though practical differentiation may often be more difficult. There are special cases where the distinction becomes blurred or breaks down, for example where previous weathering regimes have led to the formation of iron oxides in subsoil which now forms the basis for present day soil development (as in some weathered basalts) or where soil development is taking place on recent alluvium derived from both bedrock and topsoil sources.

7.3 Magnetic minerals and material flux

Iron compounds are among the most ubiquitous components of natural materials, forming some 2% of the Earth's crust. On the longest geological timescales the cycles and patterns of mountain building, denudation, geosyncline development and crustal displacement all involve large-scale movements of materials which include forms of magnetic iron oxide. Geological processes, such as sea-floor spreading, the eruption and flow of lavas, the dispersal of volcanic ashes and the transport of eroded sediments under the influences of water and wind, control the magnetic mineralogy of natural materials, sometimes by forming or transforming magnetic oxides through the influence of heat, hydration/dehydration or changes in E_H and pH, sometimes by merely transporting stable magnetic oxides to environments where they may persist unaltered for long periods in sediments and sedimentary rocks. The magnetic oxides in the crust together with paramagnetic forms of iron (which are potentially convertible to magnetic oxides by thermal or chemical action, both natural and anthropogenic) form the main primary source of the minerals considered in the present account (Fig. 7.1). The second primary source, the cosmic flux of extra-terrestrial magnetic particles, is significant only in the immediate vicinity of meteor impact sites and in certain marine environments where the flux from lithospheric sources is minimal, e.g. the central

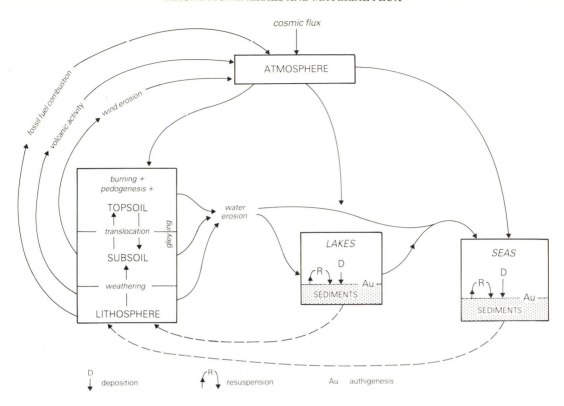

Figure 7.1 Schematic diagram of the cycle of magnetic minerals.

Pacific. Chapter 3 is concerned with the various magnetic minerals present in primary crustal and extraterrestrial materials.

As rock surfaces are exposed to atmospheric processes, to colonisation by plants and to resultant weathering and soil formation, the iron compounds present in bedrock may be subject to many processes of concentration, dilution and transformation. Secondary magnetic oxides may even be formed near the soil surface. Soil iron compounds are among the most abundant and most sensitive components of the soil system. Where chemically stable magnetic oxides are the end product of weathering and pedogenic processes, they may be diagnostic of soil horizon type or of weathering régime. The processes involved in the formation of secondary magnetic oxides and their effects in the soil are a major theme of Chapter 8.

The secondary magnetic oxides formed at or near the soil surface often differ in crystal form and size from the primary magnetic oxides present in the underlying substrate. Both primary and secondary magnetic minerals are eroded from soils and substrates, and may become incorporated in river and lake sediments. Differences between primary and secondary magnetic minerals can thus form the basis for sediment source identification in rivers, lakes and estuaries (Chs 9, 10 & 15).

Primary iron compounds are transformed to magnetic oxides, not only by weathering and soil formation, but also through the combustion of fossil fuel by man. The burning of solid fuels in particular, whether in domestic or industrial appliances, generates large volumes of magnetic spherules which are an important component of fly-ash from solid fuel fired power stations. Other industrial processes such as metal smelting and steel manufacture are of major significance in the discharge of magnetic particulates into the environment. The magnetic characteristics of virtually every type of particulate emission from industrial and domestic fossil-fuel combustion as well as many other industrial processes, provides hitherto little used scope for the magnetic monitoring of particulate pollution both atmospheric and marine (see Chs 11 & 12).

7.4 Natural remanence and mineral magnetic properties

In the following chapters we make use of two different but interrelated types of magnetic parameters. In Chapters 13 and 14 emphasis is placed on remanent magnetisation acquired in the Earth's magnetic field. Such remanence is called natural remanent magnetisation (NRM) whether formed as a result of cooling through the blocking temperature, crystal growth through the blocking volume or the deposition and 'fixing' of detrital particles. Chapters 8–12 are concerned with mineral magnetic properties. These are the magnetic characteristics of a substance which are an expression of the intrinsic magnetic properties of the constituent magnetic crystals and of their dispersion as fine particles within the substance. Such characteristics are generally measured by monitoring the signal induced by an instrumentally generated field, as in the case of magnetic susceptibility and magnetic hysteresis measurements. They may also involve monitoring the change in induced signal with temperature as in Curie temperature measurement.

Palaeomagnetic measurements of natural remanent magnetisation are important for the insight they give into the behaviour of the Earth's magnetic field through time. It is out of the results of natural remanence-based studies that geologists and geophysicists have established the magnetic polarity timescales of such enormous significance in the present theories of plate tectonics and sea-floor spreading. They are also a prime source of empirical insight into the nature of the dynamics of the Earth's fluid core.

In all palaeomagnetic measurements, the primary aim is to recapture the directional and intensity characteristics of the Earth's magnetic field for a given period and location. The signal retained in rocks or sediments will however reflect an assemblage of conditions including magnetic mineral types and concentrations, mechanisms of formation, depositional and post-depositional history and so forth. In order to compensate for these so that as much as possible may be learned about the primary concern, the Earth's magnetic field, many supplementary palaeomagnetic and mineral magnetic measurements

may be required in order to isolate and to some degree 'standardise' the palaeomagnetic signal.

From the above, we can see a distinction between magnetic properties which reflect ordering consequent on the existence of the Earth's magnetic field – natural remanent magnetisation – and magnetic properties which are solely a consequence of the magnetic crystal structures, grain sizes and shapes present. Although in practice the various magnetic properties can be closely connected as explained below. The five parts of Figure 7.2 diagramatically illustrate the palaeomagnetic and mineral magnetic characteristics of some different natural substances.

(a) The abundance of magnetic minerals in the Earth's crust varies from rock to rock. In some rocks the remanence of the magnetic minerals has been aligned with that of the Earth's magnetic field and can provide palaeomagnetic data.

(b) Magnetic crystals are also found in the atmosphere. Dust particles containing magnetic minerals may have a magnetic remanence. However, the natural remanences of the various particles in any bulk atmospheric samples we investigated will not be aligned and so cannot be detected in such assemblages.

(c) Atmospheric fallout and erosion and weathering of the Earth's crust contribute magnetic minerals to river systems. As with the atmospheric dusts, mineral magnetic studies can be performed on river sediments in order to investigate the concentration and types of assemblages of magnetic minerals contained in the sediments. Investigations of natural remanences again cannot be used because of the lack of alignment. Laboratory remanences, such as isothermal remanence are, however, readily usable in mineral magnetic studies of both river and atmospheric samples.

(d) Recent lake and marine sediments are also derived from crustal erosion and atmospheric fallout. Their constituent magnetic particles may have natural remanences   field during sediment formation. The alignment process may have been more efficient in some horizons producing a higher intensity of natural remanence (B versus G). In some horizons the remanence of the magnetic particles may not have been aligned (C and E). In other horizons

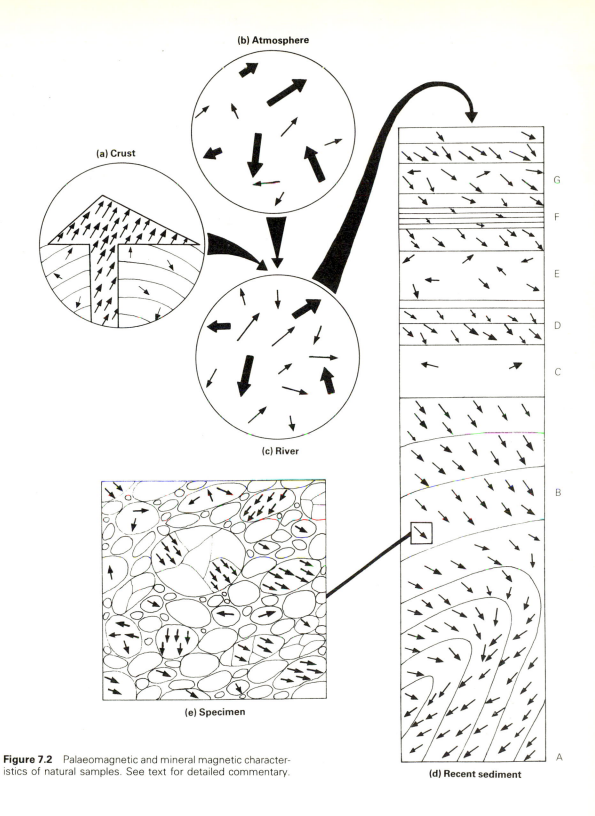

(a) Crust

(b) Atmosphere

(c) River

(e) Specimen

(d) Recent sediment

G
F
E
D
C
B
A

Figure 7.2 Palaeomagnetic and mineral magnetic characteristics of natural samples. See text for detailed commentary.

the remanence of the magnetic particles may be aligned but its direction may not coincide with that of the Earth's magnetic field (A). The types and concentrations of magnetic minerals may also vary from horizon to horizon (A to G).

(e) Variations in magnetic properties are also found on a microscopic scale. For example a 10 ml specimen from a recent sediment core will be composed of a range of particle sizes and mineral types. Some of the particles may be magnetic. The composition and natural remanence of these particles will vary and the remanence may be aligned to a certain extent so the specimen as a whole has a natural remanence. This picture may be further complicated by some of the magnetic particles themselves being subdivided into magnetic and non-magnetic crystals and by the magnetic crystals being even further subdivided into magnetic domains.

Although, we will be mainly interested in variations in the magnetic properties of bulk samples on a macroscopic scale, we need to bear in mind the microscopic complexity when making environmental interpretations from mineral magnetic and palaeomagnetic measurements. Also, for our purposes it is useful to retain a clear distinction between palaeomagnetic studies and measurements (involved with the history of the Earth's magnetic field), and mineral magnetic studies and measurements (related to the intrinsic properties of magnetic minerals).

7.5 Sampling and measurement

Magnetic measurements demand their own approach to field sampling, and to sample storage, preparation and treatment. In the case of natural remanence measurements of sediments the main requirements are:

(a) *Orientation.* A minimum requirement for inclination measurements alone is knowledge of which way is up. For relative declination measurements, constant azimuth must be established. For absolute declination, true azimuth must be known for each sample.

(b) *Preservation of structure.* Samples should not be dried out or frozen at any stage between sampling and measurement. This applies equally well to whole cores and to subsamples which should always be stored in a damp environment, e.g. in sealed plastic containers. The size of sample used will depend on the type of magnetometer available. 10 ml plastic cubes are ideal for use with fluxgate and parastatic magnetometers for 1″ cylinder rock samples. Some superconducting magnetometers (Section 6.2.4) will only accept smaller samples.

(c) *Isolation from stray magnetic fields.* Storage should be in magnetically screened environments wherever possible. This has the double advantage of isolating samples from contemporary fields which may induce artificial viscous remanences (Section 4.3.2) in the sample and establishing the samples in an environment where gradual decay of previously acquired viscous remanences will take place.

Samples for natural remanence measurement can be taken from free faces or from core tubes either split length ways or extruded from the base and many kinds of core samples provide material suitable for palaeomagnetic study. Alternative types of corer are described in a wide range of recent articles and reviews of which those by Hakanson and Jansson (1983), Goudie (1981), Barber (1976), Mackereth (1958, 1959), Davis and Doyle (1969) and Digerfeldt (1978) are especially useful.

Mineral magnetic measurements can be carried out on a very wide range of materials. Surfaces can be measured by magnetic susceptibility search loops and probes (Section 6.3.1). Here the main problem is ensuring constant geometry between surface and sensor for every location measured. Signals decline with distance following a power law, so spurious variations are easy to produce through careless field practice. Downhole susceptibility probe readings can be made using appropriate sensors from the Bartington Instruments range. Cores of rock, soil or sediment can be scanned for volume susceptibility provided, where necessary, they are retained in clean non-metallic core tubes or can be extruded into non-metallic liners without distortion and consequent volume variation (Section 10.3). A very wide range of mineral magnetic measurements can be carried out on subsamples of fresh, moist material provided it is not too loose or sloppy to retain physical coherence and magnetic moment after magnetisation or whilst spinning. Mass specific measurements on dried material pose no problems provided the material has not been heated above *c.* 60 °C and the sample is packed so as to immobilise all particles. Filter paper residues are easily measured either by placing them in 10 ml plastic sample pots or laying them on special templates (Section 11.7). The special and subtle nature of magnetic contamination poses unexpected problems especially in artificial remanence measurements where the sample holder, filter paper or packing material is inevitably magnetised along with the sample. When dealing with weakly magnetic samples, all holders and packing materials should be screened by being washed then pre-magnetised and measured empty. Unsuitable pots and packing can then be rejected. Glass fibre filter paper is best avoided (Section 11.7) but if it is used, blanks must be pre-measured and values for them subtracted. With very weakly magnetic peat, the magnetic signal-to-noise ratio can be greatly improved simply by compression. Finally, the diamagnetic properties of sample holders and packing materials must be taken into account in susceptibility measurements of weak samples.

Although magnetic measurements are thus extraordinarily versatile the wide range of environments and applications considered and the rather divergent purposes and requirements of natural remanence as against mineral magnetic measurements, have led us to make special reference to sampling and measuring procedures where appropriate in each chapter.

7.6 Summary

It follows from the processes and relationships briefly described in this chapter that magnetic measurements are potentially of great environmental interest. The contribution of natural remanence measurements establishing chronologies of sedimentation may be vital to palaeoenvironmental studies in both marine and lacustrine environments. Coupled with rapid magnetically based core correlations, they greatly enhance our capacity for more reliable quantitative studies of material flux. In lake watershed ecosystems this approach is important especially where either the terrestrial or aquatic ecosystems are mineral-limited, or where changes in land use and vegetation have been associated with major shifts in material flux. Mineral magnetic parameters are valuable stratigraphic tools in palaeoenvironmental and palaeolimnological studies since they can be used alongside more time-consuming and often destructive analyses to give additional insights into sediment types and sources. At the same time, since the combustion-derived anthropogenic forms of magnetic minerals are often associated with processes releasing substances which are ecologically damaging to the environment (e.g. SO_2, and many heavy metals), magnetic monitoring of particulate pollution may also have important ecological implications.

[8]
Soil magnetism

Even at low concentrations in a soil, iron oxides have a high pigmenting power and determine the color of many soils. Thus soil color, as determined by the type and distribution of iron oxides within a profile, is helpful in explaining soil genesis and is also an important criterion for naming and classifying soils.

Schwertmann, V. and Taylor, R. 1977

8.1 Introduction

Although the magnetic properties of soils are seldom if ever quoted in standard texts and few articles on soil magnetism have looked beyond low field susceptibility measurements, a number of authors have pointed to ways in which magnetic measurements can be used in pedology and related fields. Following Le Borgne (1955), several Western and Russian authors have studied the mechanisms whereby magnetic susceptibility (4.4) is often 'enhanced' in surface layers. Lukshin *et al.* (1968) and Vadyunina and Babanin (1972) have shown how susceptibility enhancement is related to major soil formations in the USSR and can be used to give some general insight into the processes affecting iron minerals during pedogenesis, as well as into specific effects such as gleying. The relationships between enhancement mechanisms and lithology (Mullins & Tite 1973), climate (Tite & Linington 1975) and fire (Le Borgne 1960), have received particular attention often through both field observation and experimental study, while recent articles have dealt more directly with the geochemistry of the iron transformations which lead to ferrimagnetic (Section 2.2.4) oxide formation (e.g. Taylor & Schwertmann 1974). Most of the work dealing with instrumentation for soil magnetic measurement has been addressed more or less directly to the use of susceptibility measurements in archaeological prospecting and survey work (e.g. Scollar 1965). The most useful and comprehensive summary of this range of work is the review article by Mullins (1977). In addition, Poutiers' monograph (1975) illustrates the use of mineral magnetic measurements, in loess and palaeosol studies. More recently, Maher (1984, 1986) has begun to explore the mineral magnetic properties of both contemporary and fossil soils with a view to relating them to soil forming processes.

The problems involved in producing a single interpretation consistent with both magnetic and geochemical measurements, brought into focus for marine sediments by Henshaw and Merrill (1980), are probably even less tractable in soils than in marine sediments, and there is a dearth of authoritative work in this critical area. Nevertheless, some characterisation of the magnetic properties of soils is of vital importance to most of the subsequent chapters, since it is within the regolith that the iron released in the weathering of bedrock is transformed to chemically stable magnetic oxides which may then persist in the soil, in the suspended load of rivers, in atmospheric dusts and in the historical record preserved in sediment, peat and ice cores.

8.2 Magnetic properties of soil minerals

The magnetic properties of a bulk soil sample reflect the varied magnetic behaviour of the range of soil minerals present. Studies dealing only with low field susceptibility have often given the impression that ferrimagnetic (Section 2.2.4) minerals alone determine bulk magnetic properties. Although this is often true, there are nevertheless many situations in which it is misleading. Thus at the outset, we need to consider briefly the magnetic behaviour of major soil constituents (cf. Chs 2 & 4). The diamagnetic (Section 2.2.1) components of the soil include quartz, orthoclase, calcium carbonate, organic matter and water. In most soils, these components can be regarded merely as a dilutant. Only in extreme cases, for example pure silica sands, pure limestones and ombrotrophic peats, will the diamagnetic component be magnetically significant. In these cases it will need to be recognised as an important component of any susceptibility measurement. Many soil minerals, both primary and secondary, are paramagnetic (Section 2.2.2) and in soils which are iron rich but poor in ferrimagnetic minerals, paramagnetism will make an important contribution to total susceptibility. Table 3.4 lists the magnetic susceptibility of some common diamagnetic and paramagnetic soil minerals. The relatively high magnetic susceptibility of iron-rich clay minerals is noteworthy.

Several canted antiferromagnetic (Section 2.2.4) minerals are present in the soil. Of these, goethite (Section 3.4) is the most abundant in well drained soils formed under temperate conditions and haematite (Section 3.2.2) is predominant in relatively drier and more highly oxidised situations. Most soils contain one or the other (Oades & Townsend 1963). Schwertmann and Taylor (1977) suggest that goethite is the more generally distributed and least climatically restricted of the iron oxides and hydroxides in the soil. They confirm that, even in those environments sufficiently oxidising for haematite formation, goethite will often also be present. Lepidocrocite (Section 3.4) is more restricted in its occurrence and is largely confined to gleyed soils where it occurs as bright orange mottles and coatings lining the walls of root channels.

Of the ferrimagnetic oxides (Section 3.2.1) only magnetite and maghaemite are generally important in the soil, though titanomagnetites and pyrrhotite may be significant on some lithologies. Magnetite will occur both as a primary mineral, derived from igneous and especially basic igneous rocks (Section 3.6.1) and as a secondary mineral formed within the soil by the mechanisms outlined in Section 8.4 below. Maghaemite is a secondary soil mineral formed in a similar way. Though widespread in temperate soils, it tends to be more abundant in highly weathered soils formed under tropical and subtropical conditions (Schwertmann & Taylor 1977). It can occur finely dispersed or as concretions.

8.3 Weathering and magnetic properties

Iron in igneous bedrock is largely present in the reduced state as Fe^{2+} within the silicates. Hydrolytic and oxidative weathering reactions release this as Fe^{3+} which, as a result of its extremely low solubility, is mostly precipitated as an oxide or hydroxide. However, oxygen-deficient conditions may subsequently reduce the Fe^{3+} to Fe^{2+}. As a result, the oxide becomes more soluble and the iron may then eventually migrate to zones of oxidation where it will be reprecipitated. Thus iron oxides can undergo repeated changes in which they are alternately precipitated in oxidising conditions and are chelated or in solution as Fe^{2+}, the overall conditions under which these reactions take place being set by soil pH, climate, soil moisture and organic matter content. In consequence, it is not possible to make simple generalisations about the position of iron oxides in weathering sequences. Some of the major trends in soil oxide formation are outlined below.

Where soils are developed from igneous rocks containing predominantly Fe^{2+} in magnetite and/or within the silicates, weathering is dominated by the hydrolytic and oxidative reactions already noted, and the resulting oxides will be mainly haematite and/or goethite. Haematite is largely restricted to those soils formed under conditions of high temperatures, good aeration, rapid decomposition of organic matter, and, frequently, relatively high pH (Schwertmann & Taylor 1977). Subsequent transformation can take place in the soil surface layers in the presence of organic matter (see 8.4 below).

Where soils are developed on a haematite-rich bedrock (e.g. Triassic sandstones) long periods of pedogenesis under cooler and moister régimes will lead to goethite or more locally, lepidocrocite formation. This occurs through dissolution of the haematite by reduction or chelation and subsequent reprecipitation when or where conditions are more oxidising. Many soils form on sedimentary or metamorphic parent materials in which haematite, magnetite and the primary Fe^{2+}-rich silicates make relatively little or no contribution to the total iron content. In these cases the iron may be present in a wide variety of forms. Such iron will often be finely disseminated and associated with the clay minerals or present as a cementing agent between coarser particles. Generally these forms of iron are finely divided, easily accessible and hence relatively readily mobilised by chelation or reduction. The mechanisms of magnetic enhancement considered in Section 8.4 below will, in these circumstances, operate on many forms of iron depending not only on lithology and major climatic type, but on all the other factors of soil formation which are so sensitively reflected in the soil iron system.

8.4 The magnetic enhancement of surface soils

Le Borgne (1955) was able to show, and many subsequent studies have confirmed, that the magnetic susceptibility of topsoil is often higher than that of the underlying material. Le Borgne ascribed this to the formation of secondary ferrimagnetic oxides within the clay-size fraction of the soil. The processes contributing to this may be considered under the general heading of magnetic 'enhancement'. Common to all the processes is the conversion of iron from non-ferrimagnetic to ferrimagnetic forms. Although this can occur as a natural process of chemical weathering, it is convenient to consider it separately.

Many authors have illustrated magnetic enhancement over a wide range, of lithologies and climatic régimes. It appears to be a characteristic of many soil

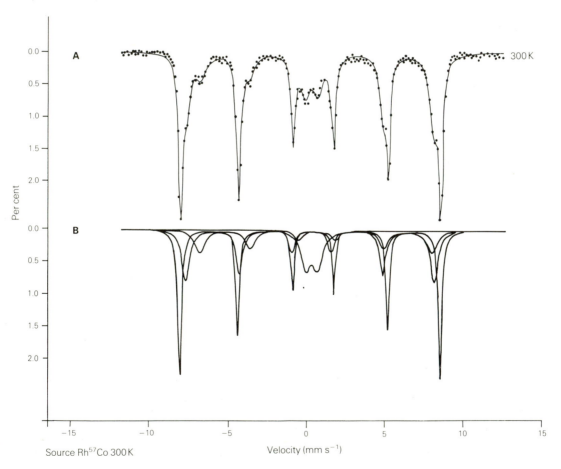

Figure 8.1 Mossbauer spectra at room temperatures of a sample of burnt soil from Caldy Hill, Merseyside, England. Plot A shows a total least squares fit to all the data and plot B subsidiary fits for individual components (see Longworth *et al.* 1979).

types under temperate conditions, although the reflection of enhancement in bulk susceptibility measurements may not always be apparent, as for example on strongly ferrimagnetic bedrocks such as basalts. It is inhibited or reversed by waterlogging and podsolisation and it may be difficult to detect in soils with low iron concentrations in the topsoil. The evidence of enhancement may be removed by truncation of the soil profile. Climates of extreme aridity or cold are also inimical to enhancement.

Mullins (1977) identifies four ways in which maghaemite, which he takes to be the product of enhancement, can be formed in the soil. The first of these maghaemite formation mechanisms, the low temperature oxidation of magnetite, is not strictly an enhancement process since no significant increase of susceptibility is likely from this alone.

Mullins' second mechanism, burning, is generally accepted as a major factor in magnetic enhancement. A wide range of observations and experimental studies show that above c. 200 °C, in the presence of organic matter, some conversion of non-ferrimagnetic iron minerals to ferrimagnetic minerals will occur. Le Borgne envisaged a two-stage process in which first, under reducing conditions, finely divided oxides and hydroxides are converted to magnetite, which is then subsequently oxidised to maghaemite on cooling under the less reducing conditions which may follow after the combustion of the soil organic matter. In practice, the processes and the final product are much more variable. Although maghaemite is undoubtedly produced by burning (Longworth & Tite 1977) non-stoichiometric magnetite may also be formed. For example Longworth et al. (1979), using Curie temperature determinations and Mossbauer spectra were able to show that at two burnt sites in Britain, Caldy Hill and Llyn Bychan, the resulting ferrimagnetic oxide was non-stoichiometric and probably impure magnetite, best approximated by the formula $Fe_{2.9}O_4$ (cf. Fig. 8.1 and also Ozdemir & Banerjee 1982).

It is likely that the fire-induced magnetic oxides formed within the soil will vary from site to site, approximating more closely to maghaemite or to magnetite depending on site conditions and the nature of the fire experienced. The finely divided nature of these oxides, their non-stoichiometric form and the frequency of isomorphous substitution within the crystal lattices makes identification complex and time consuming. Detailed studies of artificially burnt materials (Oldfield et al. 1981a) suggest that even in a single sample a range of oxides will often be present at each stage. Visually, burnt topsoil may range in colour from black through shades of grey and pink to salmon pink or bright orange. This visual gradation is often accompanied by an increase in the level of enhancement, together with a shift from 'soft' multidomain (Section 2.4.3) 'magnetite' to higher concentrations of viscous (Section 2.5) and superparamagnetic (Section 2.4.5) 'magnetite' alongside a growing haematite component.

The third process quoted by Mullins is the dehydration of lepidocrocite (γ FeOOH) to maghaemite. This takes place between 275 and 410 °C and, since lepidocrocite is limited to poorly drained soils, the mechanism is likely to be restricted to local situations in which gleyed soils are drained or subject to high temperatures.

Mullins' final mechanism is probably the most generally important. He envisages the formation of microcrystalline maghaemite or magnetite from weakly magnetic iron oxides and hydroxides via the reduction–oxidation cycles which occur under normal pedogenic conditions. The processes involved in this type of enhancement are complex and poorly understood. Organic matter is required as a substrate for the heterotrophic microorganisms which provide the reducing conditions and chelating agencies needed to bring into solution the iron formerly present in non-ferrimagnetic oxides and hydroxides. The extent to which subsequent recrystallisation as ferrimagnetic oxides is a purely chemical process or one which may be dependent on microbial action is still open to doubt. One possibility is that soil bacteria, analogous to the magnetotactic (Section 15.2) types shown by Blakemore (1975) to grow small chains of stable single-domain magnetite crystals within their cells, are responsible for some of the enhancement observed as a normal part of pedogenesis.

Measurements by Mullins (1974) and by Mullins and Tite (1973) largely carried out on cultivated soils developed on sedimentary deposits suggest that the secondary ferrimagnetic crystals so formed are a mixture of superparamagnetic, viscous and stable single-domain (Sections 2.4.4–6) types in roughly constant proportions, with the viscous component generally comprising between 5 and 10%.

In many situations, especially in areas of past or present cultivation and in fire-stressed ecosystems, it is difficult to tell to what extent surface enhancement is a reflection of burning or of the 'pedogenic' mechanism. However, the relative importance of the latter may be indicated by the high enhancement of many old forest soils and by the evenness of the enhancement level over wide areas of comparable lithology. No systematic attempt has been made to use magnetic measurements themselves as a basis for distinguishing different types of enhancement though this is clearly possible under favourable circumstances.

At sites close to major urban and industrial centres, surface soils may have a higher susceptibility as a result of fallout from the atmosphere of magnetic spherules derived from fossil-fuel combustion. In the industrial areas of northern England, the susceptibility values for recent ombrotrophic peats are within the same range as strongly enhanced forest soils in rural areas. This atmospheric component has not been recognised in the literature on soil magnetism, but it must be taken into account in site selection for studies

of true enhancement processes and products, and in any evaluation of soil magnetic properties close to major sources of spherules (cf. Ch. 11).

8.5 Particle size relationships

The ferrimagnetic crystals in soils can be seen, from the above account, to derive from both primary and secondary ('enhanced') iron minerals. The latter are most often of stable single-domain size or less and associated with the clay fraction, whereas the former are, depending on the particular lithology, usually associated with sand and coarse silt-size fractions (cf. Ch. 10). The presence of both primary and secondary ferrimagnetic minerals will often give rise to a bimodal distribution of specific susceptibility with respect to

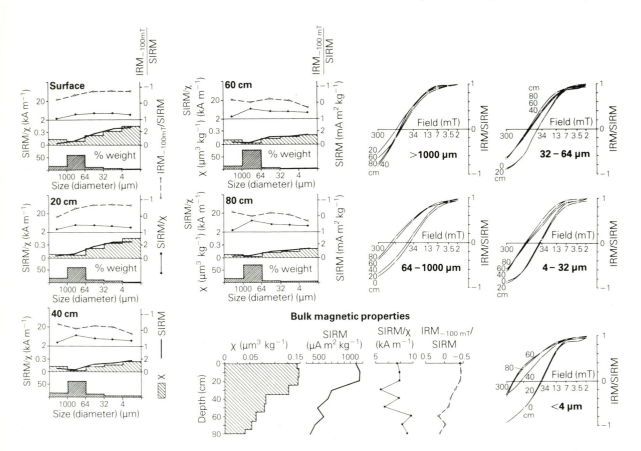

Figure 8.2 Newton Mere – Soil Pit B. Mineral magnetic parameters for bulk and particle size fractioned samples from a cultivated brown earth soil developed on haematite-rich glacial drift (Smith, unpub.). Above and to the left of the plot of bulk parameters, each diagram shows the variations in χ, SIRM, SIRM/χ and IRM$_{-100mT}$/SIRM against particle size. To the right, coercivity of SIRM curves are plotted for each particle size range and sample depth (see text).

particle size. In soils which are rich in non-ferrimagnetic forms of iron, this bimodal distribution will be superimposed on the effects of paramagnetic iron compounds and antiferromagnetic oxides (commonly including goethite) which may be associated with a wide range of particle sizes. These non-ferrimagnetic iron minerals will also include forms which are often more easily soluble than the ferrimagnetic minerals and which, along with adsorbed and dissolved iron, will be among the more readily chemically mobilised components of both soil and sediment systems.

Figure 8.2 shows an example of the relationship between magnetic properties and particle size for samples taken from a soil pit dug within the catchment of Newton Mere in central England (Smith pers. comm.). The soil is a cultivated and well drained sandy loam developed on glacial drift derived almost entirely from surrounding Triassic sandstones and red marls all rich in haematite. Igneous and meta-morphic erratics occur sparsely in the drift. The primary magnetic components are therefore largely antiferromagnetic with ferrimagnetic forms much less abundant. The bulk soil magnetic properties show marked enhancement especially within the top 30 cm, and atmospheric fallout may be partly responsible for this (cf. Chs 10 & 11). The ratio of saturation remanent magnetisation to susceptibility, $SIRM/\chi$ (see Section 4.6) peaks in the upper and lower parts of the profile. Above 30 cm, in the enhanced layers, the high $SIRM/\chi$ values are associated with high χ and strongly negative $IRM_{-100\,mT}/SIRM$ (see Section 4.6.4) values and may be ascribed to the relatively high proportion of stable single-domain grains associated with the secondary ferrimagnetic component. Below 60 cm maximum $SIRM/\chi$ is associated with minimum χ and less strongly negative $IRM_{-100\,mT}/SIRM$ indicating that at this depth below the enhanced layer, the ratio has increased as a reflection of the greater relative importance of the primary haematite, abundant in the parent material.

The magnetic measurements on the size fractions may be summarised as follows:

(a) At all depths, χ values are weakly bimodal with the peak in the coarsest material always much less important than that in the fines. The peak in the clay fraction is most strongly developed in the topmost samples within the enhanced layer.

(b) $SIRM/\chi$ peaks in the 63 μm–1 mm range in all but the surface sample. In all samples the coarsest fraction has minimum $SIRM/\chi$ values and the clay fractions have lower $SIRM/\chi$ than do the silts and fine sands.

(c) $IRM_{-100\,mT}/SIRM$ values and remanent coercivi-ties $(B_0)_{CR}$ vary little down profile in the coarsest size fraction. Above 20 cm, within the enhanced layer, $IRM_{-100\,mT}/SIRM$ values become more strongly negative with increasing particle size whereas below this, the reverse applies. The peak $SIRM/\chi$ in the 63 μm–1 mm range corresponds in each case with a less strongly negative $IRM_{-100\,mT}/SIRM$ value and higher values of $(B_0)_{CR}$ (see 4.6.3).

These variations with particle size and depth may be tentatively interpreted in terms of four magnetic components within the soil system as follows. The *primary ferrimagnetic* component is restricted largely to the coarsest fraction where it occurs as erratics in the drift. Its presence to some degree at all depths gives rise to the slight increase in χ in material coarser than 1 mm and to the fairly constant $(B_0)_{CR}$ and $IRM_{-100\,mT}/SIRM$ values observed in the same fractions. The $SIRM/\chi$ minimum in the coarsest fraction of each sample reflects a proportionally large multidomain contribution. The greater relative importance of the *antiferromagnetic* component (fine-grained haematite) is responsible for less strongly negative $IRM_{-100\,mT}/SIRM$ values and higher $(B_0)_{CR}$ with depth and, below the enhanced layer, with decreasing particle size. It also gives rise to maximum $SIRM/\chi$ in the 63 μm–1 mm range from 20–80 cm, but as particle size decrease below 63 μm within these samples, the *paramagnetic* component, becomes increasingly important producing a decline in $SIRM/\chi$ but having no effect on $(B_0)_{CR}$ and $IRM_{-100\,mT}/SIRM$. The *secondary ferrimagnetic* component dominates bulk soil and SIRM within the enhanced layer and is responsible for the peaks in the clay fraction at all depths down profile and probably reflect lessivage in the sandy soil. The picture is there-fore a highly ordered one within which the magnetic mineral assemblages shows the effect of soil forming processes on the parent material.

Although the bulk soil measurements give a less detailed picture than do those performed on particle size fractions they clearly reflect the main characteristics attributable to the nature of the parent material and the relative homogeneity of the cultivated and enhanced layer. The nature of bulk soil magnetic properties under different pedogenic régimes and on different lithologies is the subject of the next section.

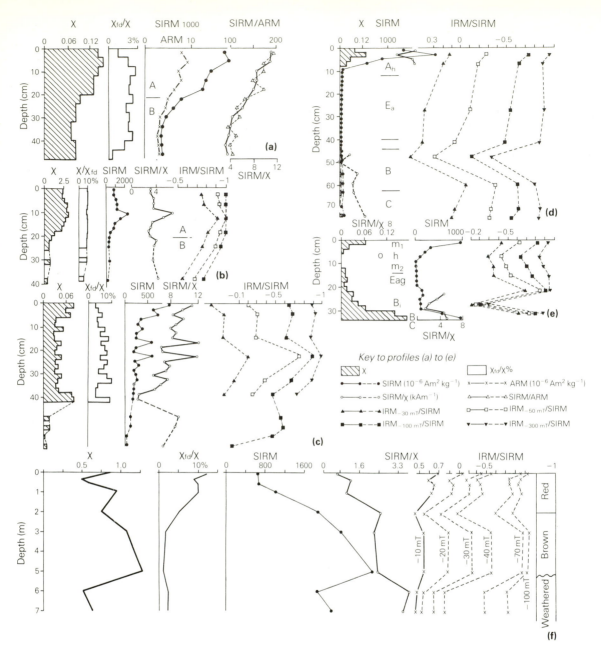

Figure 8.3 Magnetic measurements from some representative soil profiles (see text). (a) Hardwick Wood – brown earth under mature deciduous forest (Maher 1983); (b) Exmoor – unreclaimed brown earth under heathland (Yates 1983); (c) New Forest–brown earth under heathland (Yates 1983); (d) New Forest – podsol under heathland (Yates 1983); (e) Exmoor – stagnopodsol under heathland (Yates 1983); (f) North Queensland – deeply weathered tropical soil on basalt.

8.6   Some representative soil profiles

Figure 8.3 shows the magnetic properties of a series of soil profiles chosen to illustrate some of the ways in which lithology, weathering régimes and pedogenic processes combine to control the magnetic properties of soils. The profiles chosen comprise two podsols developed on tertiary Barton Sands and on middle-Devonian slates, three brown earths developed on the first two lithologies and also on cretaceous clay, and a deeply weathered humid tropical soil developed on basalt. The parent lithologies thus vary from weakly paramagnetic to strongly ferrimagnetic and a single tropical soil is included for comparison with soils developed under cool temperate conditions. The paired podsol and brown earth profiles allow comparison between weakly mixed but strongly eluviated profiles and relatively well mixed profiles which are not so strongly eluviated.

The podsol profiles (d and e), irrespective of underlying parent material, have the following features in common:

(a)   Peaks in χ and SIRM indicate shallow layers of low enhancement at or near the soil surface, within the F and H layers between the fresh undecomposed litter and the bleached underlying mineral soil of the eluviated A horizon. Where measured, SIRM/χ and IRM$_{-100\,mT}$/SIRM values are within the range typical of secondary 'magnetite'. Any or all the enhancement mechanisms, including atmospheric deposition, may be significant (Yates 1983).

(b)   Significant mineral magnetic changes coincide with the zone of iron enrichment in the illuviated B horizon. The combination of relatively low SIRM/χ, but extremely hard IRM$_{-100\,mT}$/SIRM values, especially in the Exmoor iron pan, indicates a magnetic assemblage dominated by paramagnetic forms of iron with some antiferromagnetic crystals present.

(c)   In the bleached eluviated mineral soil, χ and SIRM reach minimum values and χ is often too low to measure. As in the gleyed profiles (8.7 and Fig. 8.5) some dissolution of ferrimagnetic oxides is implied. The IRM$_{-100\,mT}$/SIRM values resemble those in the overlying organic soil, suggesting that clay translocation may be responsible for the few magnetic minerals present.

(d)   Below the B horizon the magnetic properties resemble those of the underlying parent material.

The brown earths (a, b and c) all show χ and SIRM peaks in the physically mixed A horizon where decomposing organic matter and mineral soil are in intimate contact. SIRM/χ, and the back-IRM parameters (IRM/SIRM) vary from profile to profile though they remain within the range of values compatible with dominance by stable single-domain magnetite. Where the frequency-dependent (cf. quadrature) susceptibility ratio, χ_{fd}/χ (see Section 6.3.4), has been measured it is seen to be a significant percentage of total susceptibility throughout each profile down to depths of at least, 40 cm, especially in the well drained New Forest (c) and Exmoor (b) examples, presumably as a result of the thorough mixing of the A horizons by worms. Also, in these two profiles, there are considerable variations in SIRM/χ and IRM/SIRM with depth. The horizons of maximum SIRM and 'softest' IRM/SIRM corresponding with zones of greater clay content. In the Hardwick Wood profile (a), which is developed on heavy clay, anhysteretic remanent magnetisation, ARM (4.4.2), SIRM, and SIRM/χ gently decline with depth in a manner consistent with a gradual reduction in the relative importance of the stable single-domain magnetite component (cf. Ozdemir & Banerjee 1982). The 'secondary' viscous grains at the stable single domain/superparamagnetic border which are responsible for the frequency-dependent 'quadrature' component measured here will be comparable to those grains around 20 nm identified by their higher viscous remanent magnetisation (VRM) in the upper part of the cultivated Barnes soil profile from Minnesota shown in Figure 8.4.

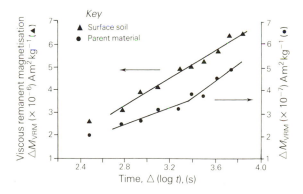

Figure 8.4   Growth of viscous remanence over time in a surface sample and a sample from the parent material of the Barnes soil formation in Minnesota (Ozdemir & Banerjee 1982). The relatively high viscous remanence in the surface soil is attributable to fine grains ~ 0.02 μm.

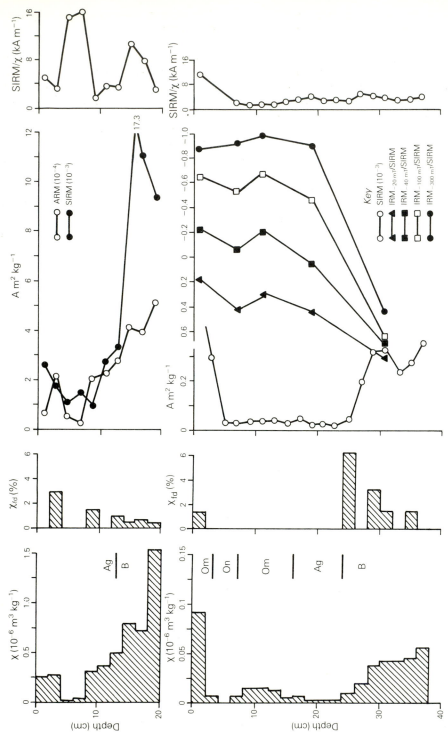

Figure 8.5 Magnetic measurements for two gleyed soil profiles from the Ardnamurchan peninsula, above, in West Scotland (Maher 1983) and, below, from Exmoor (Yates 1983). Note the minimum χ, χ_{fd}/χ and SIRM values in the horizons of most intensive gleying (see text).

The final soil type to be considered is a deeply weathered basalt based regolith from tropical Queensland (Fig. 8.3f) (Isbell *et al.* 1976). Here, χ and SIRM peak around 5 m, at the upper contact of the weathered basalt (cf. Dearing 1979), and neither parameter shows any clear evidence for 'enhancement' near the surface. However, χ_{fd}/χ rises to peak values of around 10% in the top metre of bright-red soil. SIRM/χ and the IRM/SIRM parameters are closely related, with peak ratios corresponding to 'harder' remanence in the weathered zone below 5 cm and at the base of the red soil layer *c.* 2 m. These would appear to be zones of relatively higher haematite and/or goethite concentration. The uppermost samples with maximum χ_{fd}/χ have relatively soft remanence probably as a result of the greater proportion of ferrimagnetic grains.

8.7 The effects of gleying on magnetic properties

Several authors have observed that gleyed soils have anomalously low susceptibility values. Figure 8.5 illustrates this effect on two gleys developed on basalt (a) and slate (b) respectively. Under the prevailing reducing conditions within the gleyed horizons, both the secondary and primary ferrimagnetic forms are readily dissolved and remain in solution or are leached from the profile. The low χ_{fd}/χ values and the somewhat higher SIRM/χ and ARM/χ values characteristic of the gleyed horizons suggest that, as might be expected, the finest superparamagnetic and viscous crystals are the most readily dissolved, and this is confirmed by Mossbauer measurements (Maher, pers. comm.). At first sight, the data from

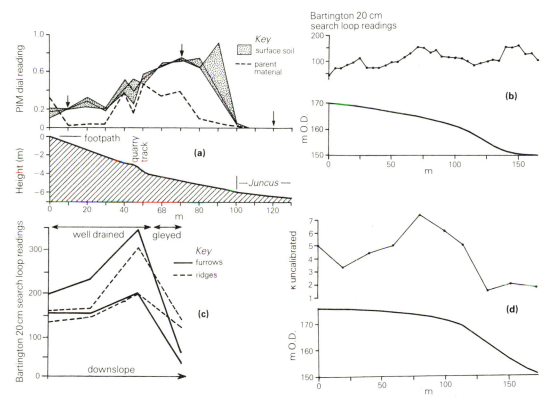

Figure 8.6 Magnetic 'catenas'. (a) Surface and bedrock pulsed induction meter (PIM) readings downslope on Ordovician slates in North Wales (see Turner 1980 and text); (b) Bartington 20 cm search loop surface readings on a Jurassic limestone slope in Oxfordshire (Sutton 1982); (c) Bartington 20 cm diameter search loop surface readings on a sloping area of former 'lazy bed' cultivation in West Scotland (Maher 1981). The effects of the slope, the lazy bed microtopography and gleying are discussed in the text; (d) Bartington 7 cm core scan loop readings on shallow drainpipe soil cores taken down a steep Jurassic limestone slope in Oxfordshire (Sutton 1982).

gleyed profiles may appear to conflict with all the evidence for the persistence of the ferrimagnetic oxides in many stream, lake and marine sediments (Chs 9, 10 & 12). Within gleyed soil however, a broad spectrum of bacteria can produce reducing conditions by using up dissolved oxygen. Dissolution of magnetic oxides may then occur as a simple chemical reaction. Iron-reducing bacteria may play a special rôle since *Clostridium* species have been shown to degrade crystalline iron oxides effectively and reduce them to non-crystalline forms (Ottow & Glathe 1971).

8.8 Soil magnetism and slope processes

Figure 8.6 plots four soil-magnetic catenas showing the relationship between susceptibility, slope processes and drainage. Transect (a) shows the envelope of values for three pulsed induction meter (6.7.3) surface readings at levelled locations down the slope of a pasture field developed on Ordovician slates in North Wales. For comparison, readings for subsoil were taken at the base of a shallow soil pit dug at each point. The subsoil readings are independent of slope and of the mostly higher surface readings, and they peak only in relation to the trackway which crosses the transect. The surface readings show a progressive downslope increase consistent with a steady enrichment in surface soil fines towards the slope foot. At the point where waterlogging gives rise to gleyed soils colonized by *Juncus* species, the surface readings fall to around zero. In (b), similar progressive downslope enrichment is shown by the Bartington susceptibility meter search loop readings plotted for the transect of a shallow soil pasture under pasture developed over Jurassic limestone in Oxfordshire.

Transect (c) subtly illustrates both the downslope enhancement trend and the effects of gleying on lazy beds, a kind of ridge and furrow cultivation associated with 'crofting' in parts of Ireland and the Scottish Highlands. The site is on the Ardnamurchan peninsula and the parent material is basalt. In the non-gleyed part of the transect, susceptibility readings from both the furrow and ridge sites show downslope enrichment, though the trend is more marked in the envelope of values for the furrow readings. This is consistent with the downslope movement of largely primary silt-sized 'magnetite' from ridge to furrow on the scale of the lazy beds themselves, as well as down the whole slope. Within the gleyed part of the transect, the surface readings for

the relatively better drained ridges are higher than those for the waterlogged furrows.

Transect (d), on the Oxfordshire Jurassic limestone, as well as illustrating the gleying effect at the slope foot, shows peak susceptibility readings at the slope crest and depletion of surface material on the steepest parts of the slope. In this transect the readings were derived from volume susceptibility scans made on shallow plastic drainpipe cores.

A more detailed evaluation of the use of mineral magnetic properties to trace soil movement on slopes has recently been published by Dearing *et al.* (1986).

8.9 The persistence of magnetic oxides in the soil

The pedologist is familiar with the sensitivity of iron compounds to changes in soil-forming conditions, as demonstrated by the changes in soil colour which occur as a result of the presence or absence of different iron minerals. At first sight, it would therefore seem unlikely that the magnetic oxides in the soil could be sufficiently stable and persistent to be of value in the recognition of earlier *in situ* events or processes such as burning or more oxidative weathering régimes. It would also seem unlikely that the magnetic oxides once released from the soil, should retain their characteristics during transport and subsequent deposition. Nevertheless, the evidence available is strongly in favour of a degree of persistence of magnetic oxides both in the soils and sediments.

The magnetic characteristics of the late Holocene buried soil from the Howgill Fells in North West England, described in Harvey *et al.* (1982) suggest that little if any mineral magnetic change has taken place during the millenium or so of burial. For the most part magnetic prospecting at archaeological sites depends on the persistence and detection of secondary ferrimagnetic oxides in burnt areas and ditch fills. On a longer timescale, Poutiers' (1975) magnetic susceptibility profiles from Pleistocene sections in terrestrial sediments from the Cote d'Azur, South East France, often show a clear correspondence between peak susceptibility and the occurrence of deep-red, allegedly interglacial palaeosols.

Within the soil, persistence depends largely on the post-formation weathering and pedogenic régimes to which the oxides are subjected. Even where conditions have changed dramatically as a result of climatic fluctuations, iron oxides from earlier régimes may persist for very long periods. As Schwertmann and Taylor (1977) note, 'transformation of initially kinetically favored metastable phases to more stable ones may be extremely slow'. Hence in Britain, we may still see the expression of 'tropical' weathering in the highly oxidised regoliths locally developed on basalt. Consistent with this is Dearing's (1979) suggestion that around L. Frisa the bright orange subsoil, rich in antiferromagnetic grains, is at least in part attributable to the persistence of weathered material of preglacial origin.

The resistance to weathering of the secondary magnetic oxides formed through fire and normal pedogenic processes (8.4 above) appears to be similar to that of the comparable primary oxides (e.g. magnetites and titanomagnetites) derived from bedrock, though both are readily broken down under gleyed conditions. In the long term, there would also be a tendency under sufficiently oxidative conditions, for all these oxides to transform to goethite or haematite. This type of long-term weathering transformation sets some limit, albeit ill-defined, on the timescales over which the weathering régimes recorded in the magnetic properties of sediment profiles can be realistically reconstructed from reference to contemporary catchment soils alone.

From the above and the range of data available so far, we may conclude that persistence of ferrimagnetic and antiferromagnetic oxide assemblages, once formed, can be accepted provided the soil has not been gleyed and the pedogenic régime has not altered too drastically. The long-term survival of these oxides in drainage systems and depositional environments is the subject of later chapters.

8.10 Soil magnetism and archaeology

The most familiar application of soil magnetism to archaeology is in prospecting and site survey (see Section 6.7). The equipment available for these purposes includes both passive and active instruments. In the first category are the proton magnetometers and gradiometers. Aitken (1974) and Mullins (1974) both give full accounts of the operation of the instruments and many authors have published studies illustrating their value. Figure 8.7, taken from Scollar (1971), shows the results of a magnetometer survey at a Roman site in the Rhineland. Active instruments are rather more varied and include the pulsed induction meter or PIM (Colani & Aitken 1966), the soil

conductivity meter or SCM (Mullins 1974), the Bartington system with both search loop and probe sensor attachments and various modifications of metal detectors and pipe finders. In general, these instruments tend to be more strongly affected by 'soil noise' and are less likely than the passive instruments to detect deep features. Pulsed Induction instruments detect both the eddy currents of metallic objects and the viscous decay of magnetisation in fine crystals. The fine magnetic viscous grains at the stable single-domain – superparamagnetic transition, also give rise to the frequency-dependent (cf. quadrature) susceptibility (χ_{fd}) component (Mullins & Tite 1973). The SCM as Mullins (1974) explains, does actually measure soil susceptibility directly as does the Bartington system which, with the addition of probe sensors for profile and borehole logging, greatly

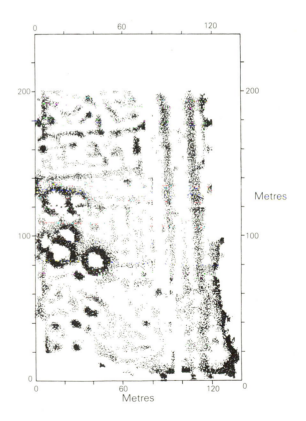

Figure 8.7 The results of a proton magnetometer survey of part of the Colonia Ulpia Triana in the Rhineland. The surveyed area is approximately 140 × 200m. The circles on the left are bomb craters. Above these are outlines of buildings and to the right are traces of two large ditches (redrawn from Scollar 1971).

increases the scope and versatility of the equipment available. The most detailed comparative evaluation of the effectiveness of proton gradiometer, PIM and SCM surveys is given in Mullins (1974) using data from Iron Age/Roman sites in southern and eastern England.

As well as using field instruments for magnetic surveys and prospecting, archaeologists have also made use of mineral magnetic measurements carried out on samples in the laboratory using various types of a.c. susceptibility bridge (e.g. Scollar 1965) and magnetometer. Krawiecki's (1982) study at Maiden Castle an Iron Age hill fort near Bickerton, Cheshire provides an interesting specific illustration of laboratory measurements (Oldfield *et al.* 1984). Charred wood occurs within the ramparts of the Maiden Castle site and visual inspection fails to reveal whether previously burnt wood was emplaced during the course of construction or the wood was burnt *in situ* during or after construction, perhaps in an

attempt at vitrifying the adjacent sandy fill. It follows from studies of the effects of fire on the magnetic properties of rocks and soils that *in situ* burning would tend to enhance the magnetic susceptibility and SIRM of mineral soil in contact with the wood, by converting the natural assemblage of haematite-dominated magnetic minerals in the Triassic sands to 'magnetite'. Figure 8.8 plots χ, SIRM, SIRM/χ and $IRM_{-100\,mT}$/SIRM for a vertical section through the rampart at a point where a charred log and layers of blackened sand are present. Clearly the sand adjacent to the log, and in the lower blackened layer, has been magnetically enhanced by one to two orders of magnitude.

Figure 8.9 is a summary plot of the same four magnetic parameters for samples taken from this and three other sections as well as for laboratory-heated sands taken from nearby at the site. Proximity to the charred wood invariably has the effect of increasing χ and SIRM, decreasing SIRM/χ and 'softening' the

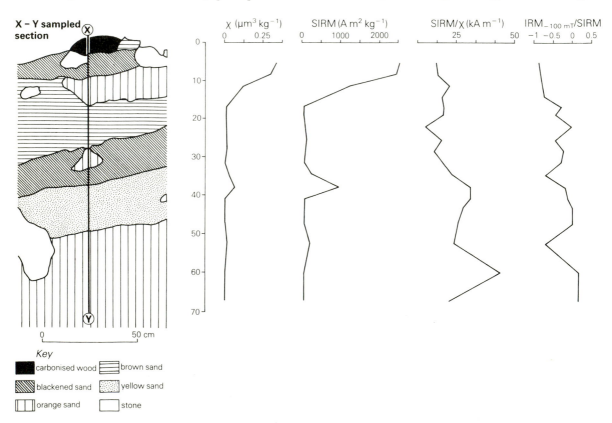

Figure 8.8 Stratigraphic and mineral magnetic profile through the outer rampart of Maiden Castle, near Bickerton, Cheshire (Krawiecki 1982).

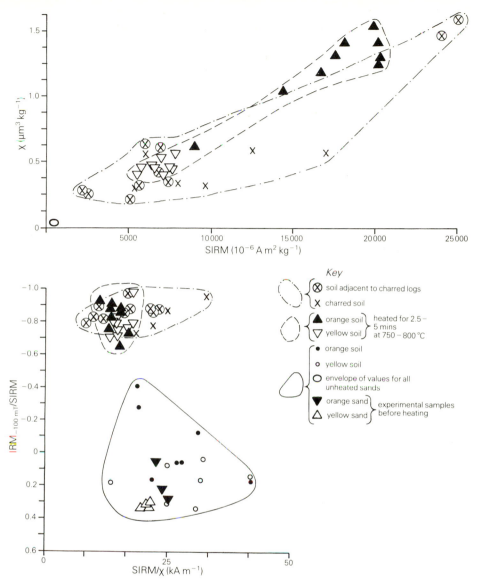

Figure 8.9 Maiden Castle hill fort, Bickerton, Cheshire. χ versus SIRM and SIRM/χ versus IRM$_{-100\,mT}$/SIRM for burnt and unburnt samples (Krawiecki 1982). Mineral magnetic parameters for samples burnt *in situ* during the construction of the prehistoric ramparts are compared with those for unheated sands and for experimentally heated sands (see text).

IRM$_{-100\,mT}$/SIRM values, all changes consistent with enhancement by fire. After many heating and cooling trials during which the parameters identified as significant in Oldfield *et al.* (1981a) were varied, closely comparable magnetic enhancement was achieved on both the yellow and orange sands abundant at the site, by heating and cooling material in a reducing environment at a peak temperature of 750–800 °C for 2.5–5 min. Subsequently, the spatial occurrence of the fire-enhanced material in exposed sections was plotted in the field using the Bartington susceptibility meter with both probe and loop sensors (Fig. 8.10).

The combination of portable magnetometer, pulse

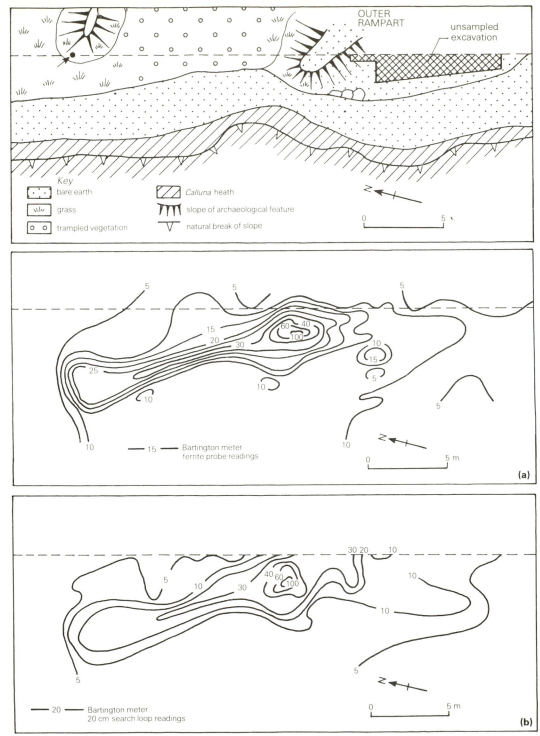

Figure 8.10 Surface susceptibility readings over part of the area of burnt timbers and adjacent magnetically enhanced sand in the Maiden Castle hill fort embankment. Contours are based on both ferrite probe (a) and 20 cm loop (b) readings using the Bartington instruments system.

magnetiser and susceptibility systems opens up new prospects in the application of mineral magnetic measurements to archaeology. Plan and section logging, detailed magnetostratigraphic description, and downhole susceptibility logging all become possible as on-site prospecting, survey and descriptive techniques. Magnetic measurements can also be used not only to locate burnt material but also to help to characterise its thermal history. It seems likely that these new opportunities herald a revival and a broadening of interest in the application of mineral magnetic measurements to archaeological contexts and problems.

8.11 Conclusions

Despite the difficulties inherent in attempting to identify finely divided iron oxides in the soil by means of either mineral magnetic or standard geochemical measurements, the categories of magnetic behaviour recognisable as a result of mineral magnetic characterisation are clearly related to soil-forming processes in a direct and coherent way. Ordered reflections of pedogenesis are apparent in variations between contrasted 'type' profiles on similar lithologies, in variations related to slope processes and poor drainage, and on a fine scale within individual soil profiles on a particle size specific basis. In consequence, the potential value of mineral magnetic measurements to the soil scientist greatly exceeds that which is apparent from the vast majority of published studies concentrating on magnetic susceptibility alone. Both as a descriptive tool in routine survey and profile description, and as an analytical technique in studies of soil-forming processes, magnetic measurements are ideally suited to complement and precede established methods. The more so since they are capable of detecting changes in magnetic mineralogy and grain size at concentration orders of magnitude below the detection limits of conventional methods. The conservation of magnetic properties and their diagnostic value also makes them of great interest in fossil soil and archaeological studies where their potential range of applications has not been fully realised. These same characteristics lie at the root of many of the most interesting applications of magnetic measurements in fluvial, lacustrine and marine systems and these are the subject of subsequent chapters. In addition, Dearing *et al.* (1985) present a general review of the role of soil magnetism in geomorphology.

[9]

Magnetic minerals and fluvial processes

> . . . he sat on the bank, while the river still chattered on to him,
> a babbling procession of the best stories in the world, sent from
> the heart of the earth to be told at last to the insatiable sea.
>
> Kenneth Graham
> *The Wind in the Willows*

9.1 Introduction

This chapter is concerned with particles in transit once they have reached a defined water course. Prior to the period of movement within a river channel, the particles will have been either released from the land surface and delivered to the channel by rainsplash, sheet erosion, rill and gully erosion or mass movement, or else removed from the channel banks or stream bed through the erosive effect of the moving water and its entrained load. Sediment within the channel is normally considered as either suspended sediment or bedload, though as flow conditions vary through space and time the distinction is more one of practical convenience than of consistent differentiation.

Many aspects of stream sediment transport are of interest to scientists in a variety of disciplines. For example channel morphology is closely related to sediment type and to rates of sediment movement, especially where the river bed includes persistent constructional features such as gravel shoals and point bars. Identifying the source of sediments is a vital aspect of erosion studies in contexts where both surface denudation and channel change are evident. Where the fine sediments in stream channels can be ascribed to soil surfaces there are often important implications in terms of soil aggregate stability in the eroded areas, and particle-associated pollution in the water bodies down river. The dynamics of sediment transport have major engineering implications where channel change reduces the effectiveness or viability of nearby structures. These engineering implications arise because the catchment, bed and banks of the river provide the mechanical load for transport. Changes in transported load often give rise to practical problems as a result of the new erosion and deposition régimes which they generate. Estimates of sediment transport are often based on hydraulic equations which assume capacity load, though this is rarely achieved in reality. There is therefore a need for better information about sediment sources and the amount of material each contributes.

In the present account, suspended sediment is examined largely from the point of view of source identification (Section 9.2) and bedload is considered experimentally with a view to improving our understanding of its transport and storage dynamics within the river channel (Section 9.3). The special case of artificial urban drainage systems is also considered in this chapter (Section 9.4) largely in relation to heavy metal concentrations and sources. Sections 9.2 and 9.3 form a link between Chapter 8 on Soil Magnetism and the succeeding chapters dealing with sediment deposited in lakes (Ch. 10) and in the sea (Ch. 12). This particular treatment inevitably implies some loss of the integrated perspective which drainage basin studies can provide (e.g. Gregory & Walling 1973, O'Sullivan 1979). To some extent this is redressed in the lake sediment based studies discussed in Chapter 10, and more specifically in Chapter 16 which is entirely concerned with a tidal river and estuarine system and the source–sediment linkages within it.

9.2 Suspended sediment sources

In most rivers the suspended sediment load constitutes the dominant mode of particulate material loss from the catchment. This is especially so where the nature of the bedrock and the weathering processes combine to supply large quantities of fine material to the channel. Macroscopic characteristics of the sediments will often give no clue as to the main source of such fine material. Geomorphological evidence in the form of active streamside erosion scars, developing gullies or truncated soil profiles may often give a clear indication of specific contributing sources (e.g. Mosley 1980), but the significance of these sources, and their relative contribution at different times may remain obscure without direct evidence from the sediment itself.

From Chapter 8, it can be seen that at least one aspect of this problem, namely the identification of topsoil erosion, is potentially tractable using mineral magnetic measurements. The processes of weathering and especially magnetic enhancement frequently give rise to mineral magnetic assemblages in the upper layers of the soil readily distinguishable from those in the underlying parent material. Where the main problem in suspended sediment source identification is one of distinguishing between surface-derived material from the catchment slopes, and subsurface material eroded from the channel and banks, then the magnetic changes associated with soil formation in Chapter 8 may be expected to aid differentiation. This particular problem has implications not only for the geormorphologist studying denudation rates, channel form and dynamics or downstream sedimentation, but also for the environmental chemist and agriculturalists interested in water pollutants, for example limiting nutrients or persistent pesticide residues, which may often move from soil surface into rivers in sediment-associated form.

9.2.1 Suspended sediments in the Jackmoor Brook

The first attempt to identify suspended sediment sources from their mineral magnetic characteristics was carried out in the Jackmoor Brook, near Exeter in South West England. The Jackmoor Brook basin has an area of 9.3 km^2 and ranges in altitude from 21.5 to 235 m. Gentle ($<4°$) slopes predominate and the soils range from well drained to poorly drained and gleyed brown earths developed on Permian red-bed desert sandstones, breccias and conglomerates rich in haematite. Mixed arable farming predominates in the area with cultivated crops and grass leys covering most of the catchment. Less than 4% of the catchment is wooded. Particulate transport within the stream is mostly as suspended sediment. Relatively high concentrations (up to 3.5 g l^{-1}) occur in storm events and the annual suspended sediment yield from the catchment is estimated as around 85 t km^{-2}. There is little floodplain development and the channel, though only rarely incised by more than 1 m, reaches bedrock in most places. Obvious actively eroding sites, whether in the channel or on the catchment surface, are rare. The basin was chosen for developing and evaluating magnetic methods of sediment source tracing for several quite independent reasons. Within the catchment, there are unresolved questions relating to sediment source. In particular the relative importance of surface soil and channel contributions is hard to establish by other methods. The well developed monitoring and sampling programme already established within the catchment provides a comprehensive hydrological framework for the study. The catchment lithology and soil types are very favourable for magnetic differentiation. Brown earth soils have developed on bedrock consistently rich in haematite. Moreover, the area lies south of the maximum limit of Pleistocene glaciation and thus lacks transported erratics. The context is ideal for testing the magnetic methodology, since *a priori* one would expect the sediment yielded from such surface soils to be relatively rich in secondary ferrimagnetic grains in contrast to channel and bankside material with a preponderance of antiferromagnetic grains derived from the bedrock. (Ch. 8).

The initial magnetic studies at Jackmoor Brook predated development of an effective portable susceptibility measuring system, and the characterisation of potential sediment sources within the catchment was achieved by sampling soil pits, and by collecting material from the plough layer of tilled soils

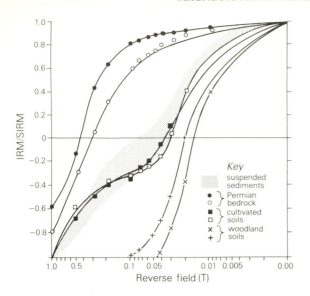

Figure 9.1 Coercivity of SIRM curves for soil, and bedrock/bankside samples from the Jackmoor Brook compared with the envelope of values for bulk suspended sediment samples from Oldfield *et al.* 1979).

and from the sites where freshy undercut exposures indicated recent erosion within the main stream channel. The first suspended sediment samples measured were obtained by centrifuging large quantities (~20 l) of stream water collected during storm events.

The results summarised in Table 9.1 from Walling *et al.* (1979) show the characteristic magnetic enhancement of surface soils developed under woodland and temporary pasture within the catchment. Figure 9.1 plots back-IRM curves for a selection of the catchment samples and the four bulk suspended sediment samples. The predicted contrast between bedrock and topsoil is confirmed, especially in the case of the deciduous woodland soils which are less vertically mixed and have been less exposed to rainsplash and rill erosion than the cultivated surfaces.

Streamside samples compare closely with parent material save at the top of river banks, within exposed surface soils. The back-IRM curves of Figure 9.1 and the very high $(B_0)_{CR}$ values confirm that the magnetic properties of samples from below the depths of active soil development are dominated by canted antiferromagnetic grains.

Gleyed soils show low susceptibility and also low coercivity of remanence values whereas the well drained cultivated soils show, for all parameters, values intermediate between those for the bedrock and the unmixed woodland topsoil. In all respects, the cultivated soils and the suspended sediment samples are directly comparable. We may therefore infer from the magnetic measurements that the dominant sources of suspended sediments are the cultivated soils of the catchment and that the mineral magnetic assemblage within these soils reflects both vertical mixing by cultivation, and some loss of surface material through rainsplash and rill erosion.

The initial mineral magnetic study opened up the possibility that distinction could also be made between the balance of sources from flood to flood and between different stages within a flood event. In order to explore this possibility further, measurements were made on filter paper residues. The suspended sediment load was sampled at the outflow gauging station using an automatic pump sampler which, during flood events, abstracted 500 ml of water at hourly intervals. The magnetic properties of the suspended sediment filtered from samples can be plotted alongside the continuous records of discharge and suspended sediment concentration (turbidity) made at the gauging station (Fig. 9.2).

Adopting the methods outlined in Walling *et al.* (1979) consistent and repeatable SIRM, $(B_0)_{CR}$ and $IRM_{-100\,mT}$/SIRM values were obtained on samples with dry weights as low as 0.02 g thus permitting over 90% of the filter paper residues to be magnetically characterised with confidence. Susceptibility was measurable on only a small proportion of these low

Table 9.1 Summary of selected magnetic properties of bulk suspended sediment samples and potential sediment sources from the Jackmoor Brook catchment.

Material	χ ($\mu m^3 kg^{-1}$)	SIRM ($mAm^2 kg^{-1}$)	SIRM/χ (kAm^{-1})	$(Bo)_{CR}$ (mT)	$\dfrac{-IRM_{-100\,mT}}{SIRM}$
woodland topsoil	>2.5	>10	≈4	<35	1
cultivated topsoil	0.2–2	1–10	5–7	24–41	0.28–0.8
poorly drained and gleyed soils	0.06–0.4	0.5–3.5	≈10	≈30	≈0.4
parent material	<0.1	1–2	>10	200–400	−0.8– −0.6
suspended sediment	0.25–0.75	2.5–9	≈10	37–60	0.06–0.4

mass samples. Figure 9.2 summarises the results obtained. Two general points emerge:

(a) The mean values and total range for all parameters strongly reinforce the conclusion previously reached that the dominant sources of suspended particles in the Jackmoor Brook are the cultivated soils of the catchment.

(b) There are both similarities and contrasts between the magnetic trends in each flood event and these are interpretable in terms of a coherent process model of sediment supply.

In floods 1 and 4 maximum SIRM values follow the peaks in discharge and suspended sediment concentration. Flood 1 was the first of the winter and followed a prolonged dry period. Flood 4 is marked by a very sharp rise in discharge to high levels. Thus in both cases channel scour may be inferred and would be expected to reach a maximum during the rising stage of the hydrograph. The effect of this has been to contribute a relatively greater proportion of bedrock-derived material up to the point of channel source depletion. Floods 2 and 3 are marked by gentler discharge rises during a period soon after the channel source depletion associated with the first event. In both cases any delay between peak discharge and sediment yield, and peak SIRM, is less apparent. Moreover, most of the SIRM values for floods 2 and 3 range from 4 to 6.5 mA m^2 kg^{-1} whereas for floods 1 and 4 the values lie between 3 and 5.5. There is therefore some indication that during the two middle flood events there is less contribution from bankside sources and a more exclusively surface-derived suspended sediment load. The tendency for the surface sources to dominate from peak discharge onwards in all floods is consistent with a situation where maximum surface yields coincide with maximum rainfall intensity and rainsplash effectiveness, and maximum surface runoff rates.

The fifth flood event portrayed was the product of snowmelt rather than rainfall. The low SIRM values are consistent with a lack of rainfall energy reducing mobilisation by splash erosion. This factor along with the persistence of freezing conditions in the soil has resulted in a relatively low contribution from the cultivated soils. Thus channel sources become relatively more significant contributors to the rather modest suspended sediment concentrations recorded.

Subsequent studies by Peart (unpubl.) have tended to confirm the above inferences whilst at the same time documenting a wide variety of mineral magnetic response to flood events and identifying gleyed soil areas close to stream channels as significant sediment sources during the early stage of some floods. His evaluation of the relative consistency of source-sediment linkages, derived from mineral magnetic measurements as well as other chemical parameters such as phosphate or carbon concentrations, suggest that only the mineral magnetic properties are conservative within the system.

9.2.2 Other studies of suspended sediment sources

Although the Jackmoor Brook site is developed on an ideally favourable lithology for the type of sediment source differentiation illustrated above, it follows from Chapter 8 that in many other contexts, magnetic differences between bedrock and the topsoil will be identifiable. Brief reference is made to two other case studies chosen to illustrate problems and possibilities elsewhere.

The Great Eggleshope Beck catchment which is instrumented by the Freshwater Biological Association covers 11.68 km^2 in the North Pennines of Britain (Carling & Reader, 1982). The channel lies within an area of colluvium head and till largely derived from upper carboniferous millstone grits which together with carboniferous limestones form the bedrock to the catchment. Shallow, acid brown earths, podsols and peaty gleys are the main soil types and these support extensive areas of rough pasture and moorland, parts of which are periodically burnt. Mineralisation upstream of the study site has led to earlier periods of mining activity which has resulted in areas of mineral spoil covering parts of the valley floor. The spoil heaps are gullied and in parts, the main channel is eroding them. In the present study (Chorlton 1981), sediment-source comparisons have been attempted largely on the basis of SIRM measurements made on bedrock, channel material, spoil and topsoil as well as on suspended sediments from the periods of maximum concentration during two flood events. The values for the suspended sediments are within the range characteristic of bedrock and channelside material and generally much lower than those of either the surface soils of the catchment or the areas of mining spoil. Similar inferences may be drawn from the results summarised by Wise (1979) for the Wadhurst Park catchment in Kent. Here, a largely pasture-covered catchment developed for the most part on Ashdown Sands yielded suspended sediments

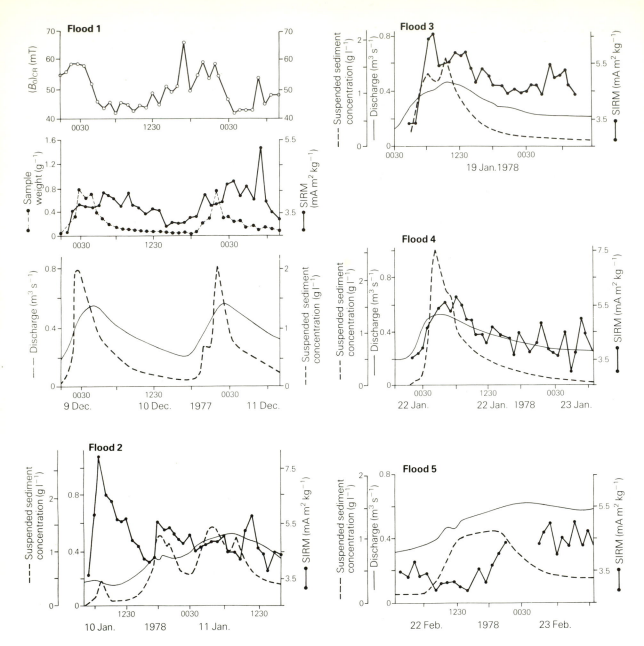

Figure 9.2 Mineral magnetic and hydrological parameters for five flood events in the Jackmoor Brook catchment (From Walling *et al.* 1979). See text for interpretation.

comparable magnetically to the exposures of eroding subsoil adjacent to the stream channel.

9.2.3 Prospects and problems

So far, we have tended to concentrate on uniform, relatively iron-rich sedimentary lithologies. The prospects for this type of approach will decline in iron-poor environments and in the situations where magnetic enhancement of topsoil is retarded or inhibited (see Section 8.4). Also, although frequency-dependent susceptibility measurements should allow differentiation between topsoil and bedrock, where the latter is rich in primary ferrimagnetic minerals, this parameter will rarely be measurable on low volume filter paper residues. Thus it will be difficult to achieve the same temporal resolution without the use of larger integrated sampling devices. In catchments of mixed lithology the added complexity may either increase or reduce opportunities depending on whether or not the variations are readily character-isable by magnetic parameters and on whether the spatial distribution of magnetically distinguishable lithological units relates coherently to the problems of source identification posed by the catchment. Glaciated catchments with varied and irregularly distributed drifts and abundant erratics are probably the least suitable for this type of study although individual till sheets and drifts of uniform provenance can be distinguished and characterised magnetically (Sugden & Clapperton 1980; Walden & Smith, pers comm.)

Rather different problems are posed by stream channels incised in and currently eroding older alluvium. Where this alluvium has been derived from eroding surface soil, then provided the eroding sites have not been gleyed, the magnetic properties are likely to compare closely with contemporary surface soils. Mineral magnetic study of this type of system must include very detailed particle size specific characterisation of the whole range of actively eroding exposures with a view to determining the range of variation present and any indications of post-depositional transformation. Sandland's (1983) and Arkell's (1984) data from the middle reaches of the Severn in an area where the river is actively eroding older alluvium suggest that there, the problem is potentially tractable.

In the studies outlined above little or no attention has been paid to the effects that changing particle size : magnetic grain size relationships will have on the magnetic parameters used to establish sedimento-logical linkages; yet discharge variations will often lead to changes in the particle size distribution of the suspended load. Where this gives rise to shifts in the proportion of grains in different magnetic domain states, we may expect changes in magnetic parameters even where no *mineralogical* variations are present. At the same time, in natural samples the mineral and domain size variations will rarely if ever be independent. For example, secondary ferrimagnetic iron oxides in the soil will always tend to be stable single domain or smaller, while equally fine-grained haematite or goethite crystals will often be associated with cemented sands. Probably the safest procedure available for avoiding either spurious or coincidental source–sediment linkages and for reducing the likelihood of invalid mineral magnetic inferences is to look in detail at the magnetic properties of particle size specific fractions for both the potential sources and the trapped sediments (cf. Ch. 16).

In catchments with substantially lower suspended sediment yields than the Jackmoor Brook, the small mass of the filter paper residues will eventually limit the effectiveness of the method. This problem can be compounded by the presence of ferrimagnetic im-purities in both the glass-fibre filters commonly used and the standard 10 ml sample pots. It has been found that 25 mm diameter Teflon filter discs without sample holders can be measured and that in practice this approach provides at least an order of magnitude better sensitivity than that described in Oldfield et al. (1979 c).

Finally, it is apparent that in their present form, the mineral magnetic methods are qualitative rather than quantitative. Oldfield et al. (1979c) attempted to make the bulk sediment characterisation at Jackmoor Brook more quantitative, but only in terms of estimating crudely the bedrock : unmixed topsoil ratio which could produce the range of magnetic parameters common to both the cultivated soils and the sediments. Quantitative estimates of the relative importance of specific sources and source types for particular episodes may however sometimes be feasible (Stott 1986). Future studies concentrating on this aspect of the technique and also on combining magnetic measurements with radiometric tracing (cf. Campbell et al. 1982) will be especially valuable.

9.3 Magnetic tagging and tracing of stream bedload

Just as the naturallly evolved magnetic properties of soils provide a basis for sediment-source identification, artificially induced magnetic characteristics can be used to provide material for use in tracing experiments. As we have seen in Chapter 8, fire can lead to a strongly enhanced magnetic signal in surface soils. In practice, most reasonably iron-rich natural materials can be magnetically enhanced by heat treatment in the laboratory, though the initial idea for magnetic tagging and tracing came from monitoring the after effects of a major forest fire. The Llyn Bychan forest fire of 1976 (Rummery 1981, 1983) in North Wales gave rise to magnetically enhanced material which persisted in the soils and the lake sediments and found its way to the Afon Abrach, the river which drains both the lake and the intensively burnt area down stream of the lake outfall. After the fire, a series of magnetic measurements were made on sieved material from successive downstream shoals beginning in the burnt area and continuing for some 2 km. In the case of the coarser clasts the downstream magnetic variation involved an order of magnitude decline. By contrast, the finest materials showed exceptionally high SIRM values close to the fire and a three order of magnitude decline downstream. These results were interpreted as reflecting the selective loss of magnetically enhanced fines from the burnt area and the gradual dilution of this material at increasing distances down stream. This observation opened up the possibility of using not only naturally but artificially enhanced material as a bedload tracer in river channels.

9.3.1 The Plynlimon case study

The Plynlimon area of central Wales was chosen for the initial testing of magnetically enhanced tracers for several reasons. One of the major concerns of the Institute of Hydrology's catchment research at Plynlimon is the large volume of bedload generated in the upper reaches of the Severn by the rapid recent erosion of forest drainage ditches. By the time the first magnetic trials began, a major programme of hydrological and sedimentological monitoring had been established by the Institute (Newson 1980) and this provided an essential framework for the trials. More recently, the work was extended downstream into the piedmont zone in response to concern expressed by

the Ministry of Agriculture Fisheries and Foods about the effects that the increased gravel yields, coupled with the water regulation policies adopted in headwater reservoirs, might be having on channel stability and possible land loss in cultivated areas.

The main reasons for using bedload tracer studies as part of the research strategy devised in response to the academic and practical problems posed by the Upper Severn are set out in Arkell (1984). Most conventional tagging and tracing techniques are limited to a particular particle size range. Pebble painting or plugging with radioisotopes are suitable only for large clasts, whereas fluorescence is more applicable to sands. Existing techniques also pose serious problems of signal persistence and particle recovery. Problems are further compounded in the study area by the very wide size range of the bedload and the preponderance in many reaches of fine gravel which is difficult to tag conventionally. Fortunately, the bedrock is a shale uniformly rich in finely disseminated paramagnetic iron giving both a consistently low susceptibility and saturation remanence in its unheated state, and a high potential for enhancement by heat treatment.

The work so far has involved developing:

(a) suitable heating procedures for enhancing the magnetic susceptibility of large quantities of gravel,
(b) instruments and techniques for magnetic measurements both in the river channels and on abstracted material in the laboratory and
(c) field trial strategies on a range of spatial and temporal scales designed to contribute both to the evaluation of the technique and to the understanding of the substantive problems of bedload transport in the area.

The heating trials leading up to adoption of a practical bulk 'toasting' method for the Plynlimon material are described and their mineralogical effects interpreted in Oldfield et al. (1981a).

Instrumentation was initially limited to a Littlemore susceptibility bridge for laboratory measurements of bulk samples. Subsequently, commercially available portable metal detectors (see 6.7) were used for location of enhanced tracer material on and down stream from 'seeded' shoals. A pulsed induction meter, a Whites Savo TR induction balance coil and an Arado VF90 acoustic loop were all capable of detecting enhanced material in the field, but not of giving more than a very crude qualitative indication of strong presence only. A 20 cm diameter search loop and both hand-held and ground-search versions of ferrite probes were finally constructed by Bartington Instruments specially for the Plynlimon bedload monitoring project.

Field trials began in two of the overdeepened drainage ditches within the afforested upland part of the Severn catchment. The ditches are essentially field flumes and together with the bedload traps at their mouths, they provide a simple and confined channel within which to test techniques for emplacement, location in transit and downstream recovery. Gravel was taken from shoals up stream from the bedload traps and replaced in the same particle size proportions by weight, with previously trapped material which had been heat treated in a large muffle furnace. These initial traces were characterised by high tracer recovery rates, and the results greatly improved insight into the rôles of storage and supply within the systems.

Encouraged by the results from these field experiments (see Arkell *et al.* 1982) and by improvements in 'toasting' techniques, the methodology was extended to three sites in the natural river channels. These were in the river Severn at Morfodion, the Llwyd, one of its tributaries, at Dolydd, and the Cefn Brwyn in the upper reaches of the Wye system. At Morfodion, the Severn is approximately 30 m wide with a 500 m flood plain. The Morfodion site provided a location for which loss from a seeded area was much more readily measurable than subsequent downstream movement. The heat-treated, magnetically enhanced gravel was placed in a trench cut normal to flow through a shoal which projected down stream into the river from the true left bank. The Dolydd trace on the Llwyd was set up beginning beneath a road bridge spanning the whole river channel. The site provided an opportunity to monitor not only loss but movement through storage locations down stream. At the Cefn Brwyn site on the Wye, the seeded site was immediately down stream of the Institute of Hydrology compound crump

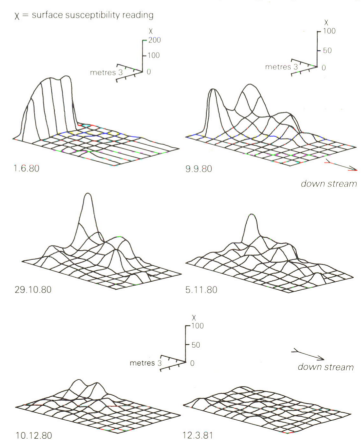

Figure 9.3 Bartington 20 cm search loop surface readings from the magnetically 'seeded' shoal at Morfodion on the River Severn. The initial isoplot of surface susceptibility readings records the effect of the magnetically enhanced gravel immediately after emplacement. Subsequent isoplots record its downstream dispersal and partial burial (see Arkell *et al.* 1982).

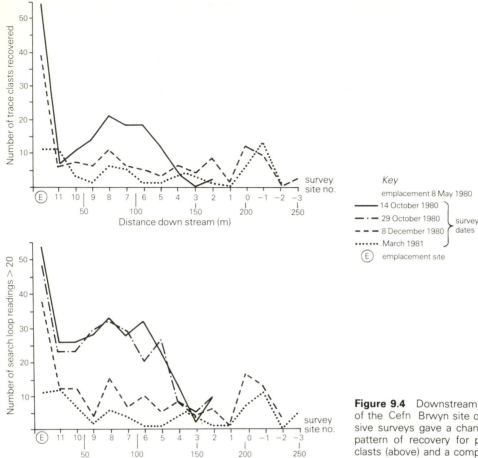

Figure 9.4 Downstream magnetic tracer recovery of the Cefn Brwyn site on the River Wye. Successive surveys gave a changing temporal and spatial pattern of recovery for positively identified tracer clasts (above) and a comparable pattern for surface search loop readings (below).

weir, allowing loss and subsequent downstream movement to be monitored without the additional complication usually found in natural channels of bedload addition from up stream. Emplacement procedures varied from site to site in response to the nature of the channel and the local bedforms, and alternative schemes of sampling and magnetic susceptibility measurement were used to tie in with different scales of topographic monitoring (Arkell *et al.* 1982).

Figure 9.3 shows some of the results obtained for the Morfodion shoal from the time of emplacement in June 1980 through to March 1981. Downstream movement of the tracer on the shoal is associated with small topographical changes. The results show that the topography on the latest survey date has been produced in part by removal of tracer along the

northern edge of the shoal and its replacement by gravel from up stream which has partially recreated the original form. Figure 9.4 is a plot of tracer movement down stream from the emplacement site at Cefn Brwyn as detected both by surface susceptibility 'search loop' readings and by the recovery and indentification of tracer clasts. The comparability of the results yielded by the two methods on each survey date reflects in part the use of the search loop to locate individual clasts for subsequent laboratory confirmation as tracer. Using the hand-held probe individual tracer particles down to 3 mm diameter can be identified. The main hydrological implications of the results obtained from all the traces so far is the overriding importance of channel storage on sediment transport even within the small forest ditch channels (Arkell *et al.* 1982, Arkell 1984).

9.3.2 The prospects for magnetic tracing

Further development of the magnetic tracing methods used at Plynlimon and their adaptation to river sediment tracing on a wide range of lithologies will be constrained by several factors, some environmental and some technical. The main ones are outlined briefly below:

(a) *Lithology.* Optimum conditions are presented by bedrock types uniformly rich in weakly magnetic forms of iron. On iron-poor lithologies the introduction and tracing of exotic enhanced material will be feasible. On strongly magnetic or magnetically heterogenous lithologies, magnetic tracing is likely to prove impossible.

(b) *Channel size, bedload flux density and shoal geometry.* The larger the channel, the greater the flux density of moving bedload through the monitored reach and the greater the probability of deep burial of enhanced material in shoals, the greater will be the logistic problems posed in downstream detection whether through field or laboratory measurements. These problems find expression both in the 'toasting' where larger volumes have so far posed problems of environmental control during heating, and in the 'post-seeding' field situation where any increase in scale of operation will tend to reduce the probability of recovering tracer material.

(c) *Heat treatment.* The optimum heat treatment of the Plynlimon material is strongly conditioned by the self-reducing nature of the sulphur-rich shales at high temperatures and by the finely disseminated nature of the iron present. Thus the experience gained so far is unlikely to be directly applicable elsewhere and each rock type will require a different combination of the main variables. Nevertheless, some general problems have emerged. Resistance of rocks to thermal stress strongly affects the possibilities of optimum magnetic enhancement in large size classes where, as in the Plynlimon studies, insertion at high temperature and a consequently rapid rate of heating were used. Moreover, heat treatment of bulky samples in large crucibles means that at any one stage in the treatment, the thermal history of material in different parts of the crucible (top v. bottom, side v. centre) will be very different. The effects of this on the shales used in the Plynlimon study are exhaustively illustrated and crudely modelled in Oldfield *et al.* (1981a). In theory, the phase-equilibrium approach should dispose of this problem though in practice it has not proved successful for large volumes of gravel. If the treatment of bulk samples can be combined with manipulation of the heating atmosphere, then reducing atmospheres (e.g. nitrogen or carbon dioxide at high temperature) should greatly increase the efficiency of conversion of all iron present to magnetite provided temperature and partial oxygen pressure can be held for long enough within the equilibrium phase.

(d) *Magnetic sensor performance.* All magnetic sensors are sensitive to variations in the geometry of the material to be measured. Thus the field search loop used in the present study is unable to detect small quantities of magnetised material *in situ* and buried beneath a thick armour layer; moreover, a small 'toasted' pebble close to the sensor rim will increase the signal by as much as a larger pebble in the centre of the sensor or some way outside its perimeter. The main response to this type of difficulty has been to develop strong, stable and sensitive probe sensors for insertion between the cobbles of an armour layer. An alternative response would be to use pipe detectors and thereby radically change the coil/sample geometry.

9.4 Magnetic measurements of stormwater-suspended solids

Heavy metal toxicity can pose serious problems in urban stormwater drainage. Many authorities regard non-point surface sources such as roads as significant primary contributors to high heavy metal loadings. Where the contribution is in a particulate, chemically relatively immobile form the heavy metals will pass through the system little modified. Where the heavy metals are in a chemically mobile form, adsorption and precipitation mechanisms in the below-ground phase of the system lead to enrichment in the suspended and benthic particulates present. Thus the heavy metals are mostly found in either a particulate or an eventually particle-associated form. In studies of heavy metal fluxes and concentrations in storm water it is therefore important to identify particle sources with a view to establishing which parts of the system are important either as heavy metal contributors or as contributors of particles with which heavy metal species become associated during transit. An initial appraisal of the possible rôle of mineral magnetic measurements in this field has been made in the separate stormwater system of a 3.5 km² catchment in Hendon, North West London.

Samples from two storms of low to medium rainfall intensity were collected on 29 November 1979 (storm 1) and 28 January 1980 (storm 2). An automatic water sampler situated at the sewer outfall, triggered by a float switch on the rising limb of the hydrograph, obtained samples every six minutes throughout the storm events. Suspended solids were isolated by filtration through $0.45\,\mu$m Millipore filter papers which were subsequently used for the SIRM measurements. Heavy metal determinations of the suspended solids were carried out by digestion of the filter papers in a nitric acid–perchloric acid mixture followed by evaporation to dryness and dissolution of the residue in 2% hydrochloric acid prior to atomic absorption spectrophotometry. Figure 9.5 plots SIRM variations against total suspended solids, and SIRM versus Pb, Zn and Cu concentrations for each flood. Generally, strong linear correlations between SIRM and heavy metals are indicated in all three cases for both flood events. Urban catchments are rich in magnetic mineral sources which are as yet poorly evaluated in terms of their type, location or spatial and temporal contributions to stormwater runoff. Atmospheric fallout and automobile emissions are known to be rich in magnetic particulates (Oldfield *et al.* 1978, Oldfield *et al.* 1979c, Hunt *et al.* 1983, Linton *et al.* 1980 & Ch. 11). Both these sources may also be expected to contribute heavy metals. Rust and attrition of vehicles, gutters, pipes and other iron and steel surfaces may be expected to yield magnetic particles to road surfaces and drainage channels. In addition, authigenic formation of magnetic oxides or sulphides in pipe and gully sediments receiving a dissolved iron input may also be important. However, some tentative conclusions on sediment and metal sources can be postulated from a study of the magnetic patterns.

SIRM values, which are here interpreted as concentration-related parameters, are reciprocally related to sediment concentrations with an exact match of sediment peaks and magnetic troughs in the first storm. The correlation between metal and SIRM values has been investigated not only in general terms but also by comparing their flow-weighted ratio through time (Revitt *et al.* 1981). Close agreements in the time distribution of these ratios is found especially

for the metals plotted in Figure 9.5. In storm 2 the flow peak lags nearly one hour behind the sediment peak. However, the SIRM values follow the pattern of the storm hydrograph with the peak discharge coinciding with the maximum SIRM values of 16×10^{-3} A m² kg⁻¹ declining to a minimum of 7.2×10^{-6} A m² kg⁻¹ on the recessional limb. The leading prime sediment peak can be explained (Ellis 1979) as an early flushing of pipe deposits lodged in the sewer system from previous low flow events. It would appear that these leading sewer sediments on the rising limb of the hydrograph possess magnetic properties different from those of the suspended sediments associated with the late flows and it is tempting to ascribe the more magnetic nature of the later sediments to scouring and transport of toxic particulate from the street surface. In storm 1 there is a general temporal coincidence of peak flow with maximum suspended solids and SIRM values although there are two subsidiary magnetic peaks associated with the rising limb of the storm hydrograph. If by analogy with storm 2, relatively low SIRM values can be interpreted as an indication of proportionally greater sediment contributions from below-ground sources, then these leading peaks may reflect early inputs from contributing surface areas, such as road gutters, located relatively close to the sewer outfall.

In terms of metal loadings and sources, the magnetic properties would therefore indicate that for storm 1 both below- and above-ground sources are contributing to outfall toxicity but at different times during the storm. In storm 2, the increase of the metal concentration well into the recessional limb of the hydrograph would point to the predominance of above-ground sources in this event. At present, the work is insufficiently developed to allow identification of magnetic characteristics and further magnetic typing of potential sources is required. The parallelism between metal levels and SIRM values would suggest nevertheless that magnetic parameters might be useful as a rapid and non-destructive surrogate method for monitoring metal patterns during storm flow events as well as for helping to identify heavy metal sources within the urban stormwater drainage system (Beckwith *et al.* 1984; 1986).

9.5 Conclusions

The three areas of application illustrated above by no means cover the whole range of potential con-

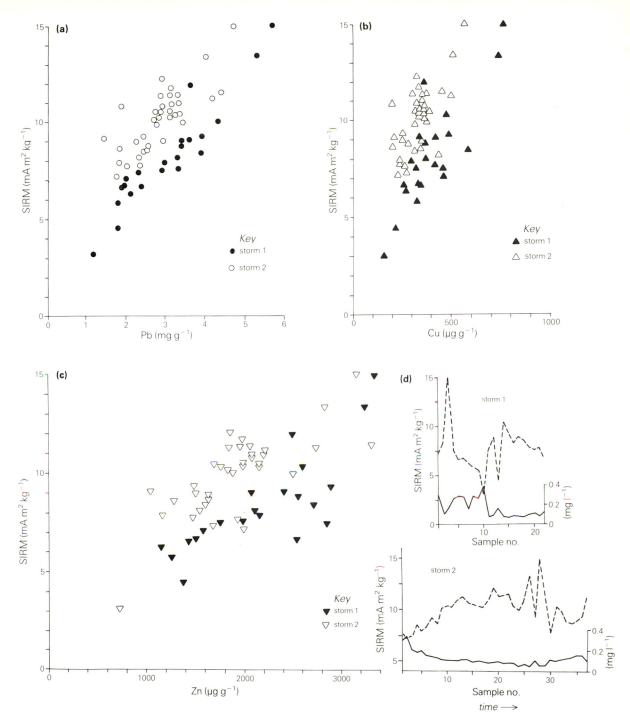

Figure 9.5 SIRM versus heavy metal concentration (a, b and c) and total suspended solids (d) for two high flow episodes in the Hendon stormwater catchment (see text).

tributions which mineral magnetic measurements may make to studies of fluvial systems. There is, in addition, scope for mineral magnetic studies of sediment transport in a much wider range of morpho-genetic régimes, on a much larger spatial scale with a focus on lithological differentiation rather than on topsoil/parent material contrasts, and on a much longer temporal scale where within valley sediment storage is a significant element in the long-term evolution of the geomorphic system. An important area of future methodological development will be the association of mineral magnetic study with complementary mineralogical radiometric and geochemical analysis. Mineral magnetic parameters that can be used to identify sediment sources may provide a basis for surrogate particulate pollution monitoring in rural as well as urban and industrial environments. The rôle of mineral magnetic measurements in integrated catchment studies is illustrated more fully in Chapter 16.

[10]

Mineral magnetic studies of lake sediments

A lake is the landscape's most beautiful and expressive feature.
It is the earth's eye; looking into which the beholder measures the
depths of his own nature

Henry David Thoreau
Walden

Lake-folk require no fiend to keep them on their toes

W. H. Auden
Lakes

10.1 Lake sediments and environmental reconstruction

In recent years, lake sediment studies have become increasingly important in many branches of environmental science. This reflects both a natural scientific curiosity in the sediment-based reconstruction of past environmental conditions, and also, especially over the last decade, a need to set studies of present day environmental processes and problems in a longer time perspective. This is often best achieved within the lake-watershed ecosystem framework (Borman & Likens 1969, Oldfield 1977, O'Sullivan 1979). Within this framework, the evaluation of human impact and the effective assessment of its present and future implications will often depend on (a) comparing present day observations and experimental results with data derived from the analysis of sediments predating anthropogenic effects (Oldfield *et al.* 1983b), (b) reconstructing in detail the history of human impact on some aspect of biosphere function, for example soil erosion (Dearing 1983), heavy metal flux (Edgington & Robbins 1976), or primary productivity (Battarbee 1978), and (c) linking present day process study and historical reconstruction so as to develop a continuum of insight into environmental processes, within which past and present states and rates can be compared (Oldfield 1977, 1983b).

Lake sediments are especially useful in historical monitoring for several reasons. Despite unresolved problems of mud–water exchange chemistry and early diagenesis, there is often a conformity of process linking past and present deposition mechanisms. Moreover, evidence both of primary productivity in aquatic communities and of material flux in the total lake-watershed ecosystem are often well preserved in the sedimentary record. Rates of sediment accumulation are usually more rapid in lakes than in marine environments. In consequence, the period of accelerating human impact on environmental processes over the last one or two centuries is usually well resolved in the upper sediments. Several new dating techniques have been developed which are applicable to recent lake sediments and allow close chronological

101

control of recent sediment-based analyses (see e.g. Oldfield 1981). Finally, the lake-watershed ecosystem is material-bounded in large measure and thus provides a convenient and spatially finite framework for study.

From the preceding chapters (8 & 9) it can be shown that variations in the type and concentration of magnetic minerals in lake catchments will often be related to soil and slope processes and land use changes. Moreover, the magnetic properties characteristic of different soils and lithologies are often highly conservative within the drainage net. The mineral magnetic characteristics of lake sediments are therefore likely to be related to and often indicative of specific sources and processes. The following sections explore the rôle of mineral magnetic measurements in lake sediment studies.

10.2 The origin of magnetic minerals in lake sediments

The magnetic minerals present in lake sediments are of varied types and origins. Interpreting the record of mineral magnetic variation in the sediments is therefore strongly dependent on evaluating alternative sources for a given lake and catchment, with a view to identifying the dominant types, sources and pathways represented. The conventional distinction between authigenic, diagenetic and allogenic sediment is useful in this respect. Authigenic magnetic minerals are those formed by chemical or biogenic processes *in situ* after deposition of the sediment. Diagenetic magnetic minerals are the result of the transformation of existing magnetic or non-magnetic

minerals to new magnetic types (cf. 12.2). Allogenic magnetic minerals are brought into the lake from outside. They may have originated within the drainage basin of the lake or have been transported (for example by wind or by man) from more distant sources beyond the immediate catchment. Table 10.1 identifies the main types and sources of magnetic minerals found in lake sediments. Although some of the least abundant of these are authigenic, and for others, such as magnetite and haematite, it is sometimes difficult to preclude entirely an authigenic or diagenetic origin, in the majority of cases studied so far, circumstantial evidence strongly suggests an allogenic origin. The circumstantial evidence includes (a) a tendency in many lakes for magnetic mineral concentrations and fluxes to peak most sharply in more marginal sediments (Thompson *et al.* 1975) and in zones close to inputs from the catchment, (b) a strong direct link between down-profile variations in magnetic susceptibility and other palaeoecological or chemical indicators of accelerated detrital mineral input, (c) clearly established linkages between sediment and catchment source in a variety of lakes (cf. Ch. 16) and (d) confident ascription of many recent mineral magnetic variations to well documented catchment events. Thus the studies completed so far have led us to regard magnetic minerals in lake sediments as overwhelmingly allogenic except where there is positive evidence to the contrary (e.g. Hilton and Lishman 1985).

ALLOGENIC MAGNETIC MINERALS FROM WITHIN THE LAKE CATCHMENT
Lithology exerts an important control on the magnetic

Table 10.1 Magnetic minerals in lake sediments: major types and sources.

Source type	Location origin	Major pathways	Magnetic mineral types
bedrock	lake catchment	streams overland flow mass movement	magnetites; haematite; pyrrhotite MD/SD
soils	outside lake catchment	wind	impure magnetite; maghemite SD/SP goethite/haematite
volcanic ash	lake catchment	streams, etc.	magnetites
	outside lake catchment	wind	
fossil fuel combustion and industrial processes	lake catchment	streams, etc.	impure magnetite haematite
	outside lake catchment	wind	
lake sediments	authigenic/diagenetic/post-depositional/*in situ*		magnetite greigite

Table 10.2 Maximum χ values from late-glacial sediments.

Site	Locality	Bedrock	χ (10^{-8} m^3 kg^{-1})
Hawes Water	Lonsdale, northern England	limestone	0.4
Loch Garten	Speyside, Scotland	schist/granite	0.4
High Furlong	Blackpool, England	marl	0.5
Roos	Holderness, England	chalk	1.0
Hornsea Mere	Holderness, England	chalk	1.1
Mellynllyn	Northern Wales	slate	1.2
Paajarvi	Southern Finland	schist/granite	1.2
Loch Morlich	Speyside, Scotland	schist/granite	1.2
Nant Ffranancon	North Wales	slate	1.7
Lough Fea	Northern Ireland	granite/gabbro	2.8
Kiteenjarvi	Eastern Finland	schist	4.0
Kingshouse	Rannoch, Scotland	granite	6.7
Vuokonjarvi	Eastern Finland	granite/gneiss	9.0
Ormajarvi	Southern Finland	gneiss	10
Geirionydd	North Wales	slate/rhyolite	20
Hjortsjon	Southern Sweden	granite/gneiss	20
Windermere	Northern England	slate/andesite	22
Loch Davan	Deeside, Scotland	schist/granite	50
Bjorkerods Mosse	Southern Sweden	gneiss/dolerite	100
Loch Lomond	Southern Scotland	schist/basalt	130
Geitabergsvatn	Iceland	basalt	160
Barrine	Queensland, Australia	basalt	200

mineralogy of lake sediments. This is most easily illustrated by reference to specific susceptibility values obtained from the earliest late-glacial sediments present in a variety of infilled lake basins (Table 10.2). Such sediments predate weathering and soil formation and are often poorly sorted. They therefore reflect and integrate the primary magnetic mineralogy of the freshly exposed parent material surrounding the lake basin at the close of the last glaciation. The full range of specific susceptibility values reflects the range of parent material from calcareous boulder clay to basalt, and confirms that lithology is a major variable controlling magnetic concentrations in lake sediments.

Modification of the primary iron compounds in bedrock during the course of weathering and soil formation is a major theme in Chapter 8 and also provides the basis for distinguishing magnetically between topsoil, subsoil and bedrock sediment sources in rivers (see Ch. 9). In the case of lake sediments, any part of the drainage basin regolith exposed to erosive processes is a potential sediment source and most allogenic sediments will be a mixture of soil-, subsoil- and bedrock-derived material in widely varying proportions. Magnetically therefore, we may expect allogenic lake sediments to reflect not only the primary magnetic minerals in the catchment but also the secondary magnetic minerals formed in the soil as a result of the processes summarised in

Chapter 8. The evidence from catchments within which secondary magnetic minerals greatly outweigh primary magnetic minerals in terms of overall importance and erodability (e.g. the Lac d'Annecy in the French Jura), confirms the persistence of some if not all of the eroded secondary minerals in lake sediments despite widespread indications of inhibition and reversal of secondary magnetic mineral formation in gleyed (waterlogged) soil horizons. The most conclusive evidence for the persistence of allogenic catchment-derived secondary magnetic minerals in lake sediments relates to those produced by fire (Rummery *et al.* 1979). Although incontrovertible *direct* evidence for the persistence of 'biogenic' or 'chemical' soil magnetic minerals in lake sediments (see Ch. 8) is more difficult to establish, the Rhode River case study (Ch. 16) clearly confirms persistence in that environment and it seems equally likely in most lakes.

Human activity within the catchment or on the lake itself may also generate magnetic minerals which pass into the lake sediments. The particulate fraction of effluent from industrial sites (12.5) and in urban stormwater drainage (9.4) is often highly magnetic. Moreover, in the recent sediments of the Grand Lac d'Annecy (Higgitt 1984, unpubl.) a striking susceptibility peak in several cores was shown to be the result of clinker from the coal barges which plied the lake in the first half of the 20th century.

ALLOGENIC MAGNETIC MINERALS FROM OUTSIDE THE LAKE CATCHMENT

Clearly, all the types of magnetic mineral encountered in the atmosphere (Ch. 11) contribute to the sedimentary record. However, only in specific circumstances will the contribution become significant in comparison with the input from the land surfaces surrounding the lake. These circumstances may result from volcanic activity, from fossil-fuel combustion and from forest fires.

Oldfield *et al.* (1980) show the impact of at least four tephra layers from 10 000 to 300 years in age, on the magnetic susceptibility of a suite of samples from three lakes in the Highlands of Papua New Guinea. Each ash layer is relatively enriched in primary magnetite and gives rise to a distinctive susceptibility maximum, (of Hamilton *et al.* 1986). There is strong evidence from the sediments of Newton Mere (Oldfield *et al.* 1983a) in the English Midlands and the nearby Whixall Moss, for the recent deposition of magnetic minerals resulting from fossil fuel combustion and industrial processes in areas lying down wind, but outside the tiny catchment (see Ch. 11). The evidence for wind erosion and for fine soil and charcoal dispersal associated with major forest and savannah fires suggests that these agencies may contribute magnetic minerals to lake sediments down wind. High ferrimagnetic concentrations have been detected in peat cores from Bega Swamp in S. E. Australia, at two horizons with high charcoal counts (Singh *et al.* 1979), and presumably were caused by atmospheric transport of fire-enhanced topsoil.

AUTHIGENIC AND DIAGENETIC MAGNETIC MINERALS

Jones and Bowser (1978) note that Fe_2O_3, presumably goethite, has been identified by electron microprobe analysis as the major mineral constituent of the iron-rich ferromanganese nodules recovered from Romahawk Lake, Wisconsin, and that both todorokite and goethite have been identified by X-ray diffraction there and at other sites including the Green Bay arm of Lake Michigan. Dell (1971) has identified greigite (Fe_3S_4) in sediment cores from Lake Superior. It is likely that magnetic phases associated with nodules and crusts, and with sulphide formation are more frequent than the current range of mineral magnetic case studies would indicate. Hilton and Lishman (1985) have shown that the volume susceptibility peak present in the fresh near-surface sediments of Esthwaite Water, in the South of the English Lake District can be reduced by aeration of the samples. These sediments are very rich in reduced forms of iron and sulphur which, at the right combination of E_h and pH appear to give rise to a rather impermanent magnetic phase. Similar results have been reported by Smith (pers. comm.) from an artificial pond receiving drainage from sulphur-rich coal waste.

The other significant authigenic magnetic phase in lake sediments arises from the production of magnetosmes by bacteria (see Ch. 16). Blakemore's observations (1975) suggest that this process may make an important contribution to the magnetic properties of sediments in a wide variety of lakes, though there is little support for this in the mineral magnetic studies completed so far.

PARAMAGNETIC MINERALS

Goddionduon, a small upland lake in N. Wales (see 10.4), has been studied in considerable detail by Bloemendal (1982). Here the peak in χ gives rise to extremely low SIRM/χ ratios and, it corresponds with maximum Fe and Mn concentrations (Fig. 10.1) and with peak Fe/Al and Mn/Al ratios. Mossbauer spectra and hysteresis loop plots for samples from the feature are typical of paramagnetic material.

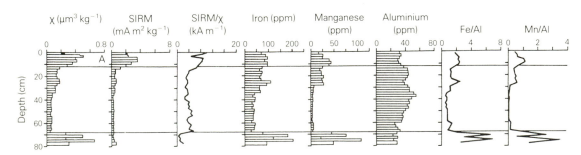

Figure 10.1 Paramagnetism in lake sediments. Mineral magnetic and chemical data from Llyn Goddionduon Core 120/-100. Susceptibility peak B (68–80 cm) coincides with minimum SIRM/χ and peak Fe, Mn, Fe/Al and Mn/Al.

Bloemendal concludes that the paramagnetic effects which dominate the magnetic properties of these samples are responsible for the very low SIRM/χ ratios. The feature coincides with an inferred lowering of lake level at the site (cf. Bengtsson & Persson 1978). At Loch Davan in north-east Scotland (Edwards 1978) and at Weir's Lough in Northern Ireland (Hirons 1983) similar χ peaks and SIRM/χ minima coincide with maximum Fe and Mn concentrations in the early Flandrian sediments. Following Mortimer (1942) and Mackereth (1966) these high concentrations are interpreted as a reflection of the acidification, leaching and podsolisation of the surrounding iron-rich soils under increasingly reducing conditions. The iron and manganese salts, once released from the catchment in solution are then deposited in the lake sediments provided conditions at the mud–water interface are sufficiently aerobic. The high susceptibility values are interpreted as a reflection of Fe and Mn accumulation as paramagnetic salts in the sediments. An increase in susceptibility and decrease in isothermal remanence observed upon cooling the Davan sediments to liquid nitrogen temperature indicates that the magnetic minerals are paramagnetic

and are not superparamagnetic ferrimagnetic grains (see Section 4.7). From Bloemendal's summary and from the published chemical analysis of many European and N. American lake sediment sites it is clear that comparable features are widespread.

10.3 Sampling and measurement

Although it is possible to carry out a wide range of mineral magnetic measurements on uncontaminated sediment samples obtained by any reliable coring method, there are major advantages in using non-metallic core tubes or liners. Moreover the whole process of initial measurement can be expedited by keeping whole cores intact so that they can be scanned for volume susceptibility variation before extrusion and subsampling. Pneumatic Mackereth (1958, 1969) corers provide ideal samples not only for palaeomagnetic studies (see Section 14.2) but also for the mineral magnetic studies considered here. Alternative types of corer are considered briefly in Oldfield (1981). Irrespective of the type of corer used, and provided the core can be retained in, laid on, or

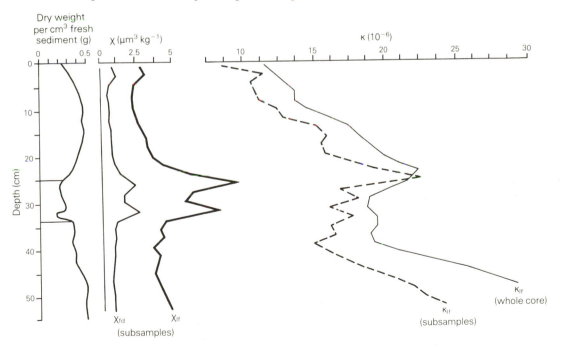

Figure 10.2 A comparison of whole core, single sample volume (κ) and single sample mass (χ) susceptibility traces for Core 16 from Lawrence Lake, Michigan. The plots show the effects of smoothing on the whole core trace and of strong variations in the dry mass/fresh wet volume ratio on the comparability of volume and mass specific measurements. The sharp rise in χ_{fd} and χ_{lf} at 34 cm is the result of early 'European' settlement round the lake and it is partly obscured in the volume-based measurements.

extruded into a rigid non-magnetic liner, whole core susceptibility scanning (Radhakrishnamurty *et al.* 1968, Molyneux & Thompson 1973) should be possible using a suitably designed sensor.

Scanning of unopened sediment cores in plastic drainpipe or perspex (plexiglass) tubes for variations in volume susceptibility (κ) provides rapid insight into down-core mineral magnetic variations. Where use of a microprocessor-controlled long core susceptibility bridge is feasible (Section 6.3.1) the rate of scan is in excess of 6 m min^{-1} with an automatic reading interval of 0.5–5 cm. Whole core scanning necessarily presents a smoothed trace of variations in magnetic susceptibility (due to instrument response) and may fail to resolve fine detail. Moreover, it is unsuitable for detecting and portraying accurately variations as the mud–water interface is approached, since the readings will be depressed both by the increase in water content which often occurs in the most recent sediments and also, very near the surface, by the fact that the sensor 'sees' both the sediment and the immediately overlying water. If after whole core scanning, a more detailed picture of variation is required, extrusion of the core and measurement of the susceptibility as well as the induced remanence characteristics of single samples on a constant volume or preferably on a dry weight specific basis, will be necessary. Figure 10.2 compares whole core and single sample susceptibility data from the same core.

10.4 Prospecting, core correlation and sediment accumulation rates

Many palaeolimnological studies are marred to some degree by the extent to which inferences rely on data from a very small number of cores and by the problems involved in correlating cores which have provided complementary but often poorly synchronised data. Both problems arise largely from practical constraints since most methods of biostratigraphic, sedimentological or chemical analysis are time consuming, sediment destructive and relatively costly. One response to this problem is to select a 'representative' or 'master' coring site and use one of several large diameter samplers now available to provide enough bulk for different types of analysis. In this case, choice of coring site is crucial and techniques which may aid site selection by speeding initial prospecting are potentially very useful. Even this approach provides

an inadequate basis for fully quantitative budget studies within a lake-watershed ecosystem framework (cf. Oldfield 1977), since the latter may call for estimates of material flux or productivity for given time intervals on a whole catchment–whole lake basis. Extrapolation from a single core or a small selection, to a whole lake is complicated by spatial variations in primary sedimentation rate, as well as by mechanisms of post-depositional sediment resuspension, redistribution and focusing (Davis & Ford 1982). Mathematical models for such extrapolations lack empirical validation (Lehman 1975), and 'whole' lake estimates based on empirical studies using data from multiple cores (e.g. Davis 1976, Davis & Ford 1982) are rather scarce. Rapid methods of correlation are therefore desirable not only for aiding the comparison and synchronisation of data from several cores, but also for developing quantitative estimates of sediment flux. In addition detailed core correlations may also aid the establishment of chronologies of sedimentation both by facilitating sample aggregation for radiometric dating (Oldfield *et al.* 1978) and by providing an independent check on the internal consistency of dates derived from different cores from the same lake. Finally, magnetically distinct horizons at or near the mud–water interface resulting from catchment events such as forest fires or land use changes, provide a natural time stratigraphic marker for subsequent follow-up studies and recurrent monitoring. The studies summarised below have been chosen to illustrate some of the above applications.

The first use of magnetic susceptibility scanning as a means of multiple core correlation arose from the realisation that in Lough Neagh in Northern Ireland, single sample volume susceptibility measurements on a range of widely separated 3 m cores from Antrim Bay provided a basis for very detailed comparison (Fig. 10.3). The correlation suggested by susceptibility measurements supported the record of secular variation previously found in the same samples (Thompson 1973). Subsequent work confirmed that similar susceptibility correlations could be established using the much more rapid whole core scanning method. Moreover, these correlations were shown to hold good in widely separated depositional environments and in sets of cores with differences in mean accumulation rate of at least three times. In order to test the validity of the susceptibility-based correlations, diatom analysis was carried out on each core. As a result, it was possible to show that the main peak in susceptibility coincided with a distinctive

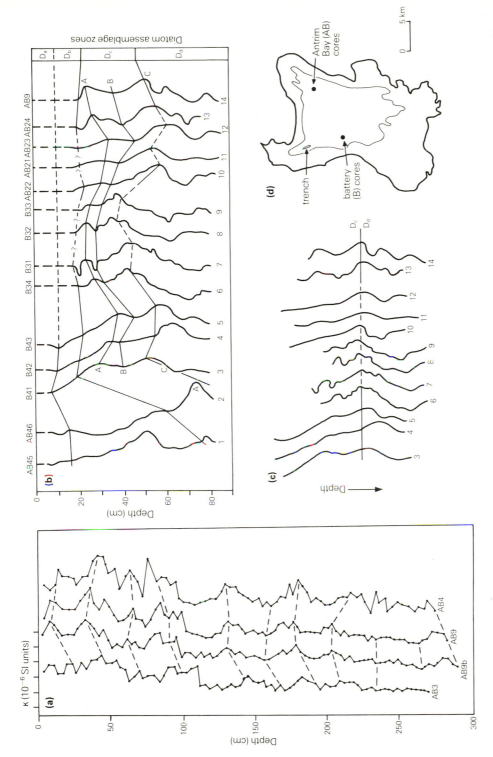

Figure 10.3 Susceptibility traces from Lough Neagh cores. Correlated single sample traces from four Antrim Bay 3 m cores are shown in (a). In (b) and (c) 14 plots from the Battery and Antrim Bay show variations in volume susceptibility measured on whole 1 m cores, compared with changes in the fossil diatom content of the sediments (Battarbee 1978). Diatom zones D_a–D_d are shown in (b) as well as several horizons of susceptibility-based correlation. In (c), the susceptibility traces are aligned by means of the D_c/D_d diatom boundary (see Thompson et al. 1975).

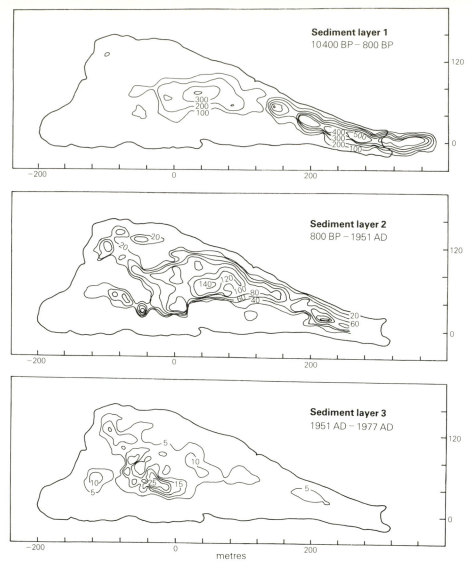

Figure 10.4 Sediment thickness for three time intervals in Llyn Goddionduon, N. Wales. For each interval sediment thickness is contoured in centimetres. The co-ordinated grid is in metres. The diagrams are redrawn and simplified from Bloemendal (1982). [210]Pb assay carried out more recently suggests that the 2/3 boundary may predate 1951 AD.

change in the fossil diatom flora preserved in the sediments (Fig. 10.3). Similar independent confirmation of susceptibility-based correlations has come from visual stratigraphy at Havgardsjon in S. Sweden (Dearing 1983) and from X-ray traces at Lake Washington, near Seattle in northwestern United States (Edmondson, pers. comm.).

At Llyn Goddionduon, a comprehensive grid of over 130 short cores was established to provide an empirical basis for estimating total sediment influx to the lake between recent dated horizons. Despite severe stratigraphic complications arising from the relatively large expanse of shallow water, high wind exposure, recent water level changes and widespread truncation and redeposition of sediments, two magnetically distinctive horizons were identified in whole core susceptibility scans and their approximate synchronism confirmed by single sample measure-

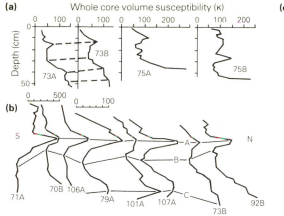

Figure 10.5 Whole core volume susceptibility logs from Lake Washington (see text). Peak A, the depths of which are plotted on the map, is a result of the lowering of lake level and the diversion of the Cedar River into the lake in 1915.

ments (Bloemendal *et al.* 1979, Bloemendal 1982). The lower feature, 'Peak B' (Fig. 10.1a) was dated to 800 BP by correlation with radiocarbon-dated peat formed at the southern end of the lake. The upper feature, 'Peak A' was ascribed to a forest fire in 1951 and the subsequent inwash of magnetically enhanced soil (see 10.7). It was thus possible to provide rough estimates of net sediment input to much of the lake bed for three time intervals: (a) from the end of the late-glacial period at the site (10 400 BP) to 800 BP, (b) from 800 BP to 1951 AD and (c) from 1951 AD to 1977 AD (Fig. 10.4). Although in this pioneer study there are uncertainties arising both from problems inherent in the method and from the particular choice of lake, it demonstrated the power of the new approach and heralded a number of similar studies during subsequent years.

One of the most useful practical applications of multicore studies is in the detection of anomalous, or disturbed sedimentation. Figure 10.5 shows volume susceptibility (κ) logs from sets of 50 cm long recent cores taken from Lake Washington. Numbers refer to coring stations, letters to replicates taken on the same day from each station. Although the features of the traces for the Station 73 cores and for those from Station 75 are replicates, the susceptibility variations are not comparable. By using susceptibility scans as a guide to core selection, and ^{210}Pb for chronology (Schell, pers. comm.), it is possible to establish the intercore correlation scheme outlined in Figure 10.5 alongside a plan which summarises the implications of the scheme in terms of the thickness of sediment accumulated since *c.* 1915 when the lowering of lake level and diversion of the Cedar River gave rise to peak

A in the traces. The same approach has been used in Lake Washington for selecting long cores for further study (Oldfield *et al.* 1983a). In Lake Washington, some of the susceptibility peaks are the result of volcanic ash layers including the Mt Mazama tephra recorded over a vast area of the N. W. of the USA. Where the main magnetic variations in a set of cores are the result of synchronous volcanic ash layers, whole core susceptibility scanning can greatly accelerate the development of tephrochronologies not only within but between lakes (Oldfield *et al.* 1980).

Dearing and coworkers (Dearing *et al.* 1981, Dearing 1983) have published two contrasted case studies of total sedimentation, one based on Llyn Peris, north Wales (53°02′ N, 4°06′ W) an oligotrophic lake with a large grazed upland catchment, and one based on Havgårdsjön (55°29′ N, 13°22′ E) a shallow eutrophic kettle-hole lake with a small low lying cultivated catchment in Scania, southern Sweden. In both cases, mineral magnetic studies were combined with ^{14}C, ^{210}Pb and ^{137}Cs as well as palaeomagnetic dating.

In Llyn Peris, rapid sedimentation associated with low susceptibility values is ascribed to slate dust input arising from quarrying in the catchment. High susceptibility values are ascribed to higher stream discharge levels and consequent channel erosion associated with periods of overgrazing by sheep. The period of recent strongly accelerating sedimentation is largely the result of erosion associated with hydroelectric plant construction.

In Havgårdsjön (Fig. 10.6), the changing erosion rates can be closely compared with the detailed record of agricultural history and land-use change available

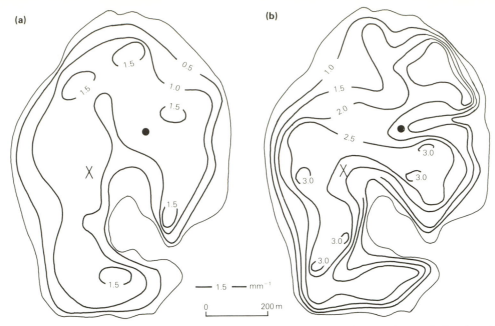

Figure 10.6 Sediment accumulation patterns for two periods in lake Havgårdsjön, Scania, S. Sweden: (a) *c.* 1550 AD – *c.* 1712 AD and (b) *c.* 1712 AD – 1979 AD. ● marks deepest zone of lake basin.

for the catchment. The period of accelerating erosion in the 19th century coincides with a major shift in animal husbandry from cattle to sheep. The overall increase in erosion rates since the late 17th century mirrors the general expansion of ploughed land over the past three centuries. Dearing has used the correlation scheme as a basis for modelling nutrient flux within the lake and catchment (Oldfield *et al.* 1983b).

Magnetic measurements combined with radiometric dating have also made possible a similar reconstruction of erosional loss for the catchment of a small crater lake in the New Guinea Highlands, Lake Egari. Intercore magnetic susceptibility-based correlation is possible here, largely as a result of two major tephra layers dated *c.* 300 and 1200 BP. The combination of susceptibility-based correlations and three ^{210}Pb-dated profiles has provided an extremely detailed picture of changing erosion rates which once more relates closely to the history of land use in the catchment reconstructed in this case from pollen-analytical evidence (Worsley 1983). Periods of forest clearance and subsistence gardening are associated with higher erosion rates, and the post-1950 decades of Australian contact record a further major increase.

Table 10.3 summarises the evidence for changing

erosion rates from Llyn Goddionduon, Llyn Peris, Havgårdsjön and Egari, and compares the rates with those derived from other recent studies. In comparing the results and in evaluating the examples outlined, attention must be paid to the limitations of this type of approach, especially in terms of the suitability of particular lakes and catchments, and the many sources of error involved in calculation (Bloemendal 1982, Oldfield *et al.* 1985a).

10.5 Sediment resuspension and focusing

One of the most widespread types of sediment redistribution is that frequently referred to as sediment focusing whereby material from marginal depositional contexts is resuspended and deposited in deeper water. The process is common in dimictic lakes where it results largely from the seasonal episodes of water 'overturn' and mixing between the periods of summer and winter stratification.

The best documented case study of sediment focusing is at Mirror Lake, New Hampshire (Davis & Ford 1982). The Mirror Lake sediments are extremely poor in ferrimagnetic minerals and the volume susceptibility variations are often masked by the

Table 10.3 Erosion rates calculated from lake sediment influx studies using mineral magnetic (A–D) and other methods (E, F). A fuller comparison with estimates derived from Reservoir Surveys and continuous monitoring studies is given in Bloemendal (1982).

	Land-use types and events	Total sed. (kg ha^{-1} a^{-1})	Organic sed. (kg ha^{-1} a^{-1})	Inorganic sed. (kg ha^{-1} a^{-1})	Watershed area (hectares)
(A) Llyn Goddionduon 1951–1977 AD	post forest fire	263–326	80–100	183–226	24.8
800 BP–1951 AD	upland grazing	126	27	99	
10 400 BP–800 BP	forest and moorland	36	8	28	
(Bloemendal 1982)					
(B) Llyn Peris 1965–1975 AD	construction	—	—	420	3800
c. 1800 AD	upland grazing	—	—	50	
(Dearing et al. 1981)					
(C) Havgårdsjön c. 1650 AD	stock farming	—	—	1000	140
c. 1850–1980 AD	and cultivation	—	—	3000	
(Dearing et al. 1981)					
(D) Egari 1950–1973 AD	subsistence horticulture and roads	2300			
c. 1600–1800 AD	deciduous forest	200			
c. 1900–1950 AD	peak subsistence	900–1200			
(Oldfield et al. 1985a)					
(E) Frains Lake 1830–1970 AD	arable and roads	907	7	900	18.4
Pre-1830 AD	oak forests	96	6	90	
(Davis 1976)					
(F) Mirror Lake (mean Holocene)	forest	—	—	36*	107
(Davis & Ford 1982)					

* Excludes estimated dissolved SiO_2 input converted to biogenic silica as diatom frustules.

diamagnetic effect of carbon, biogenic silica and water. Whole core susceptibility scans are thus unsuitable for detailed stratigraphic correlation. Instead we have used measurements of isothermal remanence, $IRM_{300\,mT}$, generated by a pulse magnetiser with a peak field of 0.3 T and carried out on single samples.

A grid of forty-six 35–50 cm long cores was obtained by working from the ice of Mirror Lake in February 1982 and 1983 using a modified Gilson benthos corer which preserved an undisturbed mud–water interface. Figure 10.7 plots IRM versus accumulated dry weight for ten of the cores. Cores 35, 38 and 36 are from the western end of the lake in 6–8 m water depth and core 1 is from 7 m at the southeastern end of the lake. The remaining cores are from central locations in greater water depths (8.5–10 m). For all or part of the period between magnetic features A and B, as for the preceding Holocene record as a whole, strong sediment focusing from shallow to deep water is recorded. During the recent post-B period, simple

focusing of this type is not apparent and both the deep central and the shallow marginal core sets include cores with light and heavy sedimentation. Cores 18, 34 and 32, with minimum post-B accumulation, lie to the north-east of the axis of the main basin and furthest away from the river-borne supply of sediments from the relatively more disturbed parts of the catchment.

The Mirror Lake IRM variations are also accompanied by changes in hysteresis ratio parameters derived from d.c. demagnetisation experiments and hence not primarily related to concentration. Figure 10.8 plots two of these for all the samples within each magnetostratigraphic subdivision for two marginal (30 and 35) and two deepwater (28 and 34) cores. From this we see that in these cores, during the period of inferred sediment focusing (pre-A and A–B) the magnetic parameters for the central and marginal cores are quite distinct. The mineral magnetic characteristics of the marginal core samples are suggestive of a relatively higher coarseparticle related haematite content, which is consistent

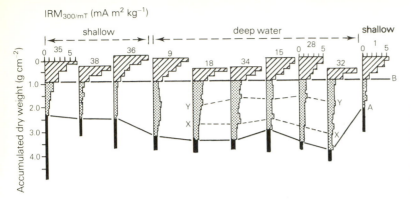

Figure 10.7 Correlated profiles of IRM_{300mT} versus accumulated dry weight for ten cores from Mirror Lake, New Hampshire. Horizon B is a late 19th/early 20th century feature, horizon A is roughly contemporary with the beginning of local colonial settlement in the late 18th century.

with the earlier results of Stober (1979) from comparable granite-dominated lithologies in Finland. By contrast, for the post-B period the parameters for all cores correspond very closely. These results are consistent with a model in which

(a) sediment focusing selectively winnows and redeposits finer/less dense material from marginal to central sites during the earlier period.

(b) either a historically documented rise in water level (Davis & Ford 1982) disrupts the focusing mechanism or an increase in sedimentation rate, coupled with possible changes in sediment type and source, gives rise to more rapid sediment burial and proportionally less effective resuspension and transfer to deeper water.

10.6 Sediment sources and ecological change

SUSCEPTIBILITY AND EROSION

The susceptibility traces from Lough Neagh, Northern Ireland, were the first to suggest a relationship between mineral magnetic variations in the sediments and the record of land-use history in the drainage basin. As Figure 10.9 shows, susceptibility tends to increase irregularly towards the surface in 3 m Core AB9. The timespan of the core is roughly the last 6000 years, beginning in the early Neolithic period characterised, in the pollen record, by reduced tree pollen representation (the 'Elm Decline') and an early increase in weeds and disturbed ground indicators. The pollen diagram from AB9 thus records vegetation history and land-use change for virtually

the whole period of significant human impact on the landscape. Forest clearance and farming are indicated by increases in the relative frequency of pollen and spore types such as ribwort plantain, bracken and the grasses. Representation of the first two tends to peak sharply during relatively brief phases of initial clearance and maximum disturbance. The grass pollen trace, which, in Lough Neagh, can be confidently interpreted as an indicator of the spread of open habitats, such as pasture and moorland (O'Sullivan et al. 1973), tends more towards a step-wise increase, each rise coinciding with bracken and plantain peaks. Progressive deforestation in a series of discrete episodes is indicated. The parallelism between the grass pollen record and susceptibility curve is remarkable and it is difficult to avoid attributing increases in the latter to the sequence of deforestation indicated by the former. The latest peaks coincide with a steep decline in carbon concentration in the sediment and both features follow the beginning of a continuous cereal pollen record, suggesting that actual tillage as well as deforestation was important in providing increasing volumes of minerogenic sediment. From the evidence summarised, Thompson et al. (1975) concluded that susceptibility may be regarded as an erosion indicator in Lough Neagh with the high values reflecting higher inputs of primary titanomagnetites from the basalt bedrock of the large drainage basin as successively more extensive deforested areas were created and the parent material exposed to erosion. More recently, by measuring the magnetic susceptibility of material collected in seston traps in Lough Neagh, Dearing and Flower (1982) have shown that susceptibility peaks correspond with monthly rainfall maxima (Fig. 10.10). They conclude that susceptibility peaks in the sediment record are

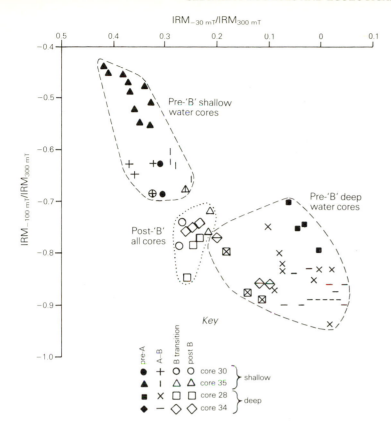

Figure 10.8 IRM_{-30mT}/IRM_{300mT} versus IRM $_{-100mT}/IRM_{300mT}$ plotted for samples from a selection of shallow and deep water cores. The variations within and between cores reflect degrees of sediment focusing within the lake (see text).

the expression of hydrological controls and that they reflect forest clearance and culvitation because these processes accelerate runoff and increase stream discharge, with a resulting increase in the proportion of coarser more magnetic sediments delivered to the lake.

A different example of the dependence of susceptibility on particle size in lake sediments emerges from Thompson and Morton's (1979) study of the Loch Lomond sediments. They show that in contrasted samples the highest specific susceptibility is in the silts below 32 μm. This gives rise to strongly particle size dependent susceptibility variations in the sediments. Peak values coincide with the highest concentrations of fines, and minima with the sandiest horizons throughout (Fig. 10.11). Magnetic crystal size varies little with sediment particle size however and the particle size related variations in susceptibility are interpreted simply as variations in the concentration of primary magnetite from parent material in the catchment. The relationship between susceptibility and pollen is reminiscent of that in the

Lough Neagh sediments, so it follows that particle size and, in consequence of this, susceptibility, vary in response to man's impact on erosive processes. In Loch Lomond, it would seem that deforestation and ploughing have provided a higher proportion of fine silt for transport to the lake bed. This theme of particle size related variation is taken up in Chapter 16 which considers sediments and sources in the Rhode River, Chesapeake Bay.

Results obtained from study of a 6 m core from Lough Fea, a small lake lying in the western part of the Lough Neagh drainage basin, cover the whole of the Holocene. Figure 10.12 plots SIRM alongside the concentration of several chemical elements and the relative frequency of *Calluna* (heather) pollen. Zones M and L (*c*.11 000 to 9000 BP) cover the end of the late-glacial and the opening of the Flandrian (Holocene) stage. K, Na and Mg which, because of their relatively high solubility, were used by Mackereth (1965, 1966) and Pennington *et al*. (1072) as indicators of the erosive input of unweathered parent material, all show parallel and synchronous

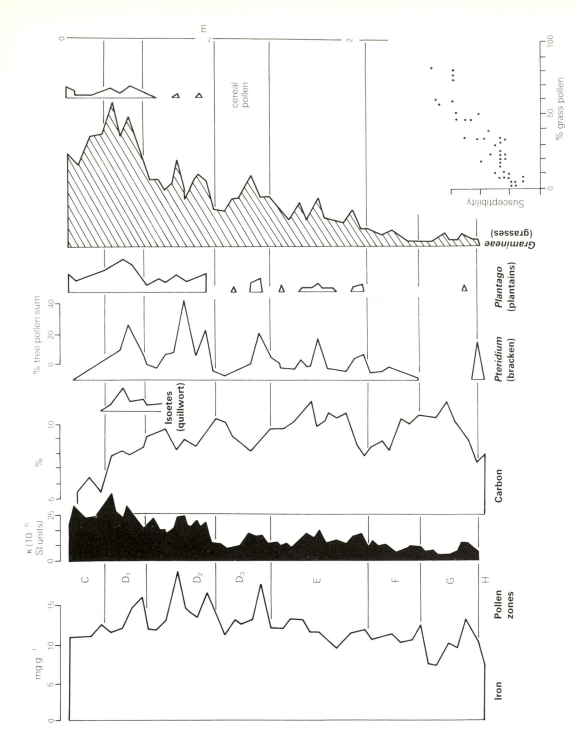

Figure 10.9 Chemical and pollen frequency data for Core AB9, from Antrim Bay, Lough Neagh compared with the magnetic susceptibility values.

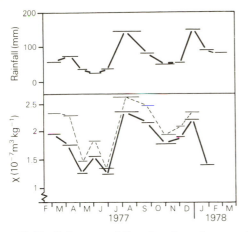

Figure 10.10 Bulk susceptibility of sedimenting material at the Antrim Bay site, Lough Neagh, calculated on minerogenic (solid line) and allogenic (dashed line) bases and compared with monthly rainfall (see Dearing & Flower 1982).

changes of concentration similar to those found in other upland sites in N. W. Britain. Peak concentrations during periods of bare soil and periglacial activity in the late-glacial period are followed by sharp declines as the land surfaces of the drainage basin become clothed in forest and in stable maturing soils. The mid-Holocene period of maximum forest cover and sustained biotic regulation of particulate loss gives rise to minimum K, Na and Mg concentrations but as soon as human activity affects the small upland drainage basin directly at the E/D_3 boundary (c.3500 BP), values begin to increase in response to deforestation and agricultural extension. At this stage, heather pollen becomes an important component in the record as a result of soil and peat erosion in the catchment. SIRM values follow the same pattern as the 'chemical erosion indicators' and the results once again indicate a direct link between magnetic mineral concentration and the erosion history of the catchment. The Lough Fea study however shows that at the opening of the Holocene the magnetic record is sensitive to climatically controlled surface processes (see Section 10.8) as well as to human activity later on (cf. Edwards & Rowntree 1980).

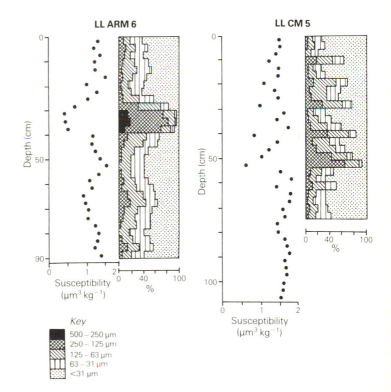

Figure 10.11 Variation in bulk specific susceptibility with depth plotted against changing particle size composition for two cores from Loch Lomond (see Thompson & Morton 1979).

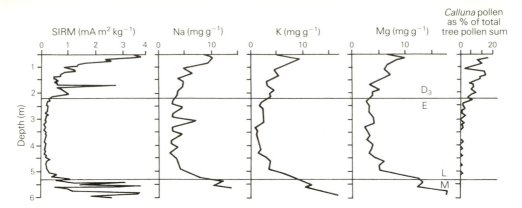

Figure 10.12 Lough Fea, N. Ireland. SIRM versus Na, K, Mg and *Calluna* pollen frequency in a 6 m core spanning the late-glacial and Flandrian sediments at the site.

SEDIMENT SOURCE SHIFTS AND LAND-USE CHANGE

The sites considered in 10.6 are all in catchments based on predominantly basic igneous bedrocks or on drift derived therefrom. In consequence, primary magnetic minerals are abundant and interpretative models of susceptibility–erosion linkage are expressed in terms of the impact, whether direct or indirect, of land-use change on variations in the accumulation rate of these primary minerals on the lake bed. In many other situations, even in some where primary magnetic minerals abound in the catchment, more complex interpretative models are required.

One of the most striking and widespread changes in land use in Britain during the last few decades has been the commercial afforestation of hills, moors and heathland areas by the Forestry Commission. The impact of extensive afforestation on lake sedimentation and especially on mineral magnetic parameters was a major theme of studies in and around Loch Frisa in northern Mull by Dearing (1979).

Loch Frisa is a deep, morphometrically complex lake lying in extensively dolerite-intruded basalts. Until *c.*1930 AD, erosion and sedimentation were dominated by the vicissitudes of crofting and latterly sheep farming in the catchment. Over the past 50 years most of the area round the lake has been planted in conifers. The three planting episodes involved extensive hand digging for drainage during the 1930s, the ploughing of a smaller area near the south-eastern shore in the 1950s and more extensive ploughing of the north-western flank in the early 1970s. Ploughing prior to planting exposed long sloping trenches of

orange subsoil to rain splash and water movement, and in the nearby Loch Meadhoin, sediment sampling in July 1976 only 6 months after similar pre-afforestation ploughing revealed a 2 cm thick layer of orange sediment in the centre of the lake. Similar orange sediments were found in the top 10–30 cm of marginal Loch Frisa cores. The Loch Frisa sediments therefore provide an opportunity for examining the impact of afforestation on mineral magnetic properties, and on sedimentation processes and patterns, as well as for comparing the effects of afforestation episodes with those of the preceding agricultural activities.

Cores were correlated by means of whole core and single sample susceptibility measurements and dated by a combination of ^{137}Cs, ^{210}Pb ^{14}C and palaeo-magnetic measurements (Appleby *et al.*, 1985). The pre-afforestation susceptibility maxima are related to peak farming activity and settlement density in the area (Dearing 1979), as would be expected from Lough Neagh and Lough Fea studies in similar lithological contexts. The effects of afforestation on magnetic mineralogy are rather the opposite. Soil profile measurements in the catchment (Dearing 1979 and Fig. 10.13) show that the orange subsoil exposed by ditching has a consistently lower susceptibility than either the basalt at the base of the regolith or the enhanced surface layers. It also includes material of much higher coercivity of remanence suggesting that, possibly as a result of tertiary weathering, some conversion to fine-grained goethite/haematite has taken place. High coercivity material can also be found within the zone of low susceptibility in the

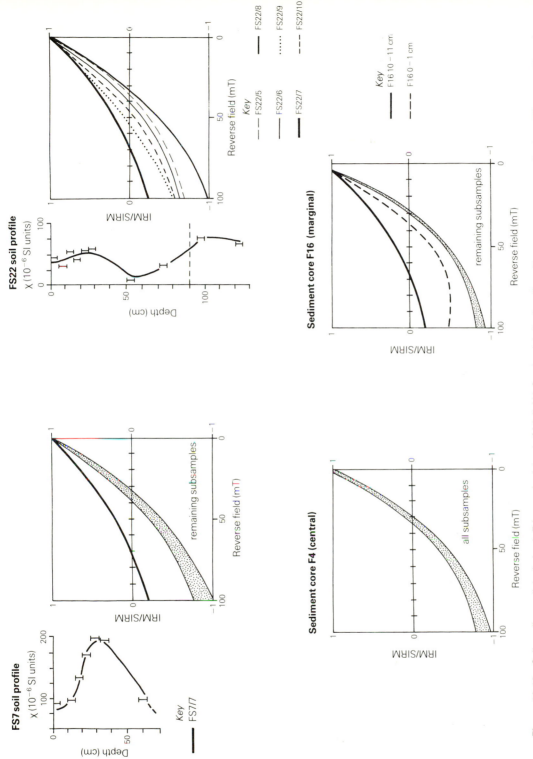

Figure 10.13 Soil–sediment links in Loch Frisa on the Island of Mull, W. Scotland. In both soil profiles one or more samples (e.g. FS7/7 and FS22/7) show high $(B_0)_{CR}$ values coinciding with susceptibility minima (at 60 cm and 50–55 cm respectively). The envelope of curves for the higher samples in FS7 and the individual curves for the remaining samples in FS22 show lower coercivities. All the samples from the central sediment Core F4 plot within a narrow envelope of curves within this lower coercivity range. In marginal Core F16 some samples (e.g. from 10–11 cm within the orange sediment layer) have high coercivities comparable to the separately plotted subsoil samples in FS7 and FS22 (see Dearing 1979).

upper lake sediments from marginal cores (Fig. 10.13). The mineral magnetic evidence thus confirms the postulated derivation from the orange subsoil exposed during the first stage of afforestation. The high value for the most recent sediments in all the marginal cores suggest that the sedimentation regime is now shifting away from dominance by subsoil input.

The Lac d'Annecy provides a strong lithological contrast to those considered so far since the surrounding rocks are almost entirely Jurassic limestones and marls extremely poor in magnetic minerals. Even the residual patches of glacial drift and a local inlier of sandstone are only minor components of the total reservoir of magnetic minerals in the catchment, most of which are present as secondary oxides in the surface soils (Dearing 1979, Longworth et al. 1979). Figure 10.14 plots mineral magnetic variations alongside chemical and pollen-analytical changes in a 6 m core from the Petit Lac, the southern part of the Annecy basin. The rapidly accumulating calcareous marls span the past 2000 years. The most distinctive change in all parameters takes place at 3 m in sediments dating from the early-mediaeval period, c. 1300 AD. Local land-use history indicates that around this time, there was a phase of expanding human activity under monastic influence. The pollen record suggests that cultivated land and orchards expanded at the expense of forested land in the catchment, the chemical changes show an increased input

of the main erosion indicators during this period and the mineral magnetic record includes a major increase in χ, χ_{fd} and SIRM as well as a decline in SIRM/χ. These mineral magnetic changes indicate both an increase in influx and a shift in mineral type towards the fine-grained assemblages typical of surface soil, as a result of erosion associated with the documented land-use change.

Section 10.6 as a whole illustrates the link between mineral magnetic variations and land-use change in various ways. In cases where bedrock completely dominates the magnetic mineralogy of the catchment (L. Neagh, L. Lomond, L. Fea), susceptibility and saturation remanence are related to deforestation and farming through the effect that these processes have on the supply of primary magnetic minerals to the sediment. The mechanism is strongly modulated by hydrological variables and their effect on particle transport. Where the vast majority of magnetic minerals available for erosion in a catchment are found in the surface soil (Lac d'Annecy) the mineral magnetic record is essentially a record of *soil* erosion in response to forest clearance and cultivation, and the magnetic properties can be used to confirm that secondary sources are dominant during episodes of intensive catchment disturbance by man. Commercial afforestation around Loch Frisa presents a special case in which, because of ditching for drainage before tree planting, magnetically distinctive subsoil is

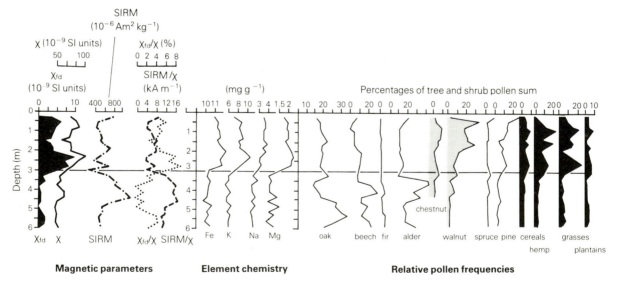

Figure 10.14 Mineral magnetic, chemical and pollen frequency data for a 6 m core from the Petit Lac d'Annecy (Higgitt, 1985). Note the major change in most parameters at c. 3 m (see text).

exposed to rapid erosion. The mineral magnetic parameters help to establish the spatial and temporal occurrence of this material and establish a source–sediment linkage characteristic of a specific kind of activity in the catchment. Comparison of these results with those emerging from continuing studies on an ever wider range of lithologies confirms that the scope for both quantifying and characterising erosion processes through mineral magnetic study is very large (cf. Dearing *et al.* 1985). However, its temporal limitations are worth noting. Since inferences of simple and direct source–sediment linkages assume that the magnetic properties used are conservative within the system, the timescales studied should always be much less than those on which the main magnetic mineral assemblage transformations are taking place in the catchment. Where this is not the case, more complex interpretative models will be required. In the next section a specific aspect of source–sediment linkage is considered, arising from the effects of fire on magnetic properties.

10.7 Magnetic measurements and fire

Many studies testify to the immense ecological significance of fire in a wide range of grassland, savannah and forest ecosystems. Not only do we find ample confirmation of man's widespread, sustained use of fire in hunting, forest clearance and land management from prehistoric times through to the present day, but we see, in many widespread ecological communities, such a complex and sensitive range of adaptations to fire incidence that it is clear that fire has been a major controlling factor in the development of many ecosystems. The impact of modern man on many fire-related communities has ranged from and often oscillated between protection and exacerbated risk. Consequently, the nature of the sustained fire–ecosystem interactions reflected in long-term population and community adaptations is difficult to reconstruct from present day observations and experiments. Nevertheless, long-term insights may be vital in formulating management and conservation policies, hence the ecologist's concern with the reconstruction of fire histories. At the same time some ecologists regard injudicious or excessive use of fire by man as a significant factor in accelerated nutrient loss from tropical and subtropical terrestrial ecosystems. Fire incidence and fire histories are also of interest in geomorphological studies. Intense fires which damage or destroy the cover of topsoil and vegetation may be important agents in accelerated erosion, especially where fire increases the exposure of surfaces to rain splash impact or strong winds. On a global scale, forest savannah and grassland fires are important processes affecting the balance, flux and storage states of carbon in the biosphere. From all these aspects, it is important to evaluate the contribution of magnetic measurements to the study of fire histories.

Fire gives rise to secondary magnetic minerals in the soil by converting paramagnetic or antiferromagnetic forms of iron to predominantly ferrimagnetic oxides (Section 8.4). The degree of conversion and hence the amount and type of magnetic mineral thus formed is related largely to the initial concentration of potentially convertible iron in the surface soil, the atmosphere during combustion and the maximum temperature reached. Field observation and laboratory experiments suggest that in most cases, enhancement of soil susceptibility begins between 100 °C and 200 °C, and peaks around 700–800 °C. It is usually greater in a reducing atmosphere such as may be created by the combustion of surface humus and litter layers.

LLYN BYCHAN

In August 1976 a ground and canopy fire on the northwestern edge of the Gwydyr forest area of North Wales destroyed the vegetation and severely damaged soil cover over a wide area including the western half of the Llyn Bychan catchment (Rummery *et al.* 1979). Bedrock comprises mostly slates and shales which are rich in iron but only weakly ferrimagnetic. Soil development is patchy and skeletal with many areas of exposed rock and scree interspersed with pockets of podsolic soils. On flat summits and in slope-foot situations, shallow peats and peaty gleys occur. As a result of the variation in soils, magnetic enhancement of erodable material during the fire proved to be localised. The sites most affected were well drained areas where podsolic soil development provided shallow readily combustible organic cover. As a result, strongly magnetic material formed at and immediately below the contact between the organic and inorganic soil horizons. Where still relatively organic, this was black in colour but where the organic material was completely ashed, the exposed soils were pink and highly susceptible. Curie temperature determinations and Mossbauer effect studies on magnetic extracts from burnt soils indicated that the main mineral formed by the fire was impure non-stoichiometric magnetite (Longworth *et al.* 1979 and Ch. 8).

Sediment sampling at Llyn Bychan began 16 weeks after the fire. Of six cores taken, only one, from the deepest central point, was affected by the influx of post-fire magnetic minerals (Rummery *et al.* 1979). Measurements made on trapped material confirmed the subsequent consistently high χ and SIRM of the sediment retrieved during 1977 and 1978. Repeat coring in 1979, showed that by then, a larger area of the total lake bed was affected by fire-enhanced material.

The work so far completed at Llyn Bychan confirms a direct link between the forest fire in the catchment and a strong distinctive peak in χ and SIRM in sediment profiles. A rapid sediment response to the fire is indicated, with a time lag of less than 16 weeks between the fire and detection of enhanced material in the central sediments. The relative importance of different pathways from the burnt areas to the lake is considered by Rummery (1981) who suggests that surface wash and overland flow have significantly supplemented stream input.

MAGNETIC MEASUREMENTS AND FIRE HISTORIES

Demonstration that the recent fire in the Llyn Bychan catchment gave rise to distinctive magnetic properties in the surface sediment encouraged attempts to determine the extent to which magnetic measurements in lake sediments might provide an additional method for reconstructing fire histories.

At Llyn Goddionduon, lying only 1 km from Llyn Bychan, the grid of cores used in the correlation exercise (10.4) included many with high χ and SIRM values in the top 10–20 cm. The increase, dated by ^{137}Cs to *c.* 1954, was ascribed to the major fire of 1951 which destroyed much of the forest in the eastern half of the catchment, though recent ^{210}Pb measurements cast doubt on this.

In the Landes, SIRM profiles of cores taken from the Etangs de Biscarrosse and Sanguinet include peak SIRM values which ^{210}Pb and ^{137}Cs dating show to be contemporary with the massive forest fires of the 1940s culminating in 1949 (Rummery 1981). In each lake, Rummery established good agreement between SIRM peaks and other indicators of fire, including charcoal content and pine pollen breakage ratios (Fig. 10.15). The above examples relate to fires within the last few decades. The first attempt to apply magnetic measurements to much earlier fire history is part of a detailed multidisciplinary study of the laminated sediments of the Lake Laukunlampi in Finnish Karelia (Rummery 1983).

It seems probable that in many situations where lake sediments are available for studies of fire history,

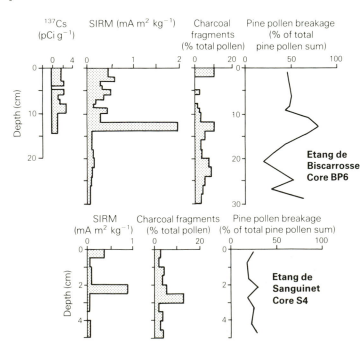

Figure 10.15 ^{137}Cs, SIRM, relative charcoal concentrations and pine pollen breakage ratios for cores from the Etang de Biscarrosse and the Etang de Sanguinet in the Landes area of southwest France (see Rummery 1981 and text).

magnetic measurements may be used to complement evidence from existing techniques such as charcoal counting. Their usefulness will depend on factors such as:

(a) *Fire location*. A fire within the lake catchment is much more likely to be recorded magnetically than one outside, since supply of minerals to the lake by streams and overland flow is likely to dominate airborne input in all but the most extreme circumstances. Magnetic minerals are thus likely to be more 'catchment specific' than charcoal fragments.

(b) *Fire type and intensity*. Intense fires affecting both ground and canopy, where the latter is present, are by far the most likely to be recorded. Fires in the canopy alone will give rise to few magnetic minerals and the low-intensity ground fires of great ecological importance in some conifer-dominated ecosystems are unlikely to give rise to high enough temperatures in surface mineral layers.

(c) *Soil and substrate type*. Where soils are very poor in iron, little fire enhancement can occur. The Landes study does however confirm that the impact on lake sediments is clearly detectable provided 'background' levels of χ and SIRM are consistently low. At the other extreme, where major sources of magnetic minerals already exist within the catchment, they may produce magnetic variations in the sediment sequence unrelated to fire.

From the above, it seems likely that magnetic measurements will contribute to fire history studies by making rapid preliminary surveys feasible and by adding insights on location, type and intensity, less readily deducible from other techniques.

10.8 Lake sediment magnetism and climatic change

Results from many sites suggest that major climatic shifts control weathering and sedimentation regimes in ways which give rise to distinctive mineral magnetic variations.

Oldfield *et al.* (1978) documented a very simple direct relationship between mineral magnetic and pollen-analytical changes in lake sediments from High Furlong, Blackpool, spanning the late-glacial

and early-Holocene period from about 1400 to 9000 years BP (Hallam *et al.* 1973). The pollen-analytical subzones record a vegetation succession from grass-herb tundra through to birch woodland during the Windermere (Allerød) interstadial, the subsequent spread of dwarf shrub and herb communities during the Loch Lomond stadial around 11 000–10 000 years BP, and the spread of birch woodland at the opening of the Holocene. Magnetic susceptibility measurements are highest during the cold stadial episodes, decline to a minimum during the interstadial and rapidly fall to around zero at the opening of the Holocene. Episodes marked by poorly developed vegetation cover, soil instability and active solifluction give rise to peaks in χ; episodes of developing

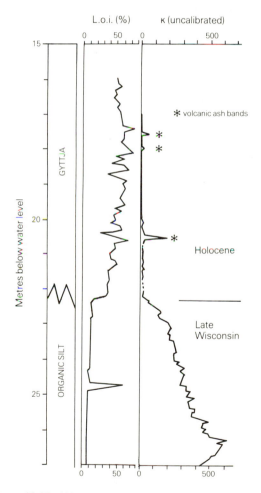

Figure 10.16 Whole core volume susceptibility and loss on ignition values for the late-Wisconsin and Holocene sediment of Battleground Lake, Washington State.

plant cover and soil maturation under milder climatic conditions give rise to minimum χ values.

Figure 10.16 plots volume susceptibility for a series of stratigraphically consecutive piston cores from Battleground Lake in southern Washington State (Oldfield *et al.* 1983a). The 12 m of sediment span late-Wisconsin and Holocene time, with a basal ^{14}C date of 14 840 $\pm$ 200 BP (QL-1539). The lake is in a closed crater basin with a small low-rimmed catchment in basalt. The major fall in susceptibility between 7.5 and 7 m clearly identifies the late-Wisconsin/Holocene boundary and reflects a

diminution in allochthonous detrital input from the catchment with the development of more stable soils and complete vegetation cover (Oldfield *et al.* 1983a).

In both cases, the magnetic mineralogy reflects the course of climatic changes by recording evidence of the associated changes in sedimentation, weathering and pedogenic regimes.

Figure 10.17 plots coercivity of SIRM profiles for sets of samples from the sediments of Lynch's Crater, northeastern Queensland, Australia. The tropical soil profile discussed in Section 8.6 was collected from a river section close to Lynch's Crater. Whereas in

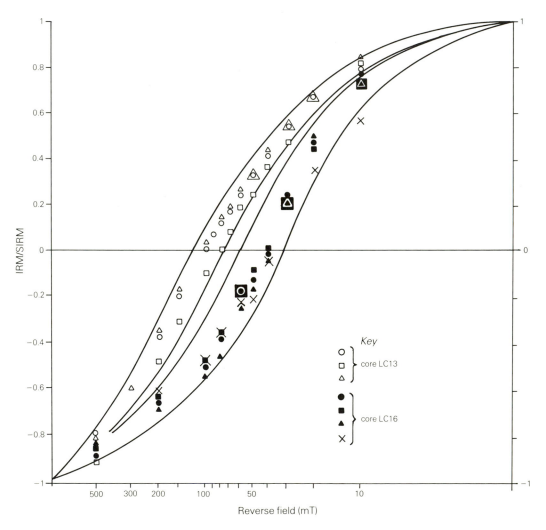

Figure 10.17 Coercivity of SIRM curves for groups of sediment samples from Lynch's Crater in northern Queensland (Kershaw 1978).

Figure 10.16 the parameter plotted is largely concentration dependent, here the values are normalised and hence not primarily a function of concentration. Virtually all the points lie within two quite separate envelopes of values, one for samples from Core section LC13, the other for samples from Core section LC16. For the LC13 samples $(B_0)_{CR}$ is just under 40 mT, but for the LC16 samples it lies between 70 and 110 mT. Core section 13 comes from Kershaw's (1978) pollen zone E1, Core section 16 from zone E2. In E1, pollen of complex vine forest taxa dominate the pollen record, and the mean annual rainfall suggested by the reconstructed vegetation type is a little in excess of contemporary rainfall at the site. In E2 higher pollen representation of scherophyll and gymnosperm elements suggests drier conditions than those prevailing at present. Here the *normalized* mineral magnetic parameters are related to palaeoclimate in ways which are not immediately apparent, though the linkage between the two persists for most of the very long 41 m deep sediment record at the site. In all probability, complex interactions between weathering, and hydrological and sedimentological regimes are involved.

From the above we may conclude that mineral magnetic measurements whether concentration dependent or not, can reflect palaeoclimatic conditions as a result of the effect that changing climate has on the environmental processes which control the concentrations and types of magnetic minerals deposited in lake sediments. A fuller discussion of these linkages is included in Oldfield and Robinson (1985).

10.9 Summary and conclusions

Mineral magnetic measurements can contribute to lake sediment studies in a wide variety of ways, ranging from initial core logging to the detailed analysis of sediment sources and sedimentological processes in the catchment or lake. The rôle of mineral magnetism in lake catchment studies and in palaeolimnology will vary with the main thrust of the investigation to be undertaken and with the nature of the lake-watershed ecosystem under study. The speed and versatility of the instrumentation now available coupled with the non-destructive nature of the measurements means that there are few areas of lake sediment based study to which this emerging methodology cannot fruitfully contribute. Further progress, as distinct from the pragmatic application of the approaches illustrated above to new sites, will depend on a closer and more quantitative specification of the implications of the measurements in terms both of sediment source (cf. Stott, 1986) and of magnetic grain size and mineralogy (cf. Parry 1965, Dankers 1978, King *et al.* 1982), especially where complex mixtures of natural magnetic oxides co-exist; on a more thorough appraisal of the relative significance of diagenetic, authigenic and bacteriological magnetic minerals in lake sediments; and on critical detailed studies relating magnetic parameters to sedimentological, granulometric and geochemical variables in a wider variety of lakes.

[11]
Magnetic minerals in the atmosphere

And Sheffield, smoke-involved; dim where she stands
　　Circled by lofty mountains, which condense
Her dark and spiral wreaths to drizzling rains,
　　Frequent and sullied.

W. G. Hoskins
Anna Seward as quoted in *The making of the English landscape*

I recall that Osvald, upon his first excursion into the South
Pennine *'Eriophorum* moors' was staggered to find that his pale-
grey flannel trousers (quite suitable for a Swedish trip) were soon
generously striped with black soot collected from the stems and
leaves of the vegetation; it was a sharp reminder of the influence
of the industrial north.

H. Godwin 1981
Archives of the peat bogs

11.1　Introduction

Less is known about the presence of magnetic
minerals in the atmosphere than about their presence
in any other major environmental system. This is a
function not only of the unfamiliarity of atmospheric
scientists with the techniques of magnetic measurement appropriate to dust and aerosol studies, but also
of severe problems involved in obtaining sufficiently
large, uncontaminated samples representative of the
full range of particle sizes present. The sections which
follow identify particular themes and outline the very
modest progress made so far in each case. No attempt
at a comprehensive review of atmospheric magnetism
is yet possible.

11.2　Sources of magnetic minerals in the atmosphere

The main sources of magnetic minerals in the present-
day atmosphere are fourfold:

(a)　*Volcanic eruptions.* Magnetic measurements of
lake and marine sediments in volcanic areas
show that tephra layers are often marked by
peaks　in　magnetic　susceptibility　(e.g.
Radhakrishnamurty *et al.* 1975, Oldfield *et al.*
1980, 1983b). This is especially the case with
basic tephra rich in ferro-magnesium minerals.
Volcanic dust makes a significant contribution
to the total particulate content of the atmosphere

as a whole and makes a proportionally greater contribution to the stratosphere especially within the Junge aerosol layer at about 20–25 km. The long residence time of submicron particles in the stratosphere coupled with the high frequency of volcanic eruptions contributing both tropospheric and stratospheric dust ensures the persistence of a volcanic dust veil, the density of which is believed to have varied markedly through time. Viewed on a timescale of centuries, Lamb (1970) sees the period 1750–1900 AD as one of high dust veil in the Northern Hemisphere. On a much longer timescale, Kennett and Thunell (1975) see the past 2 million years of the Pleistocene as marked by a peak in explosive volcanism suggestive of a link between volcanic ash concentrations in the atmosphere and climatic change. Haggerty (1970) and others have suggested that volcanic dust is a major contributor to magnetic mineral assemblages in deep-sea sediments.

(b) *Wind erosion.* As noted in Chapter 8, surface soils are often strongly ferrimagnetic as a result of enhancement mechanisms such as fire which is also a major agent in wind erosion. Even in hot arid areas, where because of the paucity of organic matter both fire-induced and 'pedogenic' enhancement are rare, and in high latitude deserts and periglacial areas, where soil development is absent, fine material deflated from the surface will rarely be less magnetic than the underlying bedrock. Soil loss by wind erosion can be a byproduct of overexploitation of marginal land as in the American Dust Bowl of the 1930s and the more recent environmental crisis in the Sahel during the 1970s. At the present-day deflation of arid areas naturally susceptible to wind erosion thus combines with the effects of marginal tillage and fire use, especially in Savannah areas, to provide a second major source of magnetic particulates in the atmosphere. Chester (1978) has summarised the evidence for heavy dust-loadings in the lower atmosphere over the Atlantic as a result of terrestrial deflation, and many other authors (e.g. Windom 1969, Prospero 1968) have documented the abundance of 'soil'-derived dust in the lower atmosphere elsewhere both on land and over the sea. It is, however, difficult to tell how representative contemporary concentrations and distributions are of the long-term situation. Increased human pressure on soils over the past few decades will almost certainly have given rise to significant changes. Moreover, we know from loess deposits and the like that in the longer term the sequence and pattern of Pleistocene climatic change have been of great importance in controlling the exposure of fine surface materials to wind erosion and the atmospheric transport of deflation products.

(c) *Industrial and combustion processes.* Until recently the abundance of magnetic spherules in the

Table 11.1 Estimates of direct global particle production for potentially magnetic components. Based on Prospero (1978) Table X, and NRC report (1979) Table 2-1. All figures are 10^6 tonnes per annum.

Direct particle emission type	Peterson and Junge (1971)		Hidy and Brock (1971)	SMIC (1971)	Other authors	Reference
	all sizes	5 m		20 m		
man-made	133.2	29.6	36.8–110	10–90	54	(1)
windblown dust (soil and rock debris)	500	250	60–360	100–500	126 60–360 128± 64 200±100	(2) (3) (4) (4)
forest fires and slash and burn debris	35	5	146	3–150		
volcanic emissions		25	4	25–150	4.2	(5)
meteoric debris	10	0	0.02–0.2		0.02–0.2 1–10	(6) (7)

* (1) Goldberg (1975); (2) Ellsaesser (1975; (3) Judson (1968); (4) Joseph *et al.* (1973); (5) Mitchell (1970); (6)Bhandari *et al.* (1968); (7) Rosen (1969).

atmosphere was ascribed to the cosmic flux of extraterrestrial particles. We know now that most spherules are the product of human activities such as fossil (especially solid fuel) combustion, metal smelting and iron and steel manufacture. The increase in spherule concentrations since the beginning of the industrial revolution is the theme of Section 5 in this chapter.

(d) *Extraterrestrial particles.* In light of the above, we may expect this source to be of significance only very locally in association with meteorites, and in those areas most remote from any form of terrigenous input such as the Central Pacific. In terms of the contemporary atmosphere and of the historical record of atmospheric particulates considered in this chapter, cosmic spherules can be ignored (but see Ch. 12).

Table 11.1 summarises various estimates of the relative importance of these sources.

11.3 Magnetic properties and aerosol modes

Atmospheric aerosols have rarely been examined with regard to their magnetic properties. Even where magnetic separation techniques have been used to differentiate between emission categories (e.g. Hansen *et al.* 1981, Linton *et al.* 1980, 1981) no subsequent magnetic measurements have been carried out to test the efficiency of the magnetic separation. Attempts to characterise more fully the mineral magnetic properties of the separated fraction are also very rare (e.g. Chaddha & Seehra 1982). In consequence, it is not yet possible to provide reliable information on the relationship between magnetic properties and atmospheric particle modes for the range of aerosol types which include a magnetic component. Such direct determinations as are available relate only to restricted samples of power station fly-ash and roadside material and these are summarised in Section 11.6. Whitby and Cantrell's (1976) scheme suggests that virtually all the magnetic material generated by the mechanisms outlined above will be within the 'coarse particle' range from *c.* 2 μm diameter upwards. In the case of urban and industrial particulates, this is consistent with all the measurements on magnetic spherules made by Puffer *et al.*

(1980) in the New York area and over the nearby parts of the N. Atlantic, as well as with studies of coal fly-ash by Hansen *et al.* (1981) and Ondov *et al.* (1979), who record insignificant amounts of magnetic iron in particle sizes below 2.2 μm and 1.6 μm respectively. Keyser *et al.* (1978) also show that auto-exhaust particles above 10 μm are rich in iron, but that the finer particle mode <1 μm contains little or no iron. At the same time it is important to note that total iron is often poorly correlated with χ and SIRM (e.g. Thompson *et al.* 1975). Consequently, until more critical magnetic measurements have been carried out on size-fractioned material, the results are inconclusive. In terms of health implications, the main concern is usually with fine particles, partly because deposition efficiency in the pulmonary and trachaeo-bronchial tracts increases with declining particle diameter and partly because some of the main toxic metals discharged for example in coal-fired fly-ash (Davidson *et al.* 1974) are enriched in the fine fractions.

11.4 Magnetic–heavy metal linkages

Although the relationship between magnetic oxides and heavy metals in fly-ash, industrial particulates and auto-emissions is poorly understood, several authors point to the possibility of close links. Theis and Wirth (1977) note that most metals in the eleven coal-fired fly-ash samples they considered were associated with specific surface oxides of iron, manganese or aluminium. Copper, chromium, arsenic and zinc they record as being associated with iron oxides in almost all cases, cadmium and nickel mostly with manganese, and finally lead with either. Hansen *et al.* (1981) show that chromium, manganese, cobalt, nickel, copper, zinc and beryllium are all significantly enriched in the 'magnetic' fraction of coal fly-ash. Linton *et al.* (1980) and Olson and Skogerboe (1975) note an association between 'magnetic iron' and lead in automobile-exhaust particulates sampled on roadways. Hansen *et al.* (1981) suggest that 'magnetite may also be a hazard to health because of its ability to occlude biologically active transition metal ions such as Mn and Ni by isomorphous substitution ... and thus act as a slow release carrier agent for toxic elements'. Lauf *et al.* (1982) have suggested that magnetic spherules generated during coal combustion may be derived directly from the conversion of pyrite framboids present in the coal. They also note that the framboids are associated with trace metal enrichment. Hulett *et al.* (1981) who identify the magnetic component in their fly-ashes as predominantly an aluminium-substituted ferrite ($Fe_{2.3}Al_{0.7}O_4$), suggest that trace elements occur as substitutions in the spinel structure. They record enrichment factors between 10 and 50 times for first-row transition elements (V, Cr, Mn, Co, Ni and Zn) in the magnetic phase of their samples. Thus despite many gaps in our knowledge and great uncertainty about the extent to which demonstrated magnetic–heavy metal linkages reflect surface association or incorporation into the crystal-line matrix of particulate emissions, it is nevertheless reasonable to explore contexts in which, in purely empirical terms, the linkage appears to occur or in which, the linkage would, if confirmed by further studies, be of major value in both historical and contemporary particulate pollution monitoring.

11.5 Peat magnetism and the history of atmospheric particulate deposition

Ombrotrophic peat bogs are those built up above the ground water table so that their surfaces of deposition and accumulation are no longer influenced by inflowing drainage. As they accumulate they often preserve a record of atmospheric deposition. Here we are primarily concerned with the magnetic record in the peat, and its value in reconstructing the history of and spatial variations in particulate pollution.

The concentrations of magnetic minerals in peats, ice and snow are usually much less than in most of the lacustrine, soil and fluvial samples discussed in the previous three chapters. As a consequence, for all but the 'dirtiest' samples, mineral magnetic character-isation must often be restricted to the most sensitive

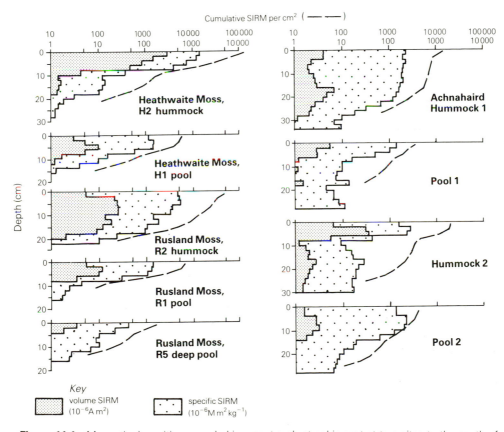

Figure 11.1 Magnetic deposition recorded in recent ombrotrophic peat at two sites to the south of the English Lake District (Heathwaite and Rusland Moss) and four profiles from the remote Achnahaird peninsula in north-west Scotland. The histograms plot SIRM values on a volume- and mass-specific basis, the dashed lines are cumulative totals for each core.

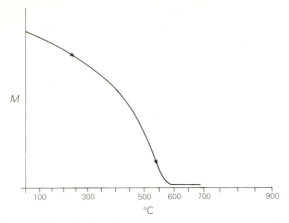

Figure 11.2 High-field Curie temperature determination for a magnetic extract from the upper 6 cm of Ringinglow Bog (see Fig. 11.12), near Sheffield, England. The results identify the dominant magnetic mineral in the extract as magnetite.

techniques of isothermal remanence and back-field ratio measurements.

Figure 11.1 shows the SIRM record in shallow, recent peat profiles from Heathwaite and Rusland Moss, which lie between the English Lake District and the heavily industrialised areas of Lancashire, and from the Achnahaird peninsula on the northwestern coast of Scotland, very remote from urban settlements and industrial activity. In all the profiles, SIRM increases upwards and invariably peaks in the top 2–6 cm. Peak values are an order of magnitude or more higher than the minima near the base of each profile. These increases have been interpreted (Oldfield *et al.* 1978) as a record of the deposition of magnetic

spherules discharged into the atmosphere as a result of fossil–fuel combustion and heavy industry. Mineral magnetic parameters in these most recent peats vary relatively little from site to site in Britain and are all consistent with an assemblage dominated by magnetite (Fig. 11.2, and cf. Hansen *et al.* 1981, Chaddha & Seehra 1982). If we take SIRM as a crude index of the volume of ferrimagnetic spherules deposited, then from the cumulative totals plotted against each profile, variations in mean flux density can be seen on three spatial scales. The southern Lake District sites as a whole have trapped, on average, between four and five times as much as the Achnahaird sites. Within the southern Lake District sites, Heathwaite which lies directly down wind of the Millom Steel Works has trapped almost twice as much as Rusland. However, at each site there is an even greater contrast when we compare hummock and pool deposition totals.

In the southern Lake District, hummock profiles contain 10–100 times more spherules than do pools. In Achnahaird, although the contrast is less marked, it is nevertheless consistently present. The difference has been interpreted as the effect of hummock vegetation filtering out particulates moving subhorizontally across the bog surface by dry eddy diffusion (Oldfield *et al.* 1979b) though it is possible that under some circumstances, magnetite dissolution below the water table may contribute to the contrast. From published evidence (National Academy of Sciences 1979) we would expect washout by rain to dominate 'pool' deposition and filtering from eddy diffusion to dominate hummock deposition. Washout is relatively

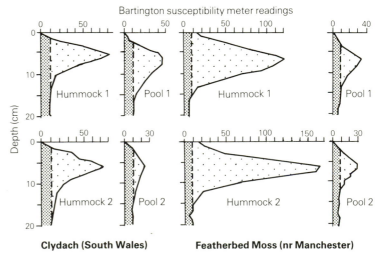

Figure 11.3 Characteristic drainpipe scans of the volume susceptibility of recent British ombrotrophic peats using the Bartington core loop sensor. The Clydach site is in South Wales, Featherbed Moss is in the south-west Pennines some 15 km down wind of Manchester. In all cores, the pre-industrial levels record low and relatively constant deposition. Above this, the contrast in total deposition between pools and hummocks is evident. Total cumulative 'industrial' fall out at each site is estimated by calculating the approximate area of the lightly stippled part of each core trace.

128

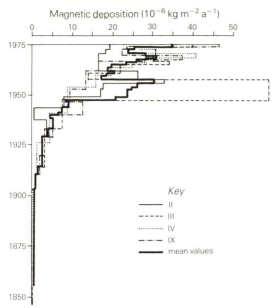

Figure 11.4 Magnetic deposition at Karpansuo Bog, south-central Finland. Dates have been determined by moss increment counting and SIRM values have been used to estimate the annual deposition of magnetite at the site since 1840 AD (see Oldfield *et al*. 1981b).

effective only for the finest and for the very coarse particles. As starting points for further study, these profiles pose a series of interesting challenges and opportunities.

Given the link between magnetic deposition and industrialisation and the relative speed with which measurement can be made, the method holds out some promise of allowing its use as a surrogate particulate pollution monitoring technique, not only historically, but also in terms of both cumulative and contemporary deposition. Significant questions and problems are evident at the outset. Peat chronology for the past 200 years is very poorly known in Britain. The link at source between spherule discharge and both heavy metal and sulphate generation may not persist through transport and deposition. Spatial patterns of variation on a regional or continental scale may be masked by the variations arising from bog microtopography. Such progress as has been made in appraising these problems is outlined below.

In most British peat bogs the point at which presumably pre-industrial values begin to increase lies in the top 25–30 cm of the profile. This has permitted development of a rapid and convenient method for identifying the industrial 'magnetic take-off point' in relatively polluted sites and for

calculating cumulative deposition per unit surface area above this. Plastic drainpipes pushed into the peat and then dug out can be scanned using the Bartington whole core sensor (Fig. 11.3). This permits rapid estimation through multiple coring and, where desirable, on-site measurement. In practice, scans from two hummocks and two pools at a given site have been used to compensate for microtopographic variation and provide an estimate of mean cumulative deposition for each locality.

Problems of chronology have so far proved rather intractable in the British context. Critical evaluation of ^{210}Pb dating in ombrotrophic peats (Oldfield *et al.* 1979a) suggests that it is not invariably reliable. However, the synchroneity of recent magnetic increases within and between sites has been confirmed by pollen-analytical study at six sites (Hughes, 1978, Richardson 1984 & 1986). Moreover the evidence from Ringinglow Bog near Sheffield shows that the dramatic rise in magnetic deposition at that site coincides with the sharp change from *Sphagnum imbricatum* to *Eriophorum vaginatum* peat believed to reflect the effects of 19th century atmospheric pollution and especially the increased concentrations of atmospheric SO_2 associated with fossil-fuel combustion.

Only in areas where, as a result of more continental

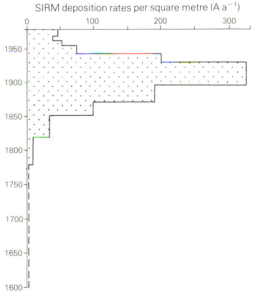

Figure 11.5 Magnetic deposition at Regent Street Bog, Fredericton, New Brunswick. The timescale has been determined from moss increment counts and bulk density calculations by K. Tolonen. The decline in SIRM deposition since World War 2 is characteristic of many north eastern North American sites.

climatic conditions, annual moss-increments can be identified and counted, has a satisfactory chronology been available. Under these circumstances magnetic deposition can be plotted in terms of both concentration and flux density versus time. Figure 11.4 shows that at Karpansuo Bog, a rather remote site in Central Finland, increased deposition begins in the mid-19th century, rises more steeply in the 20th and

peaks in the past 30 years. This mirrors the late spread of heavy industry to southern Finland during the post-war period. At Regent Street Bog, near Fredericton New Brunswick (Fig. 11.5), the increase begins earlier and the post-war period shows a steep decline. This is probably primarily an expression of the history of the local iron and steel industry which it parallels with remarkable accuracy (Tolonen & Oldfield, 1986).

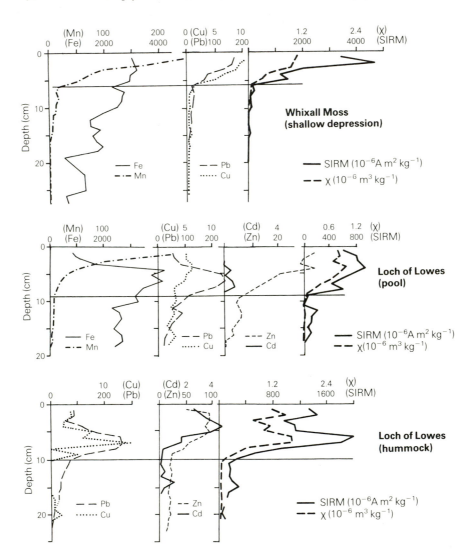

Figure 11.6 The stratigraphic record of heavy metal element concentrations in recent ombrotrophic peat from two British sites compared with χ and SIRM measurements. Whixhall Moss is in Shropshire, England. The location of the Lowes site is between Dunkeld and Blairgowrie close to the south-eastern edge of the Scottish Highlands. The horizon of rapid increase is synchronous between these metals and the χ and SIRM curves. The data have been provided by J. M. Jones.

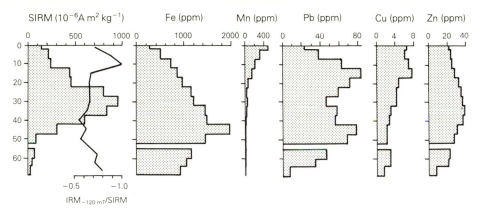

Figure 11.7 SIRM and IRM$_{-120mT}$/SIRM plotted against heavy metal element concentrations for a core from a hummock of ombrotrophic peat at Regent Street Bog, Fredericton, New Brunswick (Fig. 11.5). Compare the lack of synchronism between the increases in SIRM values and in the concentrations of Pb, Cu and Zn, with the British results plotted in 11.6.

The possible link between magnetic and heavy metal deposition has been examined both stratigraphically and spatially. Oldfield *et al.* (1981b) record such a link in a series of peat profiles taken from west to east across Finland. From the Harparliltrasket site in the west, close to the power station and heavy industry, to the remote sites of Karelia in the east, close to the Russian border, there is, in each profile, a clear parallel between magnetic and total iron deposition in the post-war period. Moreover the proportion of total iron accounted for by magnetic deposition declines from west to east with increasing distance from industrial sources (see Oldfield *et al.* 1981b). In British peats, comparisons between the stratigraphic records of SIRM or χ and copper, lead and zinc deposition show a strong similarity (Fig. 11.6), suggesting that after more detailed study, χ and SIRM may be used realistically as surrogate measurements for reconstructing the history of deposition of these elements in the peat. By contrast in N. American and in continental sites where the chronology is more readily resolvable, the levels in the peat at which the heavy metal concentrations increase, lie almost invariably well below the magnetic 'take-off' (Fig. 11.7). Critical factors in the contrast may be the relatively more open 'fluffy' character of the *Sphagnum fuscum/Polytrichum* hummocks from which the North American and continental evidence comes, as well as the much greater seasonal water level fluctuations experienced at these sites. Both factors would be conducive to greater vertical mobility of soluble metals in the peat column. Significantly, Aaby *et al.* (1979) report a convincing stratigraphic

record of lead deposition in the pool site at Draved in Denmark, but a much less satisfactory record in the hummock. At the very least, magnetic measurements in recent peat will often provide an early industrial marker horizon, as well as a stratigraphic record against which to compare that of the heavy metals when questions of their mobility in the peat column are considered.

Figure 11.8 summarises an exercise designed to compare cumulative magnetic and heavy metal deposition as they have varied spatially from very heavily to lightly polluted areas in Britain. Paired hummock and pool 'drainpipe' cores from each site were volume susceptibility scanned and the top 8 cm

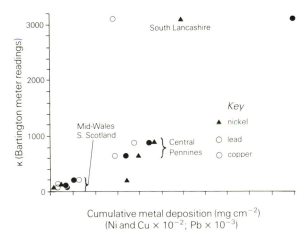

Figure 11.8 Cumulative metal and magnetic deposition at six British sites ranging from S. Scotland to S. Lancashire using the methods outlined in the text (Section 11.5).

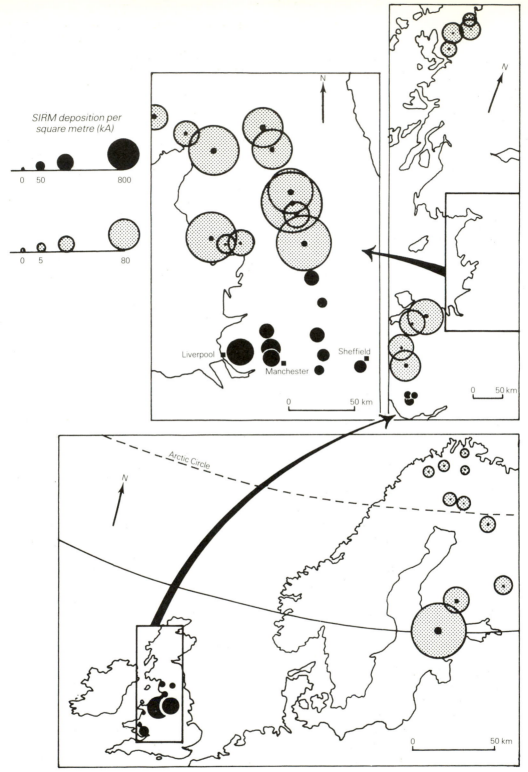

Figure 11.9 Cumulative 'industrial' magnetic deposition for British and Finnish ombrotrophic peat sites. The figures have been calculated from Bartington loop scans (cf. Fig. 11.3), single-sample SIRM measurements (cf. Fig. 11.1) and accumulated flux values (cf. Figs 11.4 & 5).

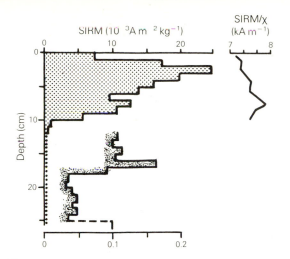

Figure 11.10 Magnetic deposition at Ely Lake Bog close to the Mesabi iron range in N. Minnesota. The profile records an initial increase at 18 cm depth coinciding with the first phase of exploitation and a later major increase at 8–10 cm. SIRM/χ tends to decline in the top 10 cm. This is one of six measured cores taken by E. Gorham.

of each sliced for heavy metal determinations on replicates at 2 cm intervals using a 'total' extraction technique (Mackereth 1969). Each co-ordinated point represents the mean of 40 susceptibility whole core readings and 16 element concentration determinations at each site. Over 80% of the points show a strong linear relationship and in the case of copper the correlation is very strong across the full range of variation.

In the light of all the above evidence SIRM measurements were carried out on samples from moss-increment dated cores from Finnish and Norwegian Lapland and these also show patterns of increased magnetic deposition, less strong than but closely parallel to the patterns from further south in Finland. Even in these remote areas the post-war period has been marked by a sharp increase in the flux density of magnetic particles. By calculating post-'magnetic take-off' cumulative deposition per cm² for all the sites studied so far it has been possible to compile Figure 11.9. From this we see that the Lancashire Plain in north-west Britain has the highest level of recent magnetic deposition while at the 'cleanest' end of the scale the far north-west of Scotland and the remotest sites in Finnish Lapland have values two orders of magnitude lower. Although this pattern must be subject to many sources of error, arising from soil contributions in remote areas, from

regional variations in bog microtopography and from the imperfections of SIRM values as a basis for estimating ferrimagnetic volumes, it provides a useful first insight into spatial variations in cumulative deposition which now require a great deal of more detailed study.

In both stratigraphic and spatial terms the patterns of variation in eastern and central North America appear to be rather different. As at the Fredericton site (Fig. 11.5), most profiles show a sharp decline in concentration towards the surface. Where, as in several moss-increment dated profiles from Nova Scotia (Tolonen, pers. comm.) chronological control is available, it is clear that this reflects a real decline in flux density, and it is tempting to see it as a possible effect of the greater American dependence on oil as against solid fuels over the last few decades (cf. Henry & Knapp 1980). Spatial variations in cumulative deposition at the sites studied so far seem dominated by 'point' sources such as the Mesabi iron range, and the Sudbury nickel smelting complex. Sharp increases in magnetic deposition within the peat profiles near each site give good chronological markers, and in both quantitative and qualitative terms, distance decay effects can be clearly seen in the magnetic measurements (Figs 11.10 & 11.11).

At all sites, whether from Europe or North America, it is clear that there are significant and readily measurable mineral magnetic variations below the main industrial 'take-off' horizon (Fig. 11.12). Moreover, the levels immediately below often have higher susceptibility and SIRM than do ombrotrophic peats of earlier historical or prehistoric age. Much work remains to be done to determine whether these variations are natural or anthropogenic, and to explore their possible implications in terms of palaeo-climates and human activity.

The foregoing account touches on many ways in which the mineral magnetic record in peat may be of value. Having set out in brief and with only a modest amount of critical evaluation the full range of possibilities so far considered, a great deal of detailed work lies ahead in the evaluation of any single aspect.

11.6 Contemporary particulate pollution monitoring

One of the most attractive possibilities opened up by the magnetic measurement of atmospheric samples is in the area of contemporary particulate pollution

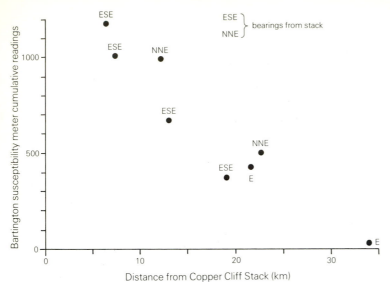

Figure 11.11 Cumulative magnetic deposition calculated from drainpipe scans of shallow peat profiles in the Sudbury region of Ontario. Volume susceptibility readings calculated in the way illustrated in Figure 11.3 are plotted against distance from the major point source. For each site two profiles have been used to give a mean value as plotted, and the approximate bearings from the main smelter are shown against each point.

monitoring. Table 11.2 and Figure 11.13 provide some preliminary results obtained in Britain by measuring material trapped using a high volume (HI-VOL) air sampler. One data set refers to measurements on fly-ash from Hams Hall power station near Birmingham, the other to measurements obtained on filters exposed for periods of 12–18 hours in the two road tunnels under the Mersey, linking Liverpool with the Wirral peninsula to the south-west (see Hunt *et al.* 1983). These latter measurements are therefore believed to reflect mostly particulate vehicle emissions. The hysteresis ratio and coercivity parameters suggest that both magnetic mineral assemblages are domi-

nated by magnetite with a proportionally higher haematite contribution in the fly-ash. There is also relatively little relationship between particle size and mineral magnetic parameters, especially between 1 and 7 μm, suggesting that in all but the coarsest spherules the mineralogical and domain size assemblages vary little and are independent of granulometry. Where heavy metal/magnetic ratios have been determined, these also often appear to be relatively constant for each source over the same range of particle sizes, though appraisal of a larger data set certainly confirms wide variations in element concentrations and ratios between coal-fired fly-ash

Table 11.2 Mineral magnetic parameters and SIRM/heavy metal ratios for fly-ash and roadside particulates. All samples are particle-size fractions obtained by a HI-VOL air sampler. The fly-ash was resuspended in the laboratory and for these samples the SIRM element ratios are relative. In the case of the roadside particulates the magnetic concentrations were too low for χ and ARM measurements. (Hunt, pers. comm.)

		ARM	SIRM	$\dfrac{SIRM}{ARM}$	$(B_0)_{CR}$ (mT)	$\dfrac{IRM_{-100\,mT}}{SIRM}$	$\dfrac{SIRM}{Fe}$	$\dfrac{SIRM}{Mn}$	$\dfrac{SIRM}{Al}$	$\dfrac{SIRM}{Pb}$	$\dfrac{SIRM}{Zn}$	$\dfrac{SIRM}{Cu}$
Ham's Hall fly ash	1	0.04	19	525	40	−0.63	—	—	—	—	—	—
	2	0.04	17	422	42	−0.53	0.19	9	—	10	1.6	121
	3	0.04	16	354	42	−0.55	0.18	8	—	9.5	1.3	133
	4	0.04	15	341	42	−0.52	0.19	6	—	7	0.9	104
Birkenhead tunnel 8/6/82	1	—	—	—	—	—	0.44	10.4	0.83	2.3	9.0	80
	2	—	—	—	—	—	0.42	15.1	0.84	1.3	8.2	39
	3	—	—	—	—	—	0.27	12.2	1.11	0.4	4.6	23
	4	—	—	—	—	—	0.39	14.6	1.32	0.6	3.0	29
Wallasey tunnel 20/7/82	1	—	—	—	—	—	0.49	16.1	1.3	9.9	16.3	86
	2	—	—	—	—	—	0.42	19.4	2.0	6.5	14.8	56
	3	—	—	—	—	—	0.34	28.5	1.6	14.3	9.5	24
	4	—	—	—	—	—	0.44	—	3.6	0.6	8.6	26

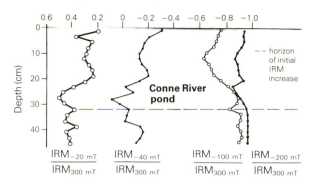

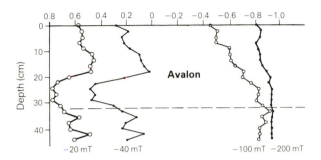

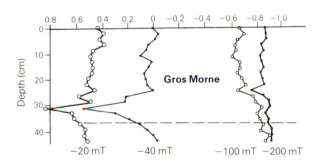

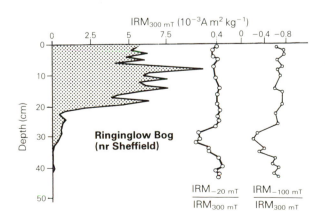

Figure 11.12 Parallel changes in IRM/IRM$_{300mT}$ plots are shown for three peat profiles from different Newfoundland sites. In each case, low reverse field values indicate 'harder' IRMs around the time of the earliest magnetic increases. The high reverse field values also show parallel trends from site to site above this level. At Ringinglow Bog (see Fig. 11.2) the −IRM/SIRM values identify a zone of relatively 'hard' IRM coinciding approximately with the first SIRM increase. Newfoundland samples supplied by E. Gorham; Ringinglow Bog data supplied by J. M. Jones.

samples of different source (Furr *et al.* 1977), as well as between these and the particulates derived from vehicle emissions (National Academy of Sciences 1979). Moreover oil-fired fly-ashes, contain on average, over an order of magnitude less iron than do the coal-fired ashes (Henry & Knapp 1980).

Provided these initial results are not unrepresentative of locations where coal-fired power generation and automobile emissions are the main sources of heavy metals and magnetic minerals, and bearing in mind both the temporal and spatial mineral magnetic–heavy metal linkages outlined in Section 11.5, and the urban stormwater and near-shore marine data summarised in Chapters 9 and 12, magnetic monitoring would appear to be feasible and to have potentially important implications.

In order to pursue this possibility further, it is necessary not only to extend detailed studies of potential sources but also to determine ways of field sampling and sample preparation which complement magnetic measurements in terms of speed, ease and economy. A recent survey by Maxted (1983) suggests that leaf measurements may provide useful results. Fifty-one sites were chosen in West Yorkshire ranging from highly urbanised and industrial areas to relatively remote open moorland. Each site corresponds to one used in the National Air Pollution Survey (1971) and at each one several leaf types were sampled. The best spatial coverage was obtained using leaves of *Chamaenerion (Epilobium) angustifolium* (rosebay willowherb) and *Acer pseudoplatanus* (sycamore). Measurements were expressed on the basis of a constant leaf surface area which was achieved by using the 2.4 cm diameter cylinders in which discs were packed, as the templates for cutting them out. Close comparability of magnetic mineral assemblage from site to site was confirmed by calculating SIRM/χ, IRM$_{300\,mT}$ χ and back-field ratios for samples across the full range of IRM$_{300\,mT}$ and χ variation. The magnetic mineral assemblage of these samples saturated in a field of 300 mT or less,

135

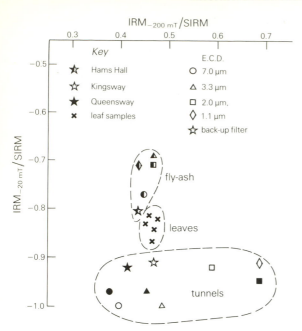

Figure 11.13 $IRM_{-20mT}/SIRM$ versus $IRM_{-200mT}/SIRM$ for particle-sized power station fly-ash and samples from the Mersey road tunnels near Liverpool (see text) (from Hunt *et al.* 1984).

suggesting that the dominant iron oxide is magnetite (cf. Fig. 11.2). Viscous loss of $IRM_{0.3 T}$ between 10 and 2000 seconds after magnetisation is low and along with the low χ_{fd} measurements suggests that the assemblages mostly lack both coarse multidomain and fine viscous grains. The range of evidence available so far indicates a rather uniform magnetite assemblage relatively independent of particle size and source or concentration. We may therefore expect χ and $IRM_{300 mT}$ to be equally useful as rough concentration indicators. Maxted records a 95% correlation between the two for his total sample set. For routine measurement $IRM_{0.3 T}$ is preferable since it obviates any need to include compensation for the diamagnetism of the leaves and moreover the high sensitivity required is more readily attainable in the remanence measurements. The results, as well as showing a general response to rural–urban gradients, conurbation size and the proximity of industrial development, suggest that local factors, for example proximity to major roads, are important. Despite the lapse of time and the fact that magnetic measurements identify only one component of 'smoke', the rank correlation between 1961–71 smoke concentrations and $IRM_{0.3 T}$ in 1982

was 90–95%. Clearly the next stage in any follow-up study will involve comparing heavy metal concentrations on leaf surfaces with the IRM measurements. The method, if validated would allow almost instantaneous on-site measurements using portable pulse magnetisers and magnetometers. The present study uses the leaves of deciduous species and is thus integrating results over a single summer growing season at most. Using conifers and winter evergreens for comparison it would be possible to determine winter levels equally effectively. The same sort of approach may also be suitable for assessing the filtering effects of trees in contexts where this process is believed to be a significant contributor to acid rain or heavy metal pollution.

11.7 Magnetic particulates in ice and snow

In view of the relative ease with which both spatial and temporal patterns of mineral magnetic variation have been resolved by measurements of ombrotrophic peat (Section 11.5), two preliminary studies have been carried out aimed at developing a comparable approach to ice and snow samples. In both studies, ice or snow was melted through a filter upon which the magnetic properties of the retained particulates were then measured.

Figure 11.14 summarises the results obtained from samples prepared by Dr A. Mannion from the Okstinden area of western Norway (Oldfield & Mannion, in prep.). The Corneluissens Glacier and Skoltbrae samples were taken from recent snow accumulated in nevee fields, the others come from cave ice and include rock debris. The parameters for the nevee field ice are comparable to all the measurements on industrial and combustion related spherules summarised elsewhere (e.g. Section 11.6) and they can be clearly differentiated from the material in the debris ice by means of their coercivity profiles. SIRM readings of 4.2 and 5.6 × 10^{-3} Am^2l^{-1} may well reflect the influence of the Mo i Rana steel works lying roughly down wind and some 40 km away on the coast. Obtaining comparable results from pre-industrial (*c.* 1600 AD) Camp Century Greenland ice has, not surprisingly, proved a great deal more challenging. The only measurements carried out so far have been obtained using the clean room facility at the Ice Core Laboratory, SUNY, Buffalo. In order to measure to the required sensitivity a cleaned pulse magnetiser and fluxgate magnetometer were used, with holders

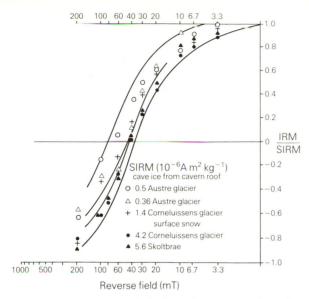

Figure 11.14 SIRM and $(B_0)_{CR}$ for surface snow and cave ice from the Okstinden region of Norway. Samples provided by A. Mannion.

and sample platforms designed to obviate the need for sample pots. Folded (0.22 μm) fluoropore filters were used without sample holders and rigorous cleaning procedures were followed at each stage. With some reservations (Oldfield, unpub.) we may tentatively infer that litre samples can provide enough material for SIRM measurements up to around 15×10^{-6} A m^2 total moment. This is about an order of magnitude above the noise level of the most sensitive portable magnetometers currently available. Comparison with the measurements from Norway suggests that concentrations in the pre-industrial Greenland ice are only about 0.3% of those in the surface snow from Okstinden. Magnetic techniques, suitably modified to contend with the low volume concentrations encountered in ice and snow samples, can be applied to historical studies paralleling those on recent post-industrial peat, to longer-term studies of dust-veil variations and to contemporary studies of local dust sources.

11.8 Global dust studies

Some dust samples supplied by D. R. Chester and collaborators from the lower atmosphere over the world's oceans have been measured. All of these were

obtained on board ship, by means of mesh samples and on HI-VOL air filters. In addition 20 samples provided by Dr Prospero from his collection of dusts obtained during the course of the Barbados Oceanographic and Meteorological Experiment (BOMEX) have also been measured. A preliminary evaluation, pending more comprehensive analysis of all these data (Hunt, pers. comm.), is set out below.

The magnetic properties of most samples are clearly dominated by ferrimagnetic crystals. The clearest indications of a significant antiferromagnetic component come from some of the samples believed to have resulted from the deflation of hot desert areas. Maximum susceptibility and SIRM per unit mass of sample or volume of air is associated with proximity to major ports, coastal conurbations and industrial complexes. Peak mass specific values may be over 10 times greater than the mean of all the atmospheric samples measured so far. The samples with peak susceptibility and SIRM are indistinguishable magnetically from those dominated by 'particulate pollution' spherules and discussed in Sections 11.5 and 11.6.

Mineral magnetic parameters are modified not only close to industrial/urban sources but also in areas of high input as a result of soil erosion. Under these latter conditions dust loading often increases dramatically and both susceptibility and SIRM

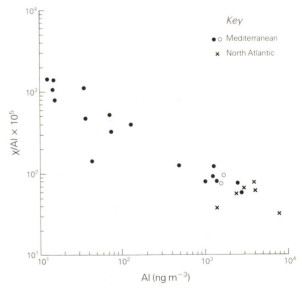

Figure 11.15 χ/Al versus Al for atmospheric dust samples from the Mediterranean and North Atlantic (from Chester *et al.* 1984).

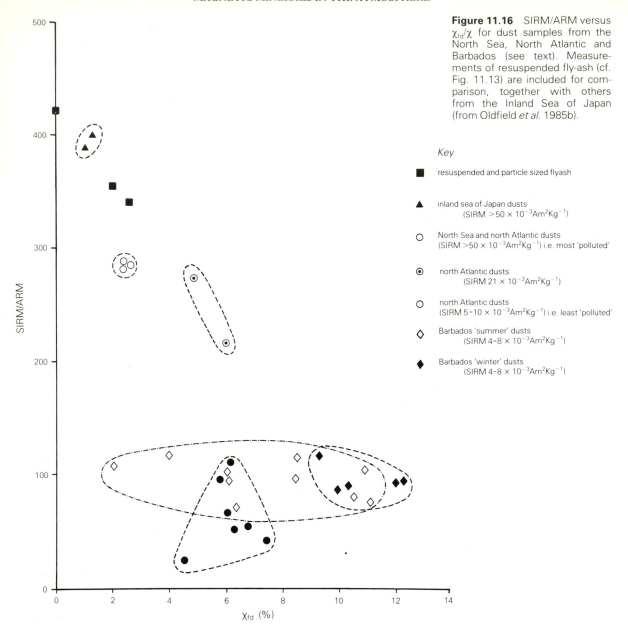

Figure 11.16 SIRM/ARM versus χ_{fd}/χ for dust samples from the North Sea, North Atlantic and Barbados (see text). Measurements of resuspended fly-ash (cf. Fig. 11.13) are included for comparison, together with others from the Inland Sea of Japan (from Oldfield *et al.* 1985b).

Key

■ resuspended and particle sized flyash

▲ inland sea of Japan dusts
     (SIRM $>50 \times 10^{-3} Am^2 Kg^{-1}$)

○ North Sea and north Atlantic dusts
     (SIRM $>50 \times 10^{-3} Am^2 Kg^{-1}$) i.e. most 'polluted'

◉ north Atlantic dusts
     (SIRM $21 \times 10^{-3} Am^2 Kg^{-1}$)

○ north Atlantic dusts
     (SIRM $5-10 \times 10^{-3} Am^2 Kg^{-1}$) i.e. least 'polluted'

◇ Barbados 'summer' dusts
     (SIRM $4-8 \times 10^{-3} Am^2 Kg^{-1}$)

◆ Barbados 'winter' dusts
     (SIRM $4-8 \times 10^{-3} Am^2 Kg^{-1}$)

decline steeply and consistently to minimum values. Chester *et al.* (1984) (Fig. 11.15) show that there is a direct correlation between χ and the χ/Al ratio in the samples collected from the Mediterranean. Moreover the high χ and χ/Al values are generally associated with low dust loadings and *vice versa*. These relationships they interpret as the result of soil derived particulates from major areas of deflation reducing

the susceptibility of a more or less ubiquitous background urban and industry-related aerosol during specific meterological episodes. High dust loadings associated with large concentrations of soil-derived magnetic minerals are associated with relatively steep viscous loss of SIRM and with higher χ_{fd} values.

Table 11.3 summarises results obtained from Barbados samples dating from 1966–68 taken by

Table 11.3 Barbados dusts mean values ± standard deviation (October and November dusts omitted).

	χ_{fd}/χ (%)	SIRM/χ (kAm^{-1})	IRM$_{LR}$/SIRM*	IRM$_{HR}$/SIRM†	$(B_0)_{CR}$ (mT)
grey dusts (Dec.–April)	10.7 ± 2.8	8.0 ± 0.23	0.24 ± 0.06	−1.44 ± 0.09	35 ± 1.7
red-brown dusts	9.5 ± 3.9	9.25 ± 0.63	0.45 ± 0.11	−1.22 ± 0.13	42 ± 4.1

$$\text{IRM}_{LR}/\text{SIRM}* = \frac{\text{IRM}_{-20\,mT}}{\text{SIRM}} + \frac{\text{IRM}_{-40\,mT}}{\text{SIRM}}$$

$$\text{IRM}_{HR}/\text{SIRM}† = \frac{\text{IRM}_{-100\,mT}}{\text{SIRM}} + \frac{\text{IRM}_{-200\,mT}}{\text{SIRM}}$$

Prospero (1968). Only normalised parameters are recorded and the mean value and standard deviation are quoted in each case. Out of a total sample set of 20, four samples from October and November, were omitted. The rest were identified as either red-brown, Sahara-derived 'summer' dusts or grey, more locally derived, South American 'winter and spring' dusts (Prospero *et al.* 1981). All the parameters used clearly distinguish the two sets. The higher coercivities, 'harder' IRM/SIRM and higher SIRM/χ values for the Saharan set are consistent with a relatively high haematite component, and in the 'hardest' samples up to 30% of the original SIRM is unsaturated in a reverse field of 0.4 T. The grey dusts have a very high χ_{fd} and are almost entirely reverse saturated at fields less than 0.4 T. All the parameters are consistent with a derivation largely from secondary ferrimagnetic grains at the surface of 'enhanced', probably burnt, soils.

Figure 11.16 plots SIRM/ARM versus χ_{fd}/χ for a larger sample set from the North Sea and North Atlantic. The Barbados dusts are grouped into seasonal sets as in Table 11.3. The cruise samples are placed in three well defined groups according to SIRM values. The gradient from highest to lowest SIRM reflects the declining relative importance of anthropogenic sources (cf. Chester *et al.* 1984) within the samples which span a large area from the coasts of Britain to low latitudes. ARM is believed to be more discriminating of true stable single-domain grain size than is SIRM, so we may expect the ratio of SIRM/ARM to decline as magnetic grain size changes, from multi- to single domain (Dankers 1978, King *et al.* 1982). χ_{fd} will increase with the greater relative importance of even finer crystals at the lower size limit of the single-domain range. Soils are thought to be the main sources of fine stable single-domain and smaller crystals. The diagram shows that the samples least affected by anthropogenic emissions have the lowest ratios and highest χ_{fd}/χ percentages, while those most

affected have values for SIRM/ARM and χ_{fd}/χ close to those for the particle-sized fly-ash samples considered by Hunt *et al.* (1983) and plotted in Table 11.2. These results reinforce the proposition that magnetic parameters may be valuable aids to dust and aerosol source identification.

11.9 Summary and conclusions

On the basis of the studies summarised above, several prospective uses for mineral magnetic measurements of atmospheric samples can be tentatively advanced. The techniques clearly have a rôle to play in the historical monitoring of particulate and possibly heavy metal pollution. It is also possible that longer-term historical studies of dust-veil variation will be feasible using older ice and peat core material, though in the latter case it is possible that solution of magnetic minerals under reducing conditions may degrade the long-term record. Applying the methods to contemporary particulate and heavy metal pollution monitoring studies will require much additional work on the relationship between the magnetic properties and element chemistry of characteristic sources, as well as pragmatic evaluation of alternative sampling strategies. Because the atmosphere lacks the largely materially bounded characteristics of lake-watershed ecosystems for example, confidence in applying the techniques to such problems as aerosol identification and tracing (Hunt, 1986), or to plotting deposition patterns and distance-dependent effects from point sources, requires a large body of contextual data with very wide spatial coverage. Initial indications are that on a variety of spatial scales mineral magnetic parameters, used perhaps in conjunction with element ratios (cf. Kleinman *et al.* 1980), will be of considerable value in helping to identify dust and aerosol sources and to plot plume dispersal and deposition. Several additional possibilities are not considered in the present account. For example, long-distance transport of sulphur compounds from power stations is believed to be partly in particle-associated form. It may therefore be possible to use magnetic properties as tracers of 'acid rain' sources. On a much smaller scale, the complex theoretical problems involved in modelling particle deposition to naturally 'rough' and heterogenous vegetated surfaces are no less daunting than the practical constraints involved in obtaining empirical data on deposition using even relatively simplified contexts (Chamberlain 1966, Clough 1973). Magnetic measurements could certainly be used in experiments designed to study particle deposition on natural surfaces.

[12]

Mineral magnetism in marine sediments

We have *first* the matters derived from the wear of coasts, and those brought to the sea by rivers . . . In oceans affected with floating ice we have land debris . . . Second – we have dust of deserts . . . In the trade wind region of the N. Atlantic we have a very red-coloured clay . . . which is largely made up of dust from the Sahara . . . Third – we have the loose volcanic materials, which have been . . . universally distributed as pumice or as ashes carried by the wind . . . While examining the deposits during the cruise I frequently observed among the magnetic particles from our deep-sea clays small round black-coloured particles which were attracted by the magnet, and I found it difficult to account for the origin of these. On our return home I entered into a more careful examination of the magnetic particles. Some of the particles are little [iron] spherules . . . that appear to have a cosmic origin.

John Murray Esq., 1876
Proc. R. Soc. Edinburgh, 247, 261

12.1   Introduction

Of all the environmental systems considered in this book, the sea is, from a mineral magnetic point of view, by far the most complex. Potentially significant sources of magnetic minerals are at their most varied and include not only all those of relevance to lacustrine and atmospheric studies but also submarine and extraterrestrial sources. Supply pathways are often more complex and extended than anywhere else. Depositional environments range from extreme high to extreme low energy, and also span a bewildering range of chemically and biologically modulated variation. Moreover marine environments include some of the least accessible on earth. In consequence it is not possible to attempt a comprehensive and integrated review. The approach adopted here is selective, thematic and empirical, focusing on a limited number of quite independent studies in contrasted marine contexts. These range from pelagic environments where our major concern is the palaeoclimatic record, to coastal situations dominated by recent human impact.

12.2   The origin and flux of marine magnetic minerals

Figure 12.1 summarises in schematic form the major sources and pathways of the magnetic minerals found in marine sediments. The present section attempts to evaluate these sources and pathways in the light of the concern with possible palaeoclimatic linkages and implications developed in 12.4 below.

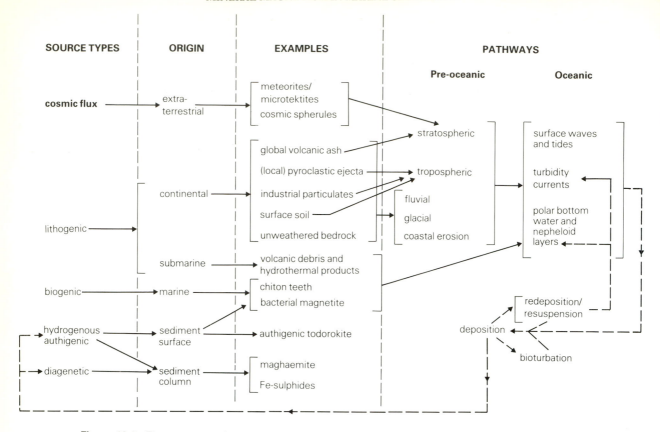

Figure 12.1 The sources and supply pathways of magnetic minerals encountered in marine sediments.

COSMIC

Due to magnetite formation during the entry of extra-terrestrial material into the Earth's atmosphere, most of the cosmic spherule component of marine sediments is strongly ferrimagnetic (Brownlee 1981). Three conditions must be met before cosmic particles can be of any real relative significance in the marine sediment record from a given area:

(a) The sediments should predate the Industrial Revolution, since as shown in Chapter 11 industrial and fossil-fuel combustion processes have, over the last century, greatly increased the atmospheric concentration of magnetic spherules, even in areas remote from urban and industrial activity.

(b) Even in pre-industrial times, the sediments must come from those areas of the sea bed most remote from any kind of particulate terrigenous input.

(c) Sediment accumulation rates must be extremely slow.

In practice these conditions are met in some abyssal sediments from open oceans, especially the Pacific. It is possible to extract cosmic spherules from this type of sediment using simple magnetic techniques and the concentrations (up to several mg kg^{-1}) may be significant in the mineral magnetic record. However, none of the cores considered in the succeeding sections fulfills criteria (a) and (b) and it is unlikely that cosmic spherules comprise more than a very minor 'background' component in the mineral magnetic record. Meteorites and microtektites are of much more limited significance spatially and temporally and can be ignored in the context of the present account.

LITHOGENIC

We can infer from preceding chapters that three types

of continental source will be important in the flux of magnetic minerals to marine sediments on the time-scale of interest in this section. These are volcanic sources, soil and bedrock. Volcanic inputs to marine sediments will range from highly local pyroclastic debris through widely distributed tephra layers, to deposition from global volcanic dust veils resulting from the injection of fine ash particles into the stratosphere. Haggerty (1970) regards the volcanic input of magnetic minerals to deep-sea sediments as a significant component in natural remanence studies.

Soil and bedrock sources contribute particulates to the oceans as a result of wind, fluvial, glacial and shoreline erosion. The figures summarised by Prospero (1981) and reproduced in Table 11.1 suggest that most authorities regard land surfaces exposed to wind erosion as much greater contributors of atmospheric dust than are volcanoes at the present day, though this may be in part a function of recent human activity. If we bear in mind the often high magnetic concentration in such surface material and add the effects of fluvial inputs (cf. Currie & Bornhold 1983), as well as glacial rafting (Mullen *et al.* 1972, Ruddiman *et al.* 1971), we must conclude that continental erosion is a major source of detrital marine magnetic minerals even in many pelagic environments. The processes of weathering and erosion which control the nature and the release of magnetic minerals in soils, as well as the atmospheric and hydrospheric pathways through which magnetic

minerals will pass from source to sediment, are all strongly controlled by climatic change.

BIOGENIC
It is rather difficult to determine the relative importance of biogenic magnetic minerals in marine sediments. In nearshore sediments, as in lakes, a wealth of circumstantial evidence points to a detrital origin for the bulk of the magnetic minerals present. However, Kirschvink (1982a) has recently inferred the presence of bacterial and possibly algal (Lins de Barros *et al.* 1981) magnetite in marine clays of Miocene age in Crete and transmission electron microscopy has revealed the presence in a variety of deep-sea sediments of several morphologically distinct magnetite crystals, which closely resemble those formed by magnetotactic bacteria (Kirschvink & Chang 1984). There is also the possibility that in the sediments of the open oceans, magnetite derived not only from magnetotactic bacteria but also from chiton teeth (cf. Lowenstam 1981) makes a significant contribution.

AUTHIGENIC AND DIAGENETIC
Recent reviews by Henshaw and Merrill (1980) and by Burns and Burns (1981) show that both authigenic and diagenetic magnetic phases are relatively common in marine sediments. Moreover, their formation is often related to variables such as accumulation rate and organic content, which are in turn often controlled by climatic variations. However, the conditions under which these phases form are becoming increasingly well understood and defined.

Authigenic maghaemite is found in 'red clay' deep-sea sediments from the north and central north Pacific, where it carries an unstable, low coercivity magnetisation (Kent & Lowrie 1974, Johnson *et al.* 1975). The maghaemite is most probably formed *in situ* by the low temperature oxidation of magnetite. Kent and Lowrie noted its widespread occurrence in sediments formed before three million years ago, possibly on account of the warmer upper Cainozoic climate and associated slower sediment deposition rates. Such low sedimentation rates increase the exposure time of sedimenting particles to oxidising bottom waters and permit oxidation near the sediment/water interface before burial occurs. Johnson *et al.* found that the most heavily oxidised magnetic minerals occurred in the non-fossiliferous cores, which displayed a slow erratic decline of magnetic mineral content towards the sediment

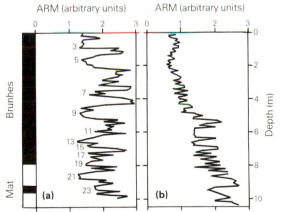

Figure 12.2 Down-core variation in intensity of anhysteretic remanent magnetisation for two North Pacific cores. (a) A fossiliferous core with a 1 m year polarity timescale; (b) a non-fossiliferous core with authigenic iron oxides. (After Johnson *et al.* 1975.) The sequence of numbers 1 to 22 in (a) indicates a possible correlation of the mineral magnetic fluctuations with Pleistocene climatic change and the oxygen isotope stages.

surface (Fig. 12.2b) and poor palaeomagnetic records. In contrast clear palaeomagnetic polarity zones and distinctive fifty thousand year variations in magnetic mineral concentration (Fig. 12.2a) were found for the fossiliferous cores. As discussed below in Section 12.4 such distinctive mineral magnetic variations can be matched to the oxygen isotope stratigraphy.

Authigenic magnetic iron sulphide minerals produced in highly reducing conditions have been identified by Kobayashi and Nomura (1972, 1974) in deep-sea sediment cores from the Sea of Japan. Unusually strongly reducing conditions are needed to grow pyrrhotite. Perhaps times of stagnant bottom waters and abundant organic matter, associated with low sea levels when the Sea of Japan may have been entirely surrounded by land, could have produced suitable conditions for the authigenic *in situ* growth of pyrrhotite and pyrite.

Selective dissolution of magnetite grains in strongly reducing hemipelagic muds on the Oregon continental slope has been demonstrated by Karlin and Levi (1983) to lead to downcore coarsening of the magnetic grain size as the smallest grains are removed. This change in the top 30 to 80 cms is accompanied by a reduction in magnetic concentration. Further downcore the magnetic grain size begins to decrease while concentration continues to decline as the remaining grains dissolve.

12.3 Core correlation in marine sediments

Several published studies (e.g. Radhakrishnamurty *et al.* 1968) point to the link between magnetic susceptibility variations and tephra layers in marine sediments, and in each case, the opportunity for core correlation using magnetic measurements is at least implicit. In the case of lake sediments (Ch. 10) detailed core correlation is often possible even where tephra layers are absent and, by analogy, we may hope to find that marine sediments offer similar opportunities despite the increases in temporal and spatial scale involved.

Figures 12.3 to 12.5 illustrate susceptibility-based core correlation in a variety of marine contexts. The cores for which whole core volume susceptibility is plotted in Figure 12.3 are part of a large set of piston and gravity cores from the eastern Mediterranean between Cyprus and southern Turkey. Even without a detailed timescale the correlation scheme highlights major spatial changes in deposition rates and patterns. The basis for the correlations is thought to be rather complex. The total time interval involved includes the later part of the last glacial interval and much of the Holocene, and part of the variation reflects fluctuations in terrigenous influx resulting from climatic change and human activity. The period of minimum susceptibility coincides with the well

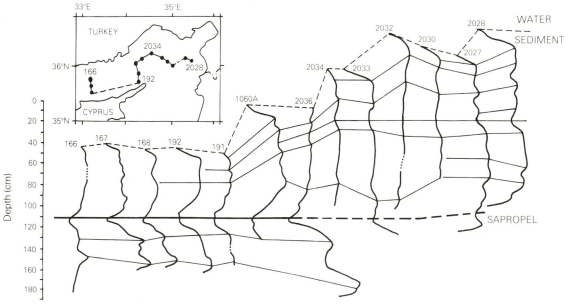

Figure 12.3 Whole core volume susceptibility logs from late-Pleistocene piston and gravity cores from the eastern Mediterranean. (All cores were scanned in the Department of Geological Sciences, Imperial College, London, UK.)

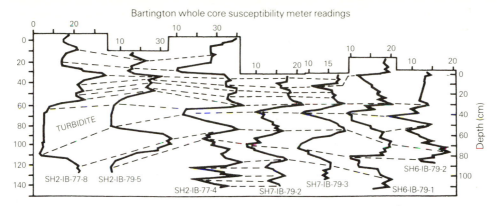

Figure 12.4 Single-sample mass susceptibility logs for cores from the Azores region of the eastern Atlantic.

known sapropel layer dated to *c.* 5000–7000 BP and may reflect a period of minimum terrigenous input between the end of the last glaciation and the beginnings of prehistoric farming and accelerated soil erosion (cf. Section 10.6). In addition, tephra layers are certainly present though not always in a recognisably continuous form from core to core, and these are responsible for some of the main susceptibility peaks.

The core traces plotted in Figure 12.4 are from a topographically varied area of the eastern North Atlantic in the region of the Azores. They include tephra layers of local origin reflecting the proximity of a region of active vulcanicity (Robinson 1982). Several cores (e.g. 79–5) show, in addition, an interval of uniformly low susceptibility corresponding with a microturbidite layer. Robinson (1982) has reported additional mineral magnetic data from cores in which the occurrence of the microturbidites has been inferred. In each core a major change in the $IRM_{-100\,mT}/SIRM$ ratio is associated with the micro-turbidite layer. The saturation remanence is clearly much 'harder' in the layer than in the samples on either side, probably reflecting some diagenetic change from paramagnetic and/or ferrimagnetic forms to antiferromagnetic as a result of post-depositional bacterially mediated organic oxidation. The parallel between the $IRM_{-100}/SIRM$ changes and the distribution of uranium and the redox-sensitive metals in these cores (Colley *et al.* 1984) suggests that the identification of this type of mineral magnetic feature has significant geochemical implications.

Figure 12.5 plots a series of single-sample susceptibility traces, compiled by Robinson (1982) from a set of cores obtained by the Institute of Oceanographic Sciences, for an area close to the King's Trough area of the North Atlantic. Here the correlations are extremely clear and largely independent of such tephra and turbidite layers as have been recognised. These cores are much more representative of relatively undisturbed pelagic environments and the degree of correlation is likely to be typical of many deep-sea sediment cores obtained from comparable environments in all the worlds major ocean floors. The basis of the correlation is believed to be climatic modulation of the magnetic mineral types and concentrations and is discussed fully in the following section.

12.4 Mineral magnetism and palaeoclimate in deep-sea sediments

One of the most striking observations to emerge from studies of deep-sea sediments has been the parallelism between palaeomagnetic parameters and palaeoclimatic indicators noted by many authors and encountered in all the major ocean basins of the world. Frequently, variations in the intensity of natural remanent magnetisation (NRM) and sometimes also changes in inclination closely parallel the palaeoclimatic signature whether this is derived from [18]O ratio determination, foraminiferal, coccolith or calcium carbonate analysis. Figure 12.6 illustrates this parallelism. Many authors have taken these and similar results as confirmation of some type of fundamental linkage between the behaviour of the geomagnetic field and changes in climate. Clearly, the

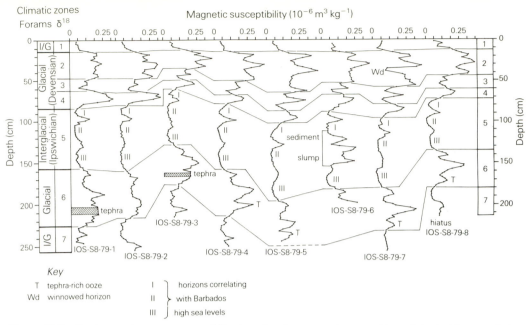

Figure 12.5 Single-sample mass susceptibility logs for eight cores from the North Atlantic. The correlations shown are related to the palaeoclimatic inferences developed in Section 12.4 of the text and illustrated in Figure 12.7.

linkage, if real, must exist on a global scale. The theories proposed have invoked a range of possible mechanisms. Harrison and Prospero (1977) proposed a direct effect of the magnetic field on the atmosphere by means of a causal link between low geomagnetic field intensities, higher effective cosmic radiative flux and warmer climates. Doake (1977) suggested that the distribution of ice may have influenced the geomagnetic field by modifying the fluid motions within the core. Wollin *et al.* (1977) and Opdyke and Kent (1977) speculated that astronomical variables, for example changes in the eccentricity of the Earth's orbit, may control both climatic and geomagnetic intensity variations. The problem of accounting for the palaeoclimatically related inclination variations has also attracted some attention. Stuiver (1972) suggested that long-term changes in planetary configuration may be responsible for these apparent correlations.

More recently several authors have either challenged the reality of a geomagnetic–palaeoclimatic link (Kent 1982) or put forward a completely different kind of explanation for the observed parallelism (Bloemendal 1980; Oldfield & Robinson 1985; Robinson 1986). Kent (1982) uses data from the southern Indian Ocean and from the

equatorial Pacific to establish two important points. He shows that not only does NRM intensity parallel calcium carbonate concentration and, where available, the ^{18}O record, but so do the mineral magnetic parameters, susceptibility and IRM. Whereas changes in NRM intensity can, in theory, reflect geomagnetic variations, changes in susceptibility and IRM cannot. Moreover, in the equatorial Pacific, the peak $CaCO_3$ concentrations and the NRM, IRM and susceptibility minima are associated with maximum ice volume and minimum temperature during glacial intervals, whereas in the southern Indian Ocean the converse applies. The parallelism between *mineral* magnetic parameters and palaeoclimatic indices observed in these cores as well as in those studied by Bloemendal from the South Atlantic and Robinson from the North Atlantic makes a global geomagnetic–palaeoclimatic linkage difficult to sustain as the primary cause of the magnetic–climatic relationship on the timescales resolved by marine sediment. This difficulty is further compounded by the opposite nature of the magnetic–palaeoclimatic relationship at Kent's two sites. Clearly an alternative interpretation is required.

Oldfield and Robinson (1985) draw together evidence from both lake and marine sediments to propose a sedimentological/mineralogical as against

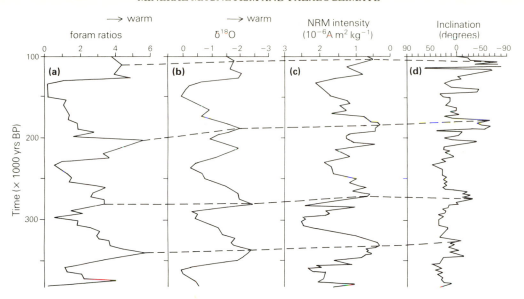

Figure 12.6 Correlation of (a) the *Globorotalia menardii* climate curve, (b) the oxygen isotope curve of *Globigerinoides ruba*, (c) magnetic intensity and (d) magnetic inclination in deep-sea cores V12-122 from the Caribbean from 380 000 to 100 000 years BP. (After Wollin *et al*. 1977.)

a geomagnetic basis for the palaeomagnetic–palaeoclimatic relationships. The essential components of the explanation are as follows:

(a) The fundamental palaeoclimatic linkage is with mineral magnetic and not palaeomagnetic properties. The linkage is reflected in variations in the intensity of NRM because these are strongly dependent on the changes in magnetic concentrations, grain size and mineralogy which are reflected in the mineral magnetic parameters (see Ch. 4). Confusion has arisen in the past from the failure of many workers to distinguish clearly between NRM intensity and geomagnetic palaeointensity, and from the fact that interest in sedimentary *palaeo*magnetism preceded any detailed *mineral* magnetic studies of deep-sea cores.

(b) Where a linkage is apparent between inclination and palaeoclimate it reflects the relationship between sediment structure and 'inclination error' noted more often in glacial varve than marine sediment studies (cf. Sections 13.2.2 & 14.5).

(c) All aspects of the magnetic–climatic linkages so far identified in deep-sea sediments are a function of the control exercised by climate over the complex of variables determining shifts in

magnetic mineral sources and types, in the flux of magnetic minerals to the sediments, and in the concentration of non-magnetic sediment components such as calcium carbonate.

Figure 12.7 shows results from IOS Core S8-79-4 in the North Atlantic. It spans most of oxygen isotope stages 1 to 7 and thus represents the last *c*. 250 000 years of the Pleistocene. The foraminiferal, calcium carbonate and ^{18}O records are set alongside plots of χ, SIRM, ARM and IRM$_{-100\,\text{mT}}$/SIRM. The non-magnetic indices provide a mutually consistent picture of palaeoclimatic variation, which is in turn reflected in each of the curves of mineral magnetic variation. Furthermore, the speed and relative ease with which detailed magnetic characterisation can be accomplished makes it feasible to resolve variations much more closely than is normally done using non-magnetic techniques. At the detailed level of resolution achieved here the minor variations in all magnetic parameters during ^{18}O zone 5 appear to reflect the sequence of subdivisions previously established by Shackleton (1977) and the reality of the palaeoclimatic signature in the mineral magnetic record can hardly be doubted.

It is especially significant that here as in Bloemendal's mid-Pliocene record from South Atlantic DSDP Core 514 (Fig. 12.8), not only do

147

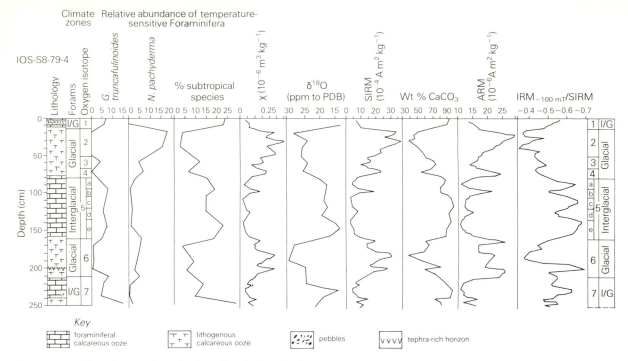

Figure 12.7 Mineral magnetic and palaeoclimatic signal in Core IOS-S8-79-4 from the North Atlantic (see Oldfield & Robinson 1982 and text).

concentration-dependent parameters such as χ, SIRM and ARM reflect palaeoclimate control but so also does the normalised, hence essentially qualitative $IRM_{-100\,mT}/SIRM$ parameter. This confirms that the mineral magnetic signature, though in part an expression of varying degrees of dilution by the non-magnetic sediment components such as $CaCO_3$, reflects also changes in magnetic grain size and/or mineralogy. In the IOS core, $IRM_{-100\,mT}/SIRM$ varies from −0.34 to −0.68, and in the 'hardest' samples up to 15% of the total SIRM remains unsaturated in a reverse field of −0.3 T. In the DSDP, core variations in the stability of the remanence are even greater and the occasional extreme sample has a coercivity of SIRM close to 0.1 T. Although recent studies by King *et al.* (1982) have stressed the importance of stable single-domain magnetite in samples with relatively high remanent coercivities, the wide range of variation found at these sites, the high coercivities of the 'hardest' samples and the presence of a significantly varying and at times highly unsaturated component at 0.3 T strongly suggest important mineralogical variations including shifts in the ratio of ferrimagnetic to canted antiferromagnetic grains. The glacial parts

of the record are marked by high magnetic concentrations and a relatively 'hard' mineral assemblage and the interglacials by the converse. Clearly the record implies changes in magnetic mineral sources and pathways as well as changes in the flux of both magnetic and non-magnetic components to the sediment. Thus in order to find an explanation to account for the full range of climatically modulated mineral magnetic variations represented here and in other deep-sea cores we need to consider the implications of Section 12.2 on the origin and flux of marine magnetic minerals.

Although tephra layers as such will give rise to changes in magnetic mineral concentration unrelated to climatic change, it is nonetheless possible that variations in the volcanic dust veil, discussed by Lamb (1970) and Kennett (1981) for example, have been of major significance in controlling both climatic change and the flux of magnetic minerals to marine sediments. Kennett, summarises many studies supporting a link between volcanic activity and climate and concludes that 'the general synchronism of the volcanic explosive episode and global climatic evolution is probably not coincidental'. However, in

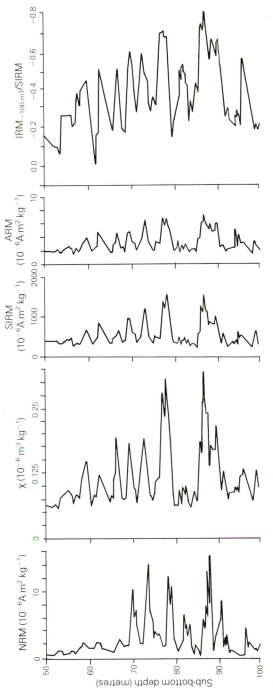

Figure 12.8 Mineral magnetic measurements from part of DSDP Core 514 from the South Atlantic (Bloemendal 1980).

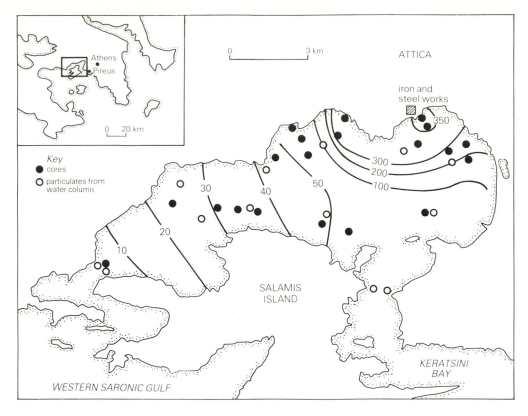

Figure 12.9 The Elefsis Gulf, Greece. The map locates both sediment coring sites and contemporary particulate sampling stations. The contours plot the decline in sediment surface mass susceptibility away from the iron and steel works on the north-eastern shore.

the light of the foregoing chapters, the magnetic evidence for the association between recurrent changes in mineralogy and climatic change is more readily interpretable as the result of shifts in weathering regimes, dust source areas and both atmospheric and oceanic pathways, than as the result of varying intensities of volcanic activity, though the evidence is far from conclusive. Moreover the importance of non-atmospheric inputs to the sediments considered here remains an open question.

Oldfield and Robinson (1985) tentatively conclude that the linkage between mineral magnetic variations and climatic change in deep-sea sediments probably arises from variations in continental sources, both volcanigenic and more especially erosive. The sedimentary expression of these variations must also be controlled by the effect of climate on the pathways by which the lithogenic minerals reach the sea bed, especially through changes in atmospheric circulation

and ocean currents. Both Ellwood (1980) and Bloemendal (1980) suggest that changing bottom current velocities and directions have been responsible for mineral magnetic variations in their cores from the South Atlantic.

It is clear that we are some way from understanding all aspects of the mineral magnetic–climatic linkage mechanisms. Moreover these are likely to vary with latitude and with core location in relation to extensive deflated arid land surfaces as well as to patterns of movement by deep benthic water masses. Even so there are strong incentives for studying the linkages since a co-ordinated view could provide valuable new insights into the nature and expression of climatic change. At the very least, the results obtained so far confirm that mineral magnetic measurements in deep-sea sediments can provide not only a convenient and informative method of logging but one which is of major palaeoenvironmental significance.

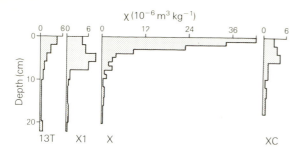

χ (10^{-6} m^3 kg^{-1})

Figure 12.10 Down-core variations in mass susceptibility at sites close to the iron and steel works located in Figure 12.9.

12.5 Particulate pollution monitoring in coastal waters

A recent study by Scoullos *et al.* (1979) illustrates the potential value of magnetic measurements in coastal pollution monitoring where major sources are discharging high particulate concentrations which include ferrimagnetic oxides. The Elefsis Gulf is a shallow, almost enclosed embayment of the eastern Mediterranean close to the main areas of industrial development associated with Athens. The major point source of particulates is an iron and steel works near the northeastern corner of the Gulf, established from 1925 onwards and first brought into production in 1948 using scrap iron in electric arc furnaces. A first blast furnace became operational in 1963 and a second in 1972. The factory complex also includes a cokery, steel making and oxygen plants, a rolling mill and port facilities.

Figure 12.9 plots the location of sixteen sample stations from which monthly samples of particulates were taken on Millipore membrane filters of 0.45 μm at depths of 0, 10, 20 and 30 m during the period March 1977 to February 1978 inclusive. Also plotted are seventeen localities from which sediment cores up to 1 m long were taken using a pneumatic Mackereth (1969) minicorer. Susceptibility and SIRM were measured on all cores and whole core volume susceptibility scans were carried out on additional paired cores taken from Station 2 in 1980. Frequency-dependent (cf. quadrature) susceptibility was measured on suites of samples from Cores 2A and 3. The same figure shows the spatial distribution of sediment surface values of susceptibility in contour form and confirms the dominance of the point source at the iron and steel works.

Figure 12.10 shows down-core variations in susceptibility in several cores taken close to the industrial site. Core X was taken only 0.3 km from the artificial lagoon which receives the plant's effluents, the others at distances varying from 1 to 3 km. The recent nature of the peak in values is clearly indicated and has been subsequently confirmed by ^{137}Cs (Baxter, pers. comm.).

Figure 12.11 plots SIRM *v.* total particulate iron for all filter paper samples taken during January 1978. This was shown by means of chlorophyll-a analysis to be the period of minimum biomass. A strong linear relationship emerges suggesting that careful calibration and use of SIRM in the Elefsis Gulf may provide a rapid index of total particulate iron concentrations. The whole core volume susceptibility plots show that rapid scanning could be used as effectively as more time-consuming single-sample measurements to monitor variations in the magnetic content of near-surface sediments in the Gulf.

The plots of frequency-dependent (cf. quadrature) susceptibility (Fig. 12.12) show that whereas coercivity curves fail to differentiate clearly between the 'magnetite' deposited prior to the major effect of the iron and steel works from that postdating its

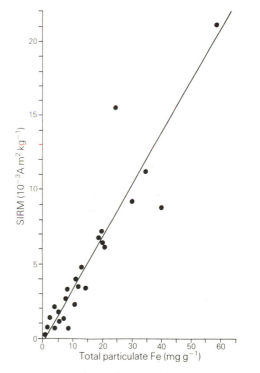

Figure 12.11 SIRM values versus total particulate iron concentration for filter samples from the water column taken from the Elefsis Gulf in January 1978.

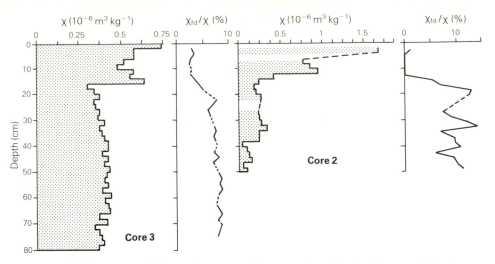

Figure 12.12 Mass susceptibility and percentage frequency-dependent (χ_{fd}/χ) susceptibility in two Elefsis Gulf cores.

development (Scoullos *et al.* 1979), χ_{fd}, expressed as a percentage of total susceptibility, declines sharply at the depth where the rapid rise in total susceptibility records the onset of the industrial development. This is consistent with a secondary soil-derived origin for the 'magnetite' in the lower part of the cores (cf. Ch. 8) and an industrial origin for the most recent material. The results as a whole, point to the value of this approach, especially in preliminary appraisals of both historical trends and current dispersion patterns. Not only is total particulate iron related to the magnetic measurements, but as in the case study noted in Chapters 9 and 11, so are other heavy metals including zinc. In considering the possible extension of this approach to other areas it is worth noting that Doyle *et al.* (1976) infer particulate pollution of the sediments of the eastern Gulf of Mexico from the identification of magnetic spherules over areas hundreds of kilometres from potential industrial sources.

12.6 Summary and conclusions

Mineral magnetic studies of marine sediments can already provide a basis for core correlation, for characterising diagenetic change, for establishing tephrochronologies, for reconstructing the course of climatic change and for near-shore particulate pollution monitoring. However, they are in their infancy and the range of illustrations and applications outlined here is unlikely to encompass anything approaching the full potential. Moreover, in each of the aspects touched on, e.g. palaeoclimatic implications and diagenetic effects, a great deal more critical work remains to be done before the crucial processes and mechanisms can be understood. This work calls for the *routine* association of mineral magnetic studies with oxygen isotope, palaeoecological, sedimentological and geochemical studies, both as precursors and complements to the traditional lines of enquiry. The speedy non-destructive nature of the techniques, the growing portability of the equipment and the development of sensors capable of scanning not only enclosed cores of any diameter but also free faces, makes the methodology ideally suited for this rôle. In addition, a broadening of the marine palaeomagnetist's concerns beyond the reconstruction and validation of palaeomagnetic directions and palaeointensities could lead to major advances in the quality and the spatial and temporal coverage of the mineral magnetic record. Only in this way will it be possible to build up a picture of the global patterns of variation upon which more general inferences of palaeoclimatic and geochemical value can be based.

[13]

Reversal magnetostratigraphy

"Footprints on the sands of time"

Longfellow
A psalm for life

13.1 Introduction

The geomagnetic field has been likened to a chronometer (Cox 1973). The magnetic clock, driven relentlessly by electric currents in the Earth's liquid core, oscillates back and forth switching between its two stable modes of a 'normal' state in which the Earth's magnetic field points north and a 'reverse' state in which the field points south. The geomagnetic chronometer lacks a regulator so it runs unevenly with switching intervals varying from a few thousand years to several tens of millions of years. The magnetic timing system of north–south flips, tags rocks as they form with its binary reversal code. Palaeomagnetists have unravelled the magnetic coding scheme so that palaeomagnetic remanence signatures can now be used to date sequences of lava flows and sediments.

Nature in her usual mischievous way has, in addition to providing such an elegant magnetic clock, laid down a rich assortment of false leads, red herrings, decoys and tricks which can bamboozle and deceive the impetuous palaeomagnetic investigator. The mechanics and workings of the magnetic chronometer are outlined in some detail below in order to alert the reader to the types of pitfalls to be encountered in magnetostratigraphy.

13.2 Geomagnetic signatures

The fidelity and accuracy with which geomagnetic polarity reversals have been recorded in rocks have varied greatly. This wide range in quality has resulted from the diversity of the geological processes which can record changes of the ancient field.

The process of remanence acquisition is generally simpler in igneous rocks than in sedimentary deposits. In some sequences of lava flows polarity reversals have been recorded in such great detail that individual reversals can be recognised by the characteristic pattern of their polarity transitions. Some sediments which have steadily accumulated, such as deep-sea sediments, have also recorded polarity changes in detail. Other palaeomagnetic records can however be rather misleading and so an understanding of the processes and mechanisms involved in palaeomagnetic remanence acquisition can be very helpful in assessing the usefulness of a palaeomagnetic record in a magnetostratigraphic study.

13.2.1 Origin of remanent magnetisation in igneous rocks

The natural remanent magnetisation of igneous rocks

is generally the residual thermoremanent (Section 4.3.1) magnetisation acquired by the rocks on cooling from their molten state. This thermoremanence is locked into the igneous material at Curie and blocking temperatures of around 500 °C, well below the temperatures of between 1000 and 1250 °C at which the material crystallised out of the igneous melt. Many igneous rocks have been found to record accurately the direction and intensity of the geomagnetic field with uniform directions within a single igneous unit. Fine-grained igneous rocks with a thermoremanence held by small iron oxide crystals have been found to be particularly suitable for palaeomagnetic work.

LABORATORY EXPERIMENTS

The stable remanence of igneous rocks can be recreated in the laboratory by heating a sample of rock to above its Curie temperature and then allowing it to cool down again to room temperature (Koenigsberger 1938). Such artificial remanences are found to duplicate both the stability and intensity of the natural remanence. These types of experiment lend strong support to the weak field cooling origin of the natural remanence of igneous material and demonstrate that a thermoremanence direction is generally coincident with the field applied during cooling. Experiments also show that the rate of cooling is not important in affecting the remanence intensity at rates slower than 0.1 °C per second (Nagata 1953, Dodson & McClelland-Brown 1980).

Experiments on remanence acquisition over restricted temperature ranges have established that the addition law (Thellier 1946) of thermoremanence applies to natural as well as synthetic samples. In essence the addition law states that the partial thermoremanence of one temperature interval is independent of the remanence produced by field cooling in other temperature intervals. This basic property of thermoremanence is exploited in two important laboratory techniques. These are first, the method of isolating more than one component of remanence by partial demagnetisation (Section 6.5.2), and secondly the method of obtaining more than one estimate of ancient field intensity by the double heating palaeointensity technique (Thellier & Thellier 1959). Both methods involve cooling the sample from some chosen temperature, below the Curie temperature, to room temperature in a well controlled zero field (Section 6.6.3) environment. This cooling procedure results in isolating the remanence of magnetic grains with blocking temperatures above

the chosen temperature, as grains with blocking temperature below the chosen temperature lose their remanence when cooled in zero field. The practical benefits of these types of experiment are that igneous rocks with quite complicated multicomponent magnetic histories, which may involve partial remagnetisation by geological events after the time of their formation, can still be used in geomagnetic and magnetostratigraphic studies.

One disturbing phenomenon observed in the laboratory is that a small percentage ($< 1\%$) of natural igneous rocks acquire a thermoremanence antiparallel to that of the applied field (Nagata et al. 1952). Néel (1955) has described a number of physical mechanisms which could cause such a reverse thermoremanence. The two major mechanisms in natural materials are first the magnetic interaction of two different magnetic constituents in a rock, and secondly the interaction of spin moments of neighbouring sites within a crystal lattice. One might expect that such reversal behaviour in a natural remanence could easily be detected by simply testing the rock in the laboratory. However, nature manages to arrange events, such as later chemical change, which can mislead or confound the investigator and in practice self-reversals of some key magnetostratigraphic igneous rocks have gone undetected for many years (Heller 1980). On the positive side the annoying complexities of self-reversals were a major challenge to early magnetostratigraphic workers and the phenomenon spurred them on to improve field collections and laboratory methods. Their endeavours resulted in the establishment of the polarity timescale (Cox et al. 1963) which played a major rôle in the emergence of the global plate tectonic hypothesis which was to revolutionise much of the Earth sciences.

BASALTS

Basalts display a wide range of oxidation and alteration states and a corresponding range of magnetic mineralogies (Section 3.6.1). Distinctive changes of magnetic mineralogy with oxidation have been documented in numerous suites of basalts from igneous provinces around the world (Ade-Hall et al. 1971). Progressive high temperature (deuteric) oxidation of natural iron oxides has been classified into six stages from no oxidation (class I) to maximum oxidation (class VI). The oxidation series begins with uniform titanomagnetite grains, with well formed outlines, which crystallised directly out of the basaltic melt. As oxidation gets under way the titanomagnetite grains begin to show exsolution of magnetite and thin ilmenite lamellae. With further oxidation titano-haematite and sphene appear. Rutile and titano-haematite become more common as the final stages of oxidation are approached and finally, at the highest oxidation state, titanohaematite and pseudobrookite or ferrorutile completely replace the original titano-magnetite. The above oxidation succession charts mineralogical alterations which can occur as a basalt cools through temperatures of between about 800 and 500 °C. The varied magnetic assemblages produced by such high temperature oxidation can still pick up a geomagnetic signal, as they cool through their blocking temperatures, and can end up carrying excellent palaeomagnetic records. Class III rocks from the middle of the range of oxidation states often display a particularly strong and stable palaeo-magnetic signal on account of the elongated shapes, small sizes and pure magnetite composition of their iron oxide grains.

Low temperature alterations, such as may arise from later hydrothermal events, can further com-plicate the high temperature sequence outlined above. Oceanic basalts extruded under water tend to be quenched and consequently have different bulk mineral magnetic properties from those of continental basalts. Many oceanic basalts display signs of extensive low temperature maghaematisation (Ade-Hall 1964, Ozima & Ozima 1971). Although the remanence of ocean basalts has been found to be predominantly stable a significant proportion also have appreciable soft magnetic components.

MAGNETISATION OF OCEAN CRUST

Following the invention of the proton precession magnetometer (6.6.1), oceanographic ships and aircraft exploring the ocean basins charted the intensity of the Earth's magnetic field by towing magnetometers behind them. Typical cruise profiles reveal magnetic anomalies with amplitudes of up to a few hundred nanotesla and widths of tens of kilometres. The anomalies have been found to have lengths of several thousands of kilometres and to be broken only when intersected by large oceanic fracture zones. The anomalies thus form a series of remarkably regular and continuous magnetic stripes which are found in all the major oceans, some stripes having been traced half way around the world (Sclater & Parsons 1981).

The explanation for the origin of the neat, regular pattern of the magnetic stripes came from combining the ideas of continental drift, sea-floor spreading and geomagnetic reversals. The idea is that the sea floor, as it forms continuously at ocean ridge crests becomes magnetised in the Earth's field, and the steady spreading of the ocean crust away from the ridges leads, as the Earth's field reverses polarity, to a regular magnetic pattern (Vine & Matthews 1963). The positive magnetic stripes formed when the field was normal, as it is today, and the negative stripes formed when the field was reversed. This link of oceanic anomalies with geomagnetic reversals allows ocean crust to be dated by magnetostratigraphic methods and conversely it allows continuous oceano-graphic magnetic field profiles to form the basis for a polarity reversal sequence (Heirtzler et al. 1968, Larson & Pitman 1972). The remarkable simplicity and symmetry of the movement of rigid plates on the Earth's spherical surface (Morgan 1968) has facilitated the magnetostratigraphic interpretation of oceanic magnetic anomalies, allowing the age of all of the oceanic crust to be deduced, together with surprisingly accurate and detailed reconstructions of the past positions of the world's major plates.

Curiously, despite the simplicity and beauty of the tectonic insights derived from oceanic magnetic stripes, the actual physical source of the magnetic anomalies remains in some doubt. The magnitude and shape of the anomalies most commonly lead oceano-graphers to model the magnetic source as being the uppermost 0.5 km of basaltic lavas and dykes with an average remanent magnetisation intensity of $10–20$ A m^{-1}, although early block models tended towards a $1–2$ km thick layer with an intensity of 5 A m^{-1}. Reasonably high Koenigsberger ratio (see Fig. 4.1) measurements from oceanic dredge and drill samples confirm that the magnetic anomalies result from a remanent rather than an induced magnet-isation. However, the remanence intensity of recovered oceanic basalts as measured in the laboratory tends to be only around 3 A m^{-1}, that is roughly half to a quarter of the modelled remanence intensity (Kent et al. 1978, Lowrie 1979). The dispersion of the remanence directions and the presumed restriction of even those moderate magnetisations to the pillow basalts (defined seismically as the upper 0.5 km (layer 2A) of the oceanic crust) have led to suggestions that the source of the stripe anomalies may lie deeper in the ocean crust. Rock magnetic and palaeomagnetic investiga-tions indicate that the gabbroic rocks of layer 3 form possible important source regions (Dunlop & Prevot

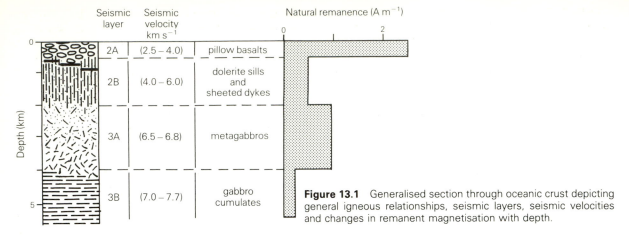

Figure 13.1 Generalised section through oceanic crust depicting general igneous relationships, seismic layers, seismic velocities and changes in remanent magnetisation with depth.

1982). Figure 13.1 illustrates the general magnetic zonation of the ocean crust envisaged as a result of these rock magnetic and palaeomagnetic investigations.

13.2.2 Origin of magnetisation in sediments

Many types of sediment have been found to contain excellent records of the past behaviour of the geomagnetic field. The exact mechanisms by which sediments acquire their natural remanent magnetism are, however, still poorly understood. This lack of understanding has arisen partly because of natural sedimentological complexities but also partly because of difficulties in interpreting the natural remanence acquisition in terms of laboratory studies.

Laboratory studies suggest that there are two basic mechanisms by which sediments can acquire a natural remanence (Section 4.3.1). One of these mechanisms leads to a remanence called a detrital remanent magnetisation. The other mechanism by which sediments can acquire a remanence is of a chemical nature and produces a remanence called a chemical remanent magnetisation. Two forms of chemical remanent magnetisation have been recognised. In one of these forms, named authigenic chemical remanent magnetisation, the remanence is produced by the growth of new minerals from solution. In the other form, named diagenetic chemical remanent magnetisation, the remanence results from the chemical alteration of pre-existing minerals.

LABORATORY EXPERIMENTS
Numerous experiments have been carried out in order to determine the properties of detrital remanent

magnetisations, but very little laboratory work has been performed in connection with chemical remanent magnetisations (Verosub 1977).

Experiments using a range of sedimentary materials, such as crushed basalt fragments (Nagata *et al.* 1943), dispersed glacial clays (Johnson *et al.* 1948) and synthetic sediment (Irving & Major 1964), have shown that depositional detrital remanent magnetisation correctly records the declination of the applied field. However, the inclination recorded by this depositional process is consistently too shallow. This biasing effect to low inclinations has been named the 'inclination error' (King 1955). It can be explained in terms of the interaction of the settling magnetic particles as they touch the substrate. For example, the actions of nearly spherical particles rolling into depressions or disc-shaped particles rotating into the horizontal plane as they come to rest at the sediment/water interface are two processes in which the action of gravity can overcome the geomagnetic field alignment and so cause an inclination error (Griffiths *et al.* 1960). A further biasing effect has been observed in the laboratory to occur on dipping beds. Particles rolling down slope at the time of deposition distort the remanent inclination in an effect called the 'bedding error' (King 1955, King & Rees 1966). Finally, laboratory deposition in flowing water has been observed to cause errors in both declination and inclination. This 'current rotation effect' (Rees 1961) has been studied in experiments using flume tanks.

Other laboratory experiments such as stirring reconstituted sediments have been performed in order to study the properties of post-depositional detrital remanent magnetisation (e.g. Kent 1973). In contrast to the experiments on depositional detrital remanences these experiments have revealed no systematic deviations of the remanence directions from the laboratory field directions. In addition to this encouraging result concerning the ability of sediments to record field directions, post-depositional detrital remanent magnetisation intensities have been found to be linearly proportional to the intensity of the applied magnetic field (Kent 1973, Tucker 1980). So these laboratory experiments, coupled with theoretical calculations based on the alignment of magnetic particles perturbed by Brownian motion, indicate that both the direction and intensity of the ancient geomagnetic field can be accurately recorded by the process of post-depositional detrital remanent magnetisation.

The time involved in the acquisition of a post-depositional detrital remanent magnetisation and the effect of sediment consolidation have been investigated in the laboratory by the gradual redeposition of long columns of sediment. Experiments have been carried out involving steady day by day redeposition, by continuous addition of sediment, for periods of up to half a year. In one experiment using organic lake muds and a redeposition rate of 2 m a^{-1} (Barton & McElhinny 1979) an artificial reversal of the ambient magnetic field was quickly locked into the accumulating sediment. The remanent acquisition process certainly took place within a period of less than 2 days. In a similar experiment using deep-sea foraminiferal clay and a redeposition rate of 3 m a^{-1} (Lovlie 1974) the field reversal was not recorded in the redeposited sediment until 10 days after the reversal occurred at a burial depth of 10 cm.

Although it must be remembered that these experiments involve deposition rates some tens to thousands of times more rapid than those in nature, they indicate that in non-organic sediments consolidation of the sediment matrix is important in fixing the magnetic particles in position. In organic-rich sediments, on the other hand, it seems that the mobility of the magnetic particles can be restricted soon after deposition, possibly within months in some sediments, by the action of organic gels (Stober & Thompson 1977, 1979).

NATURAL REMANENCE OF DEEP-SEA SEDIMENTS

The widespread correlation of faunal zonations with palaeomagnetic reversal sequences in deep-sea sediment cores demonstrates that the acquisition of remanence by many deep-sea sediments is more or less contemporaneous, at least in terms of the resolution of dating, with deposition (Opdyke 1972). However, as many deep-sea sediment cores which exhibit a coherent pattern of magnetisation are bioturbated near the surface, their remanence has probably resulted from a post-depositional mechanism with its accompanying magnetisation delay. The similarity of the mean inclination of the remanent magnetisation of recent marine sediments to the local geomagnetic field inclination further suggests that the predominant remanence in oceanic materials is a post-depositional remanence rather than a depositional detrital remanence.

The depth of active bioturbation has been estimated in deep-sea sediments by studying the redistribution of ash layers and microtektite horizons and by investigating the mixing of radioactive contaminants in surface samples (Ruddiman et al. 1980). These various techniques show that surface material can be mixed with subsurface materials to depths ranging from 10 to 60 cm. Possible effects of variable bioturbation on a polarity subzone are illustrated in Figure 13.2. These effects can range from boundary displacements to the elimination or spurious addition of short polarity subzones.

The range of magnetic iron and manganese oxides, hydroxides and sulphides found in marine sediments has been described by Haggerty (1970) and by Henshaw and Merrill (1980), and Chapter 12 outlines their sources and significance in marine sediment

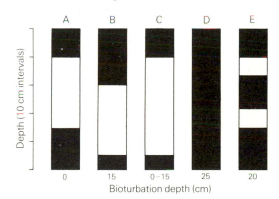

Figure 13.2 Possible effects of bioturbation on a single polarity subzone. A perfect palaeomagnetic recording of the subzone is shown in sediment column A. Distorted recordings of the same subzone are shown in columns B to E. The minimum depths of faunal activity required to create each effect are listed beneath each column. (A) no bioturbation; (B) constant bioturbation; (C) intermittent surface bioturbation; (D) constant bioturbation after a period of inactivity; (F) intermittent subsurface bioturbation. (After Watkins 1968.)

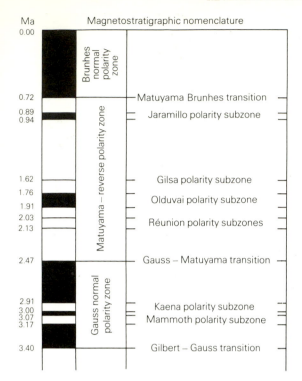

Ma

Magnetostratigraphic nomenclature

Ma		
0.00	Brunhes normal polarity zone	
0.72		Matuyama Brunhes transition
0.89		Jaramillo polarity subzone
0.94		
1.62	Matuyama – reverse polarity zone	Gilsa polarity subzone
1.76		Olduvai polarity subzone
1.91		
2.03		Réunion polarity subzones
2.13		
2.47		Gauss – Matuyama transition
2.91	Gauss normal polarity zone	
3.00		Kaena polarity subzone
3.07		Mammoth polarity subzone
3.17		
3.40		Gilbert – Gauss transition

Figure 13.3 Magnetostratigraphic nomenclature as applied to the past 3.5 million years. Ages of boundaries derived by McDougall (1979) by combining K–Ar age determinations from many localities around the world using the direct timescale approach. Some debate still remains over the existence and age of short subchrons such as the Gilsa subchron and possible reversed polarity subchrons in the Brunhes zone. Normal polarities are depicted by solid shading.

stratigraphy. The loss of quality and resolution of the magnetic stratigraphy at depth in some cores, particularly from the north central Pacific, is attributed to a chemical remanence replacing the primary post-depositional detrital remanence (Kent & Lowrie 1974).

PARTICLE REALIGNMENT

The original palaeomagnetic record of a sediment can be distorted by later physical movements of the micron-sized particles which carry the natural remanence. Particle realignment can take place rapidly if the sediment is disturbed or more slowly over hundreds or thousands of years. On the thousand year timescale, it can gradually lead to the resetting of reversal patterns or the reduction of secular variation features while on a timescale of years, prolonged core storage can lead to marked changes in remanence direction. Magnetic particle realignments occur within minutes when the sediment is disturbed, for example by drying (Granar, 1958). Such physical rotations of magnetic particles and their associated remanence realignments can be difficult to detect because the magnetic coercivity and character of the new remanence is often very similar to that of the original remanence.

The effect of drying is surprisingly important and rather complicated (Johnson *et al.* 1975). Drying inorganic sediments with an original high water content generally fixes their remanence (Henshaw & Merrill 1979). In contrast drying organic-rich lake sediments can lead to a reduction of up to 50% of their remanence (Stober & Thompson 1977, 1979). In further contrast, drying of some sediments can lead to the acquisition of a new remanence. These three distinct effects are all thought to be related to the physical mobility and realignment of magnetic carriers. The first effect, of remanence stabilisation, can occur naturally due to either evaporation or compaction processes. Some laboratory experiments (Verosub *et al.* 1979) suggest that stabilisation can be produced by a remarkably small change in water content through a critical limit. The precise critical water content appears to depend both on the size of the magnetic particles and the sediment matrix (Payne & Verosub 1982), but to be about 75% for many deep-sea sediments with compaction being the controlling factor. The second effect, of remanence loss, has been found in some lake sediments steadily to reduce the natural remanent intensity as the water content is decreased. A likely mechanism is that during dehydration the small magnetic carriers are physically pulled towards the larger sediment particles by surface tension effects. The net remanence is thus

Table 13.1 Recommended terminology for magnetostratigraphic polarity units.

Magnetostratigraphic polarity units	Approximate duration in years	Geochronological equivalent	Chronostratigraphical equivalent
polarity subzone	$10^4 - 10^5$	subchron	subchronozone
polarity zone	$10^5 - 10^6$	chron	chronozone
polarity superzone	$10^6 - 10^7$	superchron	superchronozone

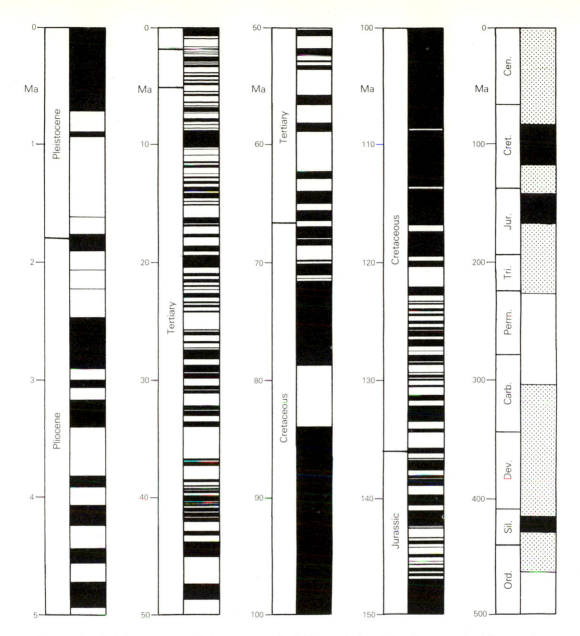

Figure 13.4 Polarity sequence of the geomagnetic field extended to 150 million years by inference from marine magnetic anomalies (modified from Lowrie & Alvarez 1981, Channell *et al.* 1982). The right-hand column divides the last 500 million years of geological time into normal and reverse polarity superchrons and disturbed (mixed) superchrons (light shading). Normal polarities are depicted by solid shading.

reduced in intensity, but the random surface tension forces do not lead to any change in remanence direction. A similar effect occurs in freezing lake sediments when disturbance of the magnetic carriers reduces the remanence intensity while the remanence direction remains largely unaltered. The third phenomenon, of the production of a new stable remanence by drying, has been attributed to sediment fabric disturbance which allows the magnetic particles to continue to rotate, at unusually low water contents, into the new field direction.

Two additional effects of drying sediments are that chemical or biochemical change associated with slow drying can reduce magnetic susceptibility (e.g. Bloemendal, 1982) and that the drying of small samples can distort their shape. This drying distortion may also change the direction of their remanence. The distortion of large blocks of sediment during drying or compression has, however, very little effect on the net remanence, because large blocks tend to crack along fracture sets which leave the remanence direction unaltered (Kodoma & Cox 1978).

13.3 The geomagnetic polarity timescale

TERMINOLOGY

The polarity timescale, although it has evolved to an extent where it is undoubtedly capable of being the foundation of an understanding of diverse global

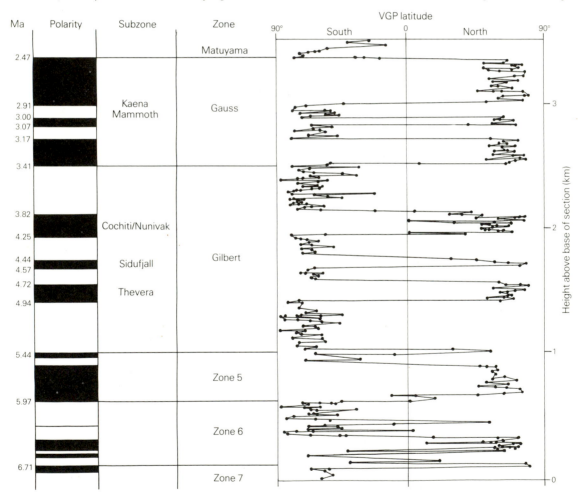

Figure 13.5 Palaeomagnetic results and polarity timescale recorded in a 3400 m sequence of lava flows in Iceland. Normal polarities are depicted by solid shading (after McDougall *et al.* 1977). The probability of a deep sea sediment sequence being datable by palaeomagnetic study is assessed at 20% by Stupavsky and Gravenor (1984).

phenomena, is still being developed and refined (Watkins 1972, Irving & Pullaiah 1976). A polarity nomenclature system is thus needed which incorporates the early magnetostratigraphic conventions and usages but which can also be developed in a coherent manner, incorporating modifications and resolving inconsistencies, confusion and controversies. A sub-commission of the International Commission on Stratigraphy has accordingly recommended a system of terms and hierarchies for magnetostratigraphic use (Table 13.1). The sub-commission discourages the use of the previously applied magnetostratigraphic terms of epoch, event and interval on account of conflicts with established stratigraphic terminology, and because the terms are not strictly physically appropriate. Furthermore, it is

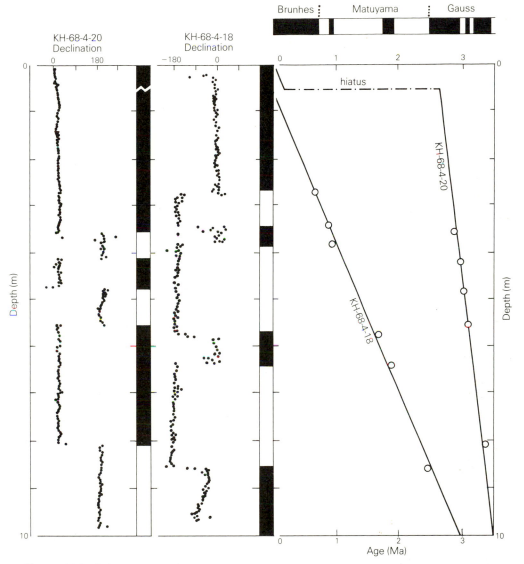

Figure 13.6 Palaeomagnetic NRM declination directions in two west-central equatorial Pacific sediment cores. The polarity zonation of each declination record is shown and correlated (open circles at right) with the polarity timescale of Figure 13.3 to give sedimentation rate estimates. A long hiatus covering the whole of the Matuyama reversed polarity chron and a large proportion of the Brunhes normal polarity chron is inferred for Core KH-68-4-20 (data from Kobayashi *et al.* 1971).

now realised that there is no fundamental geophysical difference between a polarity event and a short polarity epoch. The preferred terms for magnetic intervals are polarity zone, subzone and superzone (e.g. Harland *et al.* 1982). Figure 13.3 illustrates the use of this revised magnetostratigraphic nomenclature for the past 3.4 Ma. Geographically derived names are preferred for major units although names, in use, derived from distinguished contributors to the science of geomagnetism, are also acceptable.

Magnetic polarity units have been established in two quite different ways: (a) by combining laboratory determinations of the remanent magnetisation direction of rock samples or cores with radiometric or biostratigraphic age determinations, and (b) through the shipboard magnetometer profiles of the intensity of oceanic magnetic field anomalies (e.g. Irving and Pullaiah, 1976). The first approach fits neatly into conventional stratigraphic classification procedures such as the use of type localities and type sections. The second ocean survey approach is not handled so easily. The practical oceanic anomaly numbering

system, which has proved so useful for reconstructing plate motions and the development and history of ocean basins, although falling outside conventional stratigraphic practice, is likely to remain as an extremely useful quasi-stratigraphic method. Figure 13.4 illustrates a polarity timescale for the last 150 Ma erected by combining the two approaches.

0–5 Ma DIRECT RADIOMETRIC APPROACH

This is the original and most elegant method of establishing a polarity timescale. It uses radiometric ages and magnetic polarities of rock samples from any part of the world. The ages and polarities are simply plotted next to each other in order to form the polarity scale. The polarity timescale was developed and refined through the 1960s principally by the work of two research groups based in California and Australia. Cox (1973) described the timescale developments as 'a rather lively competition, resembling a long-distance chess game in which the two sides communicated via letters to the journals *Nature* and *Science*'.

The direct radiometric method is limited by the

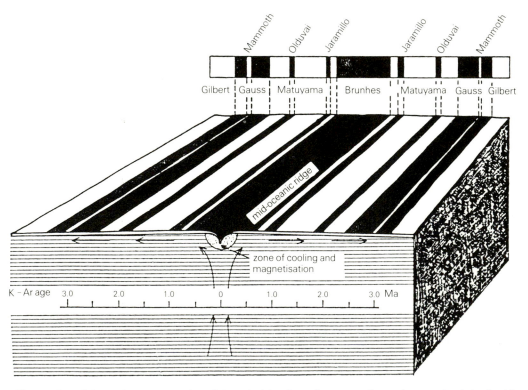

Figure 13.7 Schematic representation of the principle of sea-floor spreading and reversals of the Earth's magnetic field as proposed by Vine and Matthews (1963) (after Allan 1969).

accuracy of dating techniques. Currently the K–Ar dating precision is around 3% for rocks a few million years in age. As the average length of time between polarity inversions, during the last 5 million years, has been about 200 000 years it works out that the possibilities of recognising individual polarity sub-zones earlier than four or five million years ago by K–Ar dating are extremely limited, even with prodigious amounts of data (Dalrymple 1972) because the resolution of the dating is inadequate.

0–20 Ma STRATIGRAPHIC APPROACH

The polarity timescale of the past 5 million years can be extended further back in time by studying strati-graphic sections rather than isolated rocks. Figure 13.5 displays the palaeomagnetic results from a thick sequence of lava flows on Iceland. The reversal pattern can easily be seen to match and extend the older part of the polarity timescale of Figure 13.3. Although radiometric dating of the older lavas presents some difficulties, analyses from the whole sequence allow an overall age scale to be assigned to the polarity transitions (McDougall *et al.* 1977). Deep-sea sediments present a further geological sequence which can be used in the stratigraphic approach to erecting an extended polarity timescale, although there are some difficulties in obtaining long, old, unbroken records (Hammond *et al.* 1974, Opdyke *et al.* 1974). Figure 13.6 exhibits palaeomagnetic results from Pacific cores illustrating the value and quality of the magnetostratigraphic information which can be found in ocean sediments (Kobayashi *et al.* 1971). A very clear reversal magnetostratigraphy has also been found preserved in Chinese loess (Heller and Liu, 1982).

0–170 Ma MARINE ANOMALIES APPROACH

By dating magnetised ocean crust a polarity timescale can be generated from the oceanic linear magnetic anomalies (Fig. 13.7) back to the time of formation of the oldest ocean crust of around 170 million years. The age of the magnetised ocean crust layer 2A can be assessed by: (a) assuming that sea-floor spreading has occurred at the same steady rate as during the past few million years (Heirtzler *et al.* 1968), (b) equating the age of the magnetised crust with that of the biostrati-graphic age of the immediately overlying sediments (Larson & Pitman 1972, Winterer 1973) and (c) matching the magnetic anomaly sequence to dated magnetostratigraphic-type sections on land (La Brecque *et al.* 1977, Lowrie & Alvarez 1981, Channel

et al. 1982). By using these three dating methods marine magnetic anomalies have been employed in constructing the polarity reversal sequence of Figure 13.4 which stretches back continuously from the present day to 150 million years ago. The marine magnetic anomalies have been invaluable in reconstructing elements of the geological history of the oceans, such as the evolution of the Atlantic ocean depicted in Figure 13.8 (Phillips & Forsyth 1972).

0–600 Ma QUIET ZONE APPROACH

The key to magnetostratigraphy before 170 Ma lies in the long, quiet polarity superchrons. Old periods of frequent field reversal are difficult to use in magneto-stratigraphy because the geomagnetic details are difficult to resolve reliably through either the direct radiometric or the stratigraphic approaches. The pattern of quiet and disturbed magnetic superchrons of Figure 13.4, however, has the potential for magnetostratigraphic correlation on an interconti-nental scale. Correlations between the rocks of six continents have been proposed on the basis of the most spectacular of these quiet zones, the Permo-Carboniferous reversed superchron (previously known as the Kiaman reversal).

13.4 Polarity transitions

Occasionally palaeomagnetists have managed to obtain records of the magnetic field when it was actually in the act of changing polarity (e.g. Fig. 13.9). Some of these polarity transitions have been studied in sufficient detail to permit mathematical models of the behaviour of the field during the polarity switch to be developed (Fuller *et al.* 1979). Estimating the time taken for the field to change polarity is extremely difficult, but it is generally taken, on the basis of the thickness of deep-sea sediment layers recording intermediate palaeomagnetic directions across reversal boundaries and on the proportion of inter-mediate directions found in lavas, to be around 5000 years. Clement and Kent (1984) in a very detailed study of the Matuyama–Brunhes polarity transition in seven deep-sea cores from the Pacific Ocean have found evidence that the duration of this transition at mid-latitudes was twice that at equatorial latitudes. During a polarity transition the field intensity typically drops to one-fifth of its pretransition value. On the basis of relative field intensity measurements

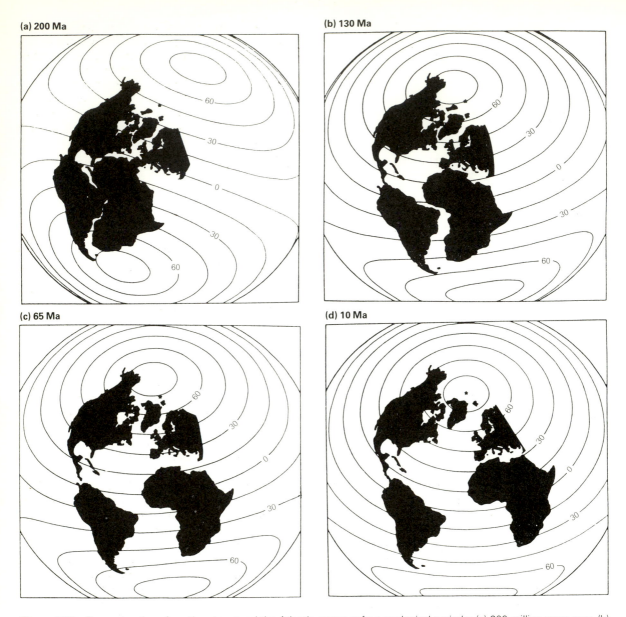

Figure 13.8 Reconstruction of continents around the Atlantic ocean at four geological periods: (a) 200 million years ago; (b) 130 Ma; (c) 65 Ma; (d) 10 Ma. Relative displacements deduced from the pattern of oceanic magnetic anomalies; latitudes and orientations derived from the mean palaeomagnetic remanence of continental rocks (after Phillips & Forsyth 1972).

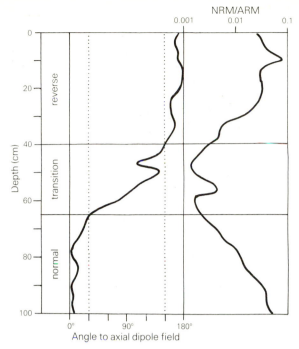

Figure 13.9 Diagrammatic representation of a polarity transition captured by the remanence of steadily accumulating sediments. Transitional directions are defined as vectors which deviate from the local axial dipole field directions by more than 30°. Low NRM/ARM ratios largely indicate times of low geomagnetic field intensity, but may also reflect low NRM intensities caused by unresolved dual component remanences.

modelling the more complicated field structure during polarity changes are (a) by axial quadrupole and octupole field configurations and (b) by transitions through standing non-dipole fields. More observations of polarity transitions are needed to evaluate the applicability of each model and in particular more transition records from the southern hemisphere are needed to distinguish between transition fields dominated by quadrupole terms and those dominated by higher order axisymmetric terms (Fuller *et al.* 1979).

13.5   Summary

Since the first polarity timescales were constructed in the early 1960s, reversal magnetostratigraphy has developed into an important surrogate dating technique. Use of the polarity timescale played a key rôle in the discovery of sea-floor spreading and the development of plate tectonics. Probably all polarity intervals younger than 150 million years which lasted for longer than 0.1 million years have been discovered. Additional shorter polarity intervals are continually being found as more detailed geological records of particular time periods are discovered and investigated. The most recent firmly established global polarity reversal was the Matuyama to Brunhes transition which took place 700 000 years ago.

Geomagnetic reversal sequences are recorded in the thermal remanent magnetisation of igneous rocks, the chemical remanence of red beds and loess, and the detrital remanence of fine-grained sediments such as clays and limestones. This broad range of remanence and rock types makes reversal magnetostratigraphy a widespread chronological and correlation tool.

Details of geomagnetic behaviour during the reversal process are beginning to be unravelled. The characteristics of individual reversal transitions can be seen to be varied. Some reversal transitions display exceedingly complex changes with pronounced non-dipole, non-axisymmetric behaviour, while other transitions have performed simpler more axisymmetric switches between polarity states.

in sediments it has often been stated that such field intensity changes persist for a longer time than the field direction changes. More recent palaeointensity data on lavas, however, suggest that the direction and intensity change more or less synchronously (Coe *et al.* 1983) and the early drop and late recovery of intensity commonly observed in sediment transition records is more likely to be a remagnetisation effect. Palaeomagnetic records of certain polarity transitions have now been obtained from different parts of the world. These records show that the predominantly dipole nature of the stable geomagnetic field is not preserved during the transitions. Two approaches to

[14]

Secular variation magnetostratigraphy

Among the great geophysical enigmas, whose unravelling promises to unlock many another of natures secrets . . . the cause of the secular variations of terrestrial magnetism plays a prominent part.

L. A. Bauer 1895

14.1 Introduction

This chapter concentrates on secular variation magnetostratigraphic dating applications within the past 10 000 years. It describes collection and measurement techniques which have been found useful in investigating lake sediments and presents type palaeomagnetic secular variation records of both declination and inclination, from seven regions of the world. These patterns of secular change can be used as master curves for dating newly acquired palaeomagnetic records.

Current interest in the use of sedimentary secular variation records as a dating tool largely derives from Mackereth's (1971) studies in the English Lake District. Mackereth demonstrated that repeatable palaeomagnetic declination records were held by the sediments of Lake Windermere and that the uppermost sediments revealed secular changes which could be matched with the known historical declination fluctuations of London. The first studies of the remanence of recent sediments were carried out in North America by McNish and Johnson (1938) and Johnson *et al.* (1948) on varved clays from New England and in Europe by Ising (1943) on late-Glacial varved sediments from Sweden. These early investi-gations were followed up in the 1950s by further studies of Swedish varved clays by Granar (1958) and Griffiths *et al.* (1960). Such varved sediments did not turn out to be ideal materials for palaeomagnetic work, as a number of quite complicated sedimento-logical effects such as inclination error, bedding error and the effects of bottom water currents, were associated with the remanence acquisition process of the varves (Section 13.2.2). These processes needed to be taken in account before a geomagnetic field signature could be discerned. It remained for Mackereth (1971) using his pneumatic corer (Mackereth 1958) to demonstrate the more repro-ducible, more straightforward secular variation records of organic-rich lake sediments and to open up the subject of secular variation magnetostratigraphy. The potential value of palaeomagnetic declination correlations within a lake can be clearly seen in Mackereth's results. The speed of whole core measurements (Molyneux *et al.* 1972) makes such correlations extremely attractive, first, as a reinforce-ment for the more traditional approaches to core correlations and, secondly, as a potential dating tool. Within- and between-lake palaeomagnetic declina-tion and inclination correlations were presented by Thompson (1973) on sediments being investigated as part of a wider environmental study of Lough Neagh in Northern Ireland (O'Sullivan *et al.* 1973). Diatom (Battarbee 1978), pollen (O'Sullivan *et al.* 1973) and magnetic susceptibility (Thompson *et al.* 1975) correlations demonstrated the quality of the within-lake remanent direction correlations and the potential of secular variation magnetostratigraphy for lakes such as Lough Neagh with severe ^{14}C dating problems. Older Quaternary sediments, recovered from dried-out lake basins, have also been subjected to palaeo-magnetic study (Denham & Cox 1971, Liddicoat & Coe 1979, Verosub *et al.* 1980, Negrini *et al.* 1984). The pattern of geomagnetic changes at the time of deposition of these sediments has proved somewhat difficult to elucidate, but the original site at Mono Lake, studied by Denham and Cox, still provides some of the best evidence available for unusually large secular change. The Mono sequence contains palaeo-magnetic direction changes of over 30° which are thought to have lasted some 500 years and to date from 25 000 years BP. Even older Pleistocene sediments contain palaeomagnetic direction fluctuations reminiscent of geomagnetic secular changes (Opdyke *et al.* 1972, Thompson *et al.* 1974, Kent & Opdyke 1977, Doh and Steele 1983), but in terms of magnetic dating the most widely used geomagnetic phenomena in the Pleistocene are the polarity reversals discussed in Chapter 13 which provide a remarkable global magnetostratigraphy.

14.2 Experimental methods

CORE COLLECTION AND WHOLE CORE MEASUREMENT

Standard palaeomagnetic methods can be used in magnetostratigraphic secular variation studies. However, as the palaeomagnetic signals under investigation have amplitudes of only a few degrees rather than the one hundred and eighty degrees found in polarity reversal studies, particular care needs to be taken at the sediment coring and sampling stages. Fortunately, for magnetostratigraphic purposes absolute sample orientation is not mandatory. A knowledge of the way up of sediment samples is sufficient for relative changes in remanence direction to be investigated. Constancy of azimuth permits relative declination to be used in addition to inclination. With this relative declination method the average palaeomagnetic declination is used as an estimate of the north direction.

A most convenient magnetostratigraphic technique has been found to be that of collecting long (>3 m) sediment columns using pneumatic (Mackereth 1958) or gravity corers and then measuring the natural remanence of the sediment while it remains undisturbed in its core liner. This technique permits the remanence direction changes to be logged quickly and avoids the sometimes severe problems of aligning short, overlapping core sections. The resolution of whole core measurements is almost as good as that of subsamples taken every 3–4 cm. In principle the best whole core measuring method is that of the three-axis open-ended cryogenic magnetometer. In practice, however, the instrument most commonly used for whole core remanence measurements is the fluxgate magnetometer, although it cannot provide whole core inclination data. This means that when using fluxgate instrumentation subsamples have to be taken to yield full palaeomagnetic direction data. Two palaeo-magnetic coring problems which repeatedly arise are core twist and core warp. If the coring system in use cannot be modified to remove twist and warp then their effect can be partly taken into account by mathematically detrending the palaeomagnetic data. If there is a choice of sediments to be investigated in any study then sites more likely to provide good material for magnetic work can be selected. Research to date supports a preference for lakes that have (a) a minerogenic input with a magnetite component derived from either basic igneous material in the bedrock/drift or from enhanced top soil, (b) flat (slopes <2°), current-free sites below the level of wave action (>8 m depth), and (c) stiff, moderately organic-rich sediments.

SUB-SAMPLING

Thin-walled cuboid plastic boxes of around 10 ml volume are handy for extracting and holding sub-samples and they fit easily into magnetometers. Commercial boxes are available with an arrow, to show orientation, a small hole, to allow air escape during subsampling and thin walls, to minimise sediment sampling disturbance. Duplicate sub-samples can be very useful for assessing palaeo-magnetic reproducibility, although much more information is to be gained by examining duplicate cores. An efficient sampling scheme is to subsample contiguously at least two cores, while occasionally (e.g. every metre) taking duplicate subsamples to check within-horizon reproducibility. No palaeo-magnetic secular variation data from sediments should be trusted unless reproducibility has been clearly demonstrated. Remanence should be measured while the sediment is still fresh, i.e. before it has dried out. Freezing of samples is likely to distort remanence directions.

STABILITY TESTING

An important weapon in the armoury of the general palaeomagnetic worker is that of partial demagnetisation. In the great majority of palaeomagnetic studies of older rocks, partial demagnetisation studies of multicomponent remanences play a most crucial rôle. Stability tests are also of great importance in reversal magnetostratigraphy studies. However, for very recent sediments, with their comparatively short, quiet histories, such stability studies turn out to be of lesser importance. Alternating field cleaning can be helpful in removing viscous remanence components, often acquired through drying during prolonged horizontal core storage, as illustrated in Figure 14.1. However, as only one significant stable remanent component is invariably found in fresh Holocene sediment, partial demagnetisation studies can, somewhat perversely, degrade rather than improve natural remanences. Such impairment arises from occasional small false remanences produced during the partial demagnetisation experiments. Zero field storage can be a very effective alternative method of removing small viscous drying components and many published Holocene palaeomagnetic results are of natural remanence measured after a period of zero field storage. Resistance to alternating magnetic fields is often used to boost claims of unusual Holocene and Weichselian remanence directions being reliable reflections of the ancient geomagnetic field. However, the magnetic stability of recent sediment holding a detrital or post-depositional remanence mainly reflects the stability of primary source minerals and not the propensity of the sediment to record faithfully the geomagnetic field. If sediment can be analysed while still fresh, before any significant drying out has begun, then secular variation magnetostratigraphic studies can be quite adequately performed with the portable fluxgate magnetometer (described in Section 6.3.2) and a mu-metal shield for zero field storage.

The intensity of the natural remanence of lake sediments is found to decrease on cooling. This phenomenon was originally interpreted as being due to the effect of the Morin transition occurring at $-10\,°C$ in haematite (Section 3.2.2) (Creer *et al.* 1972,

Barton 1978). It is now recognised to be due to the randomising effects of grain rotations associated with the growth of ice crystals (Stober & Thompson 1979).

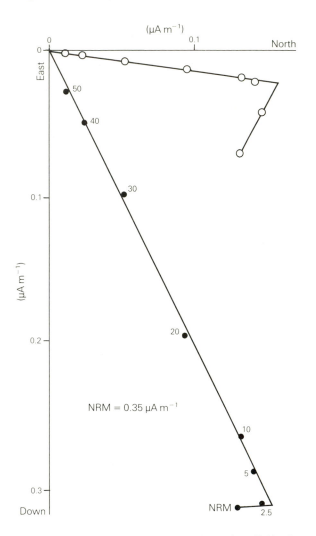

Figure 14.1 Orthogonal plot of alternating field demagnetisation of the natural remanence of sediment 6000 years old from Gass Lake, Wisconsin (sampled by S. Webb). Open and solid circles mark remanence changes projected on the horizontal and north/down planes respectively with partial demagnetisation in fields up to 50 mT. Two remanence components are seen. The high coercivity component points 8° east of north with a dip of 63° and is taken to be an original remanence, marking the direction of the ancient field. The low coercivity component is taken to be a 'viscous' drying remanence formed during 18 months' core storage. These preliminary results suggest that fresh Gass Lake sediments would have high potential for secular variation magnetostratigraphic work.

SEDIMENTOLOGICAL INVESTIGATIONS

The only sedimentological analyses commonly used in palaeomagnetic secular variation studies are estimates of particle size. Such granulometry studies are valuable, as it is widely recognised that coarse sediments make poorer geomagnetic recorders than fine sediments. Reconstitution and more rarely redeposition experiments, in beakers and tubes respectively, have been tried in connection with estimating palaeointensities (Section 13.2.2). Unfortunately, despite considerable effort, little progress has been made towards establishing a practicable palaeointensity method for lake sediments. However, reconstitution experiments at least show that lake sediments are capable of acquiring a post-depositional remanence (Section 13.2.2) in the ambient field direction. Fabric measurements, particularly through the use of anisotropy of susceptibility instrumentation (Section 6.3.3), are relatively easy and rapid to perform. Again, however, they are infrequently used in magnetostratigraphic studies as the magnitude of the anisotropy of most recent lake sediments is extremely low, well below instrumental noise levels. Laboratory procedures based on such fabric, granulometry or reconstitution experiments are badly needed to help distinguish those sediments capable of accurately recording the ancient field from those in which sedimentological effects have masked or surreptitiously mimicked geomagnetic fluctuations.

STATISTICAL AND TRIGONOMETRICAL METHODS

The two main trigonometric manipulations used by palaeomagnetic workers are (a) rotation on the sphere and (b) conversion of field directions to pole positions. Both calculations are straightforward, but they are quite time consuming without the aid of a computer. Descriptions of the manipulations are to be found in standard texts such as Mardia (1972) and McElhinny (1973). Two common frames of reference used in presenting palaeomagnetic directional data are with respect to (a) the vertical direction and (b) the local direction of a geocentric axial dipole field.

The chief palaeomagnetic statistical methods were formulated by Fisher (1953). They are based on his spherical distribution, which can be thought of as the equivalent of a Gaussian distribution on the sphere. Fisher (1953) gives the relevant formula for calculating the mean, its standard error, the variance and other useful statistics of palaeomagnetic data.

Watson (1970) has been prominent in the further development of statistical aspects of palaeomagnetic data. Mardia (1972) summarises the wealth of statistical literature dealing with circularly and spherically distributed data. Computer programs for carrying out such trigonometric rotations and statistical calculations are widely available.

Hyodo (1984) has suggested that deconvolution using a digital filter can improve secular variation records in marine cores by correcting for amplitude attenuation and phase shifts occurring during the post depositional defrital remanence acquisition process.

14.3 Magnetic dating and magnetostratigraphy

THE HISTORICAL PERIOD

Sediment sequences spanning the past 400 years can be dated directly from their magnetic remanences by matching their palaeomagnetic secular variation signature with historically documented geomagnetic field fluctuations. Three examples of such field changes, observed over the last few hundred years, are shown in Figure 5.5 for London (Malin & Bullard 1981), Rome and Boston. The earliest known written record of the value of the local geomagnetic field comes from Hartmann's report concerning the 4° difference in magnetic declination between Nurnberg and Rome in 1510 AD as found in connection with the use of portable sun dials (Mitchell-Crichton 1939). The direction of magnetic declination appears to have been inscribed onto some sun dials and this permits the historical declination record to be extended back at least to 1492 AD when sun dials manufactured in Nurnberg and Augsburg by master craftsmen were marked with a declination of 11° to the east of north. Compass maps unfortunately do not appear to be of sufficient accuracy to be of use in establishing even earlier ancient field changes. Navigators' declination observations made since the sixteenth century provide the earliest geomagnetic field measurements in many parts of the world and have been collected together, most notably by Sabine (see Malin & Bullard 1981) and Veinberg (Veinberg & Shibaev 1969). Inclination measurements began in London in AD 1576 with the work of Norman (1581), while intensity measurements had to await the genius of Gauss (1833). All these historical records have been combined using spherical harmonic analysis with a few archaeo-magnetic measurements to produce Figure 14.2 which

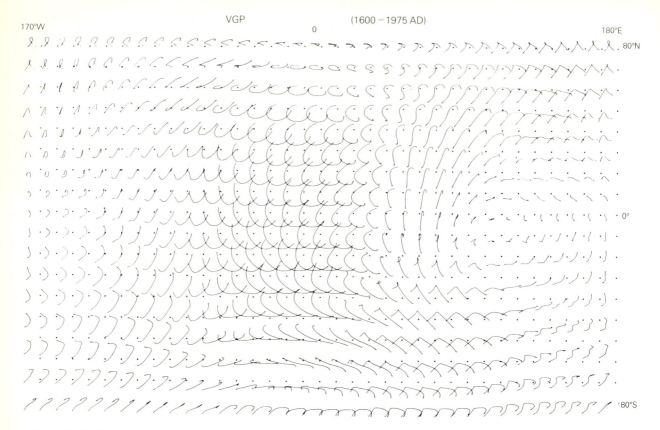

Figure 14.2 Historical virtual geomagnetic pole plots on a 10° latitude and longitude grid. Each curve charts the motion of the local virtual geomagnetic pole from 1600 AD to 1975 AD. The sense of motion is clockwise except in the region of the Indian ocean. Note the large geomagnetic direction changes that have occurred over Africa and Europe compared with the low geomagnetic changes of the Pacific region.

is a plot of the local changes in declination and inclination over the whole world on a 10° by 10° grid (Thompson & Barraclough 1982). The direction changes of Figure 14.2 form the basis of the magnetic dating method. The rather complex global pattern of Figure 14.2 is described mathematically by 120 parameters (24 two-piece cubic splines). The easiest historical field features to recognise in palaeomagnetic secular variation logs are the declination and inclination turning points. For examples the 1810 AD westerly declination maximum and 1710 AD inclination maximum observed at London (Fig. 5.5) can be seen in the recent sediment record from Loch Lomond (Fig. 5.8). Note in the grid of Figure 14.2 how the direction turning points occur at different times in various parts of the world and how the amplitudes of the features also vary from place to place. The most promising localities for historical magnetic dating are

those with rapid high amplitude secular changes and clear turning points.

THE PAST 10 000 YEARS
Magnetic age determinations are necessarily indirect for sediments older than 400 years. The approach followed is one of correlating a new palaeomagnetic remanence log with that of a nearby type sequence which has been dated independently, for example by [14]C. The palaeomagnetic records of seven such special type sequences are shown in Figure 14.3 and the ages of their turning points are tabulated in Table 14.1. These seven type records (Thompson 1983) are almost exclusively dependent on the radiometric ^{14}C method for their dating. Radiometric dating of the carbon content of lake sediments is more problematical than the dating of the carbon of many other materials such as charcoal, bone, ombrotrophic peat

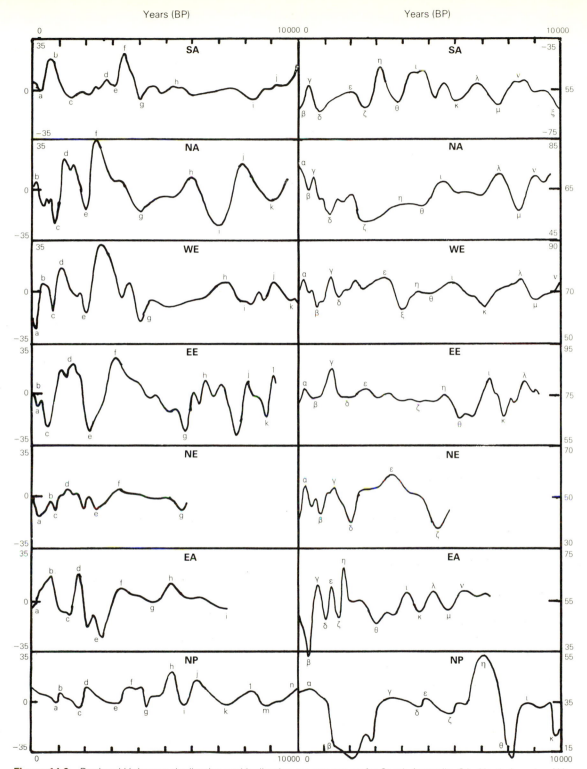

Figure 14.3 Regional Holocene declination and inclination master curves for South Australia, SA; North America, NA; western Europe, WE; eastern Europe, EE; Near East, NE; East Asia, EA and North Pacific, NP. Tree ring timescale calibrated in calendar years BP. Curves based on data from Barton and McElhinny (1981), Banerjee *et al.* (1979), Mackereth (1971), Turner and Thompson (1979), Huttunen and Stober (1980), Tolonen *et al.* (1975), Thompson *et al.* (1985), Horie *et al.* (1980) and McWilliams *et al.* (1982).

Table 14.1 Ages of magnetostratigraphic features.

	SA*	NA	WE	EE	NE	EA	NP
Declination							
a†	300	—	140	160	220	0	900
b	680	100	450	300	700	700	1100
c	1300	750	600	600	850	1200	1800
d	2000	1200	1000	1400	1300	1650	2150
e	2800	2000	2000	2200	1900	2200	3200
f	3500	2400	2600	3100	2100	3100	3900
g	4500	4000	4900	5700	2400	4400	4400
h	5500	5900	7100	6500	3200	5100	5300
i	8300	7000	8300	7600	5600	7300	5600
j	9000	7900	9100	8000	—	—	6000
k	—	9000	10000	8700	—	—	8350
l	—	—	—	9000	—	—	8900
Inclination							
α	—	50	240	300	300	—	200
β	—	420	650	600	550	400	2150
γ	400	750	1150	1300	700	760	3500
δ	900	1200	1650	1900	900	1000	4700
ε	1900	2300	3100	2600	1400	1300	5100
ζ	2600	2900	3800	4600	2000	1550	5800
η	3200	3700	4300	5500	3600	1750	7000
θ	3600	4400	5000	6400	5300	2800	8200
ι	4600	5300	6000	7200	—	4100	8950
κ	6000	6600	7100	7800	—	4600	9800
λ	6800	7700	8300	8600	—	5100	—
μ	7900	8400	8800	—	—	5600	—
ν	8600	9600	9700	—	—	6600	—
ξ	10000	—	—	—	—	—	—

* SA South Australia (35°S 140°E) based on Barton and McElhinny (1981).
NA North America (45°N 90°W) based on Banerjee *et al.* (1979).
WE Western Europe (55°N 05°W) based on Turner and Thompson (1981).
EE Eastern Europe (60°N 30°E) based on Huttunen and Stober (1980).
NE Near East   (30°N 35°E) based on Thompson *et al.* (1985).
EA Eastern Asia (35°N 140°E) based on Horie *et al.* (1980).
NP North Pacific (20°N 155°W) based on McWilliams *et al.* (1982).

† a to l declination turning points. α to ξ inclination turning points. Ages tabulated in calibrated ^{14}C years BP. The EA ages are rather poorly known, based here on a linear interpolation between the basal tephra layer and the archaeomagnetic features preserved in the upper sediments. Errors in ^{14}C ages at all the sites possibly amount to several hundred years. Labelling of the palaeomagnetic features is purely for convenience of reference. Any likenesses in ages or in shapes of similarly labelled features are probably chance occurrences, unlikely to be duplicated in other parts of the World.

and wood. The recent sediment isotopic difficulties are rather subtle involving natural contamination and hard-water effects (Olsson 1974). The main problem is that the recent sediment carbon being dated is often older than its stratigraphic position, which results in erroneously old radiocarbon ages. Kidson (1982) notes that 'Despite caveats in the literature ... many authors still appear to believe that the age of a sample lies within one standard deviation given by the dating laboratory ... No allowance is made for non-counting errors. Many potential inaccuracies widen the possibility that the true date lies outside the one sigma limits'. The problem of old carbon contamination due to stable organic residues, such as mor humus or peat, and geologic carbon, such as graphite, being washed into lake sediments from eroding soils in their catchments is a widespread dating difficulty. Additional dating methods such as tephra chronologies, pollen zoning and lamination counts are consequently highly desirable for enhancing the reliability of the chronology of type palaeomagnetic sequences. The geographic range over which palaeomagnetic types records can be used is still under investigation. They can certainly be used without significant loss of

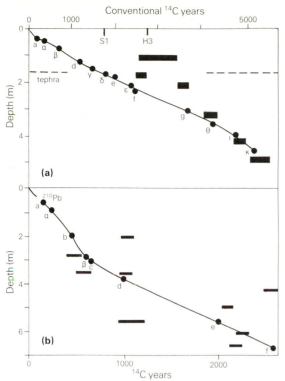

Figure 14.4 Time–depth curves comparing ¹⁴C and magnetic chronologies for: (a) Vatnsdalsvatn, north-west Iceland and (b) Rostherne Mere, England. Horizontal bars mark ¹⁴C age determinations and their laboratory counting precision. Solid circles mark magnetic turning points dated by correlation with the WE master curves of Figure 14.3. Additional dating information comes from the tephra at 1.7 m depth in Vatnsdalsvatn and ²¹⁰Pb dating of the uppermost Rostherne sediments. Rostherne data after Nelmes (1983).

magnetic direction changes repeat well from core to core and show some similarities to the British master sequence of Figures 5.8 and 14.3, 1500 km distant. If the Icelandic and British magnetic features are correlated by matching their turning points then a quite different time–depth curve from that of the radiocarbon age determinations is found (Fig. 14.4a). The magnetostratigraphic dating method indicates that the lower Vatnsdalsvatn ¹⁴C dates are reliable but that the upper dates have been contaminated by old carbon. Furthermore, the magnetostratigraphy suggests that the Vatnsdalsvatn tephra layer is only 1700 years old, presumably originating from the other nearby explosive volcano Snaefells and that by and large the sediment accumulation rate has remained constant over the last 6000 years.

The second example compares the ¹⁴C and palaeo-magnetic chronologies of Rostherne Mere, England. Figure 14.4b plots nine radiocarbon ages as bars and the palaeomagnetic turning points as a time–depth profile labelled a to f. The radiocarbon age determinations were made as part of a detailed diatom study (Nelmes 1983) in order to provide estimates of accumulation rate. Such a need for accumulation rate data rather than just the ages of isolated horizons is common in many palaeolimnological studies. The investigator faced only by the isotopic ages, (i.e. the solid bars, of Fig. 14.4b) would be hard pressed to calculate useful accumulation rates for any part of the 7 m sequence. The Rostherne Mere palaeomagnetic record showed many similarities to that of the Western Europe type record (see Fig. 14.3) allowing the turning-point profile to be constructed as far back in time as declination turning point WE f of around 2600 years BP. This Rostherne Mere palaeomagnetic time–depth relationship fits in well with the short-lived ²¹⁰Pb isotopic information and can be seen to provide a basis on which diatom accumulation rates can be calculated. The magnetic results suggest that the Rostherne Mere sediments are most unsuitable for conventional ¹⁴C dating having both excessively young and excessively old radiocarbon ages.

It must be made clear that by no means all lake sediments are suitable for palaeomagnetic studies. During the past 20 years the sediments of over 100 lakes in Europe have been analysed magnetically and less than one in five have yielded useful results (in the sense that both the declination and inclination data were repeatable in two or more cores).

One statistical approach to analysing the scattered palaeomagnetic data of secular variation logs is to fit

accuracy over distances of tens and even hundreds of kilometres. The upper practical limit for Holocene palaeomagnetic pattern matching seems to be working out at around 2000 km.

Figures 14.4a and b illustrate two examples of the use of secular variation magnetostratigraphy in helping to tie down the chronology of recent lake sediments. The first example shows a series of six radiocarbon age determinations from Lake Vatnsdalsvatn (Lochglenlochy) in northwest Iceland. The radiocarbon dates, drawn as solid bars in Figure 14.4a, appear to be internally fairly consistent. The tephra at 1.7 m depth appears to have an age of around 3000 years BP, quite close to the 2900 years BP age of the Hekla-3 eruption. Also there appears on the basis of the ¹⁴C ages to have been a dramatic fall in sediment accumulation rate in recent years. The palaeo-

them with a smooth curve. The idea is to estimate the proportion of the scatter that can be adequately explained by random orientation and measurement errors and the proportion of the palaeomagnetic signal likely to have been caused by geomagnetic fluctuations. Several smoothing methods have been tried, of varying degrees of mathematical complexity. The simplest methods tend to produce ill-fitting jagged curves containing spurious high frequency components. The most complicated methods can be very time consuming in their application. A routine method employed to calculate the curves of Figure 14.3 is that of fitting least squares cubic splines on the sphere (Thompson & Clark 1981). The method uses a robust biweighting procedure (Mosteller & Tukey 1977) in order to remove outliers (the cause of much high frequency contamination) and a cross-validation procedure (Clark & Thompson 1978) in order to assess the appropriate degree of smoothing. Such least squares methods permit confidence bands to be placed about the smooth best fitting curves. An alternative approach to investigating secular variation logs is to assess the scatter by calculating a single statistic. A particularly simple but handy statistic is the median solid angular difference (ξ) between adjacent declination and inclination measurements. Good quality lake sediments have ξ values below 6°. The high quality Loch Lomond data of Figure 5.8 have an ξ value of 2.5°. Epp *et al.* (1971) discuss a more comprehensive statistic based on this approach to assessing data consistency.

Core correlation methods using palaeomagnetic secular variation data have been suggested by Denham (1981) and Clark and Thompson (1979). Other more general correlation methods (e.g. Rudman & Blakely 1976, Gordon 1973, 1982) have also been applied to secular variation data. In practice, however, none of these procedures has proved to be workable and subjective trial and error methods are still used for the magneto-stratigraphic task of matching a palaeomagnetic log to its nearest master curve. Averaging of logs by stacking is a very useful approach in helping to confirm the reliability of palaeomagnetic fluctuations, a good example being provided by Barton and McElhinny's (1981) work on three South Australian lakes.

14.4 Origin of palaeolimnomagnetic secular variation

Spectral analyses of palaeolimnomagnetic logs (Thompson 1973, Denham 1975, Turner & Thompson 1981) reveal a concentration of power at periods of around 2000 to 3000 years. This dominant thousand year timescale of the palaeomagnetic direction fluctuations can be clearly seen as the major oscillations in the raw data plotted in Figure 5.8 and can be seen in all the spline plots of Figure 14.3. The concentration of spectral power at thousand year periods does not of course mean that the data are periodic nor does it mean that the same signal is being recorded at all the sampling sites. From a uniformitarian point of view the geomagnetic *processes* operating at present are likely to have been operating in the past and so present geomagnetic secular change is likely to be a useful first guide to gaining an understanding of ancient secular variation and palaeo-limnomagnetic logs.

As described in Section 5.1.2 the present features of the Earth's non-dipole field are varying on timescales of tens and hundred years. Hide and Roberts (1956) put the average timescale of secular variation at round 40 years on the basis of kinematic changes of the secular variation field. Thompson (1982) found an average lifetime of around 500 years for the major non-dipole field foci (Fig. 5.4) which have evolved since 1600 AD through a global analysis of historical field observations and archaeomagnetic data (Thompson & Barraclough 1982). Despite the great difference in the characteristic timescale of non-dipole field change (100 years) and the characteristic timescale of palaeolimnomagnetic secular variation features (2000–3000 years), it has been suggested (Thompson 1983) that the dominant geomagnetic cause of the lake sediment secular variation features is the growth and decay of non-dipole foci that we can see taking place at the present day. The thousand year palaeolimnomagnetic fluctuations are viewed as resulting from chance successions of several non-dipole foci with 500 year lifetimes evolving almost contemporaneously near the sampling sites and producing apparently persistent geomagnetic features. Higher frequency components of geomagnetic change are viewed as having been lost from the palaeolimnomagnetic records through (a) the averaging effects of post-depositional remanence acquisition, (b) the slow viscous degradation effects of particle realignment and (c) the introduction of sedimentological noise by processes such as bioturbation and micro-slumping. On this uniformitarian model the differences between long period secular variation patterns of various continents (Fig. 14.3) are explained as arising from the effects of local, short-lived geomagnetic sources, while the few similarities which are to be found between such widely spaced secular variation records are thought to have little geomagnetic significance and to have arisen by chance, as would be expected for such a comparison of field patterns generated by local magnetohydrodynamic processes.

Holocene palaeomagnetic secular variation records can be mimicked well by random processes such as random walk or autoregressive processes (author's unpublished calculations). Historical main field spherical harmonic coefficients can be used through their autocovariance functions to judge the order and variance of the autoregressive processes while the ratio of the present day main field to secular change coefficients allows the remaining parameters of the autoregressive processes to be calculated for spherical harmonic degree two to six. With some smoothing, to imitate signal attenuation associated with remanence acquisition effects, declination and inclination time series indistinguishable from those presented in Figure 14.3 can be generated by such autoregressive random processes based around historic field data.

The lack of reliable palaeointensity records from lake sediments means that only angular magnetic information is available from palaeolimnomagnetic studies. This fundamental drawback severely limits the chances that specific Holocene geomagnetic models of lasting value can be constructed with present data. As an alternative approach to that of geomagnetic modelling, an empirical description of Holocene secular change records has been built up in order to try to answer certain questions about Holocene field behaviour (Thompson 1982, 1983). Such an empirical description can potentially demonstrate periods of quiet and noisy secular changes, times of tilted as opposed to axial dipole fields, and times of average eastward or westward longitudinal drift. Figure 14.5 summarises an attempt at obtaining these features from the seven secular variation records plotted in Figure 14.3. The mathematical procedures followed in establishing the empirical description are described by Thompson (1982, 1983). In particular a quantitative method of assessing the average global sense of longitudinal drift, through the use of the curvature of the spline functions fitted to secular variation records, has been developed; Halley's (1692) approach of timing the longitudinal drift rate of geomagnetic features, which worked so well for 17th and 18th century changes, is unfortunately not appropriate with lake sediment

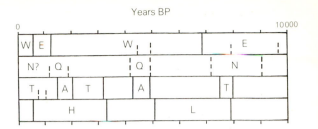

Figure 14.5 Summary of Holocene global averages: (a) E, W, eastward and westward drift; (b) Q, N, quiet and noisy secular variations; (c) A, T, axial and tilted dipole orientation; (d) H, L, high and low dipole intensity.

records, as Denham (1974) has pointed out, since ^{14}C dating is not sufficiently accurate to date the rate of movement of short-lived geomagnetic foci. The main features indicated by Figure 14.5 are that during the past 10^4 years (a) there were periods of both westward and eastward drift (b) that quiet periods of low secular change occurred and (c) that these were times when the dipole axis was more closely aligned with the spin axis than during the last 400 years.

14.5 Palaeomagnetic pitfalls

The magnetic dating method outlined above requires careful interpretation of the palaeomagnetic record. A serious stumbling block in magnetostratigraphy is the confusion of sedimentologically induced palaeomagnetic signals with geomagnetic signatures and the consequent proposal of quite mistaken correlations and chronologies. Sedimentological influences, particularly the well documented 'inclination error', make many sediment types unsuitable for magnetostratigraphic investigations. Two guidelines we have found particularly useful in differentiating between valuable geomagnetic records and stratigraphically worthless sedimentological influences are (a) the repeatability between sequences of both palaeomagnetic declination and inclination and (b) the lithological independence of palaeomagnetic direction data.

A palaeomagnetic difficulty of some philosophical interest concerns the 'reinforcement syndrome'. In 1971 Watkins discussed many aspects of the reinforcement syndrome and its connection with the misinterpretation of short geomagnetic polarity events. He summarised how the first published description of a geomagnetic phenomenon can have great influence over future studies as 'it enables workers pondering their own "curious" data to realize its real(?) meaning', and he pointed out that 'a substantial trap will have been laid . . . if the behaviour described in the initial publication is in fact erroneous' as it will lead to the generation of spurious data. An important human element can serve to reinforce an initial (erroneous) discovery for, as pointed out by Watkins, 'it is far more reasonable to generate both the energy and belief (faith?) required for publication of data confirming a discovery than to publish much negative data of a pedestrian nature'. He concluded that 'the case for super critical evaluation of data pertaining to polarity events is both obvious and strong'. Secular changes are naturally also ripe for abuse through the action of the reinforcement syndrome. A likely example can be found in the approach of correlating secular variation features having little or no dating control with previously published features several thousand kilometres distant, and then of concluding that the correlations show that similar geomagnetic field behaviour has occurred over large regions.

14.6 Excursions and the reinforcement syndrome

Geomagnetic excursions are taken to represent magnetic field behaviour intermediate between that of polarity reversals and large secular change. They may be unsuccessful or aborted reversals and as such possibly expected to be at least as common as successful reversals and could be of global extent (Verusob 1982). Naturally they are of great potential magnetostratigraphic significance (Denham 1976). Recent igneous rocks carrying unusual remanence directions which have been taken to be records of geomagnetic excursions have been found in France (Bonhommet & Babkine 1967), Iceland (Peirce & Clark 1978, Kristjansson & Gudmundsson 1980) and Indonesia (Sasajima et al. 1984). Unfortunately the development of the subject of excursions within the Brunhes normal polarity chron has led to a very confused situation and the likelihood of Brunhes excursions being able to play a major rôle in Quaternary magnetostratigraphy is receding.

The most debated recent excursions are the Gothenburg (Morner et al. 1971, Noel & Tarling 1975) in Europe and its North American equivalent the Erieau (Creer et al. 1976). Both excursions are claimed to be of late-Weichselian age and to provide detailed information about the pronounced changes in direction of the late-Weichselian geomagnetic field. Following the original proposals, numerous reports of excursions or unusual geomagnetic field behaviour have been made. The sediments involved have included those from lake (Nakajima et al. 1973, Noltimier & Colinvaux 1976, Anderson et al. 1976, Vitorello & Van der Voo 1977), cave (Kopper & Creer 1976), loess (Bucha 1973), till (Stupavsky et al. 1979), shallow marine (Vilks et al. 1977, Abrahamsen & Knudsen 1979, Abrahamsen & Readman 1980, Stoker et al. 1983) and deep-sea (Clark & Kennett 1973, Freed & Healy 1974) environments. These reports should not necessarily be taken at face value as they could well have been encouraged by the snowballing action of the reinforcement syndrome. Our interpretation of most of these unusual Brunhes palaeomagnetic directions is that they have been strongly influenced by sedimentological effects in high energy environments, such as the action of water currents or by slumping or by the inclination error of deposition remanence (Thompson & Berglund 1976).

It is interesting to reflect that an important observation used in the original identification of the late-Weichselian excursions was that the Laschamp/Olby lava flows of the Auvergne district, France had been found to have approximately reversed palaeomagnetic directions (Bonhommet & Babkine 1967) and to have an age which could conceivably have been late Weichselian. Since the initial dating studies the ages of these lavas have been revised and they are now taken to have been erupted between 35 000 and 45 000 years ago. Also it has since been discovered (Heller 1980) that the lavas self-reverse (Section 13.2.1) their remanence on reheating and cooling in the laboratory. The self-reversal is caused by the intimate association of different magnetic phases produced by both high and low temperature titanomagnetite oxidation. Although the laboratory self-reversal of many of the Laschamp/Olby lavas cannot be taken to prove unequivocally that the Laschamp geomagnetic event did not occur, these new findings coupled with the revised Laschamp ages destroy the crucial original observation used in justifying a late-Weichselian excursion and emphasise the need for an understanding of the rôle of the reinforcement syndrome in prompting the publication of spurious palaeomagnetic data.

14.7 Summary

Secular changes of the geomagnetic field can be used for correlating and dating some kinds of recent sediment. The geomagnetic field continually changes, producing an irregular but characteristic magnetic pattern which can form the basis of a magnetostratigraphic core correlation method. Magnetic direction fluctuations have been dominated by local magnetohydrodynamic effects in the Earth's fluid core. These processes give rise to a regional distribution of geomagnetic changes at the Earth's surface.

The secular variation magnetostratigraphic method is straightforward in principle, involving matching the palaeomagnetic remanence record of a new sediment sequence with a nearby, previously dated type palaeomagnetic section. The accuracy of the secular variation magnetostratigraphic method largely depends on the accuracy of the dating of the type palaeomagnetic record. At best the dating of the type records is probably only good to within one or two hundred years. However, this uncertainty does not prevent secular variation magnetostratigraphy from competing with alternative dating methods.

[15]

Biomagnetism

My mind is in a state of philosophical doubt as to the origin of
animal magnetism.

Coleridge 1830

15.1 Introduction

The study of the weak magnetic fields originating in biological systems has been made possible by the production of cryogenic magnetometers (Section 6.2.4) and high quality magnetic shielding (Section 6.6.3). The subject of biomagnetism can be subdivided into three main topics according to the origin of the biomagnetic fields (Williamson & Kaufman 1981). These three topics of (a) magnetic precipitates formed through biochemically controlled processes, (b) magnetic contaminants and (c) electric currents arising from ion flow in various organs such as the heart and the brain are all discussed below.

Biochemically precipitated magnetite has been found in the body tissues of organisms as diverse as bacteria, algae, insects, birds and mammals; indeed magnetite is probably the fourth most common biochemically precipitated mineral. In many of the magnetic organisms the magnetite precipitates can be used for detecting the Earth's magnetic field and hence can be used for orientation and navigation. Recent advances in biomedical aspects of biomagnetic fields and magnetic contaminants have been summarised in the excellent review article of Williamson and Kaufman (1981).

15.2 Magnetic navigation

The clearest demonstration of magnetic navigation by living organisms is Blakemore's (1975) discovery of a diverse group of bacteria which orientate in the Earth's magnetic field and swim along its magnetic field lines. These magnetotactic bacteria possess a biomagnetic compass in the form of a chain of magnetite particles which they have synthesised from soluble iron. Many organisms, besides bacteria, have also been found with magnetite precipitates. The list of organisms includes pigeons, bees, algae, chitons, turtles, tuna, dolphins and butterflies. The quite extraordinary navigational and homing abilities of many of these organisms are well known and it has often been suggested that the geomagnetic field may play a rôle in aiding their perception of direction. Unambiguous demonstration of the use of a biomagnetic compass in higher organisms is rather difficult as it must involve both behavioural and physiological experiments. Nevertheless laboratory experiments on the location, mineralogy and mineral magnetic properties of the biochemically synthesised magnetic particles provide a wealth of information about the likely sensitivity, value and possible biophysical methods of geomagnetic field sensing.

MAGNETOTACTIC BACTERIA
The response of magnetotactic bacteria to a magnetic field is easily demonstrated by observation of the bacteria under low power magnification. Viewed under one hundred times magnification magnetotactic bacteria, in a droplet of clean water, appear as moving points of light. They naturally swim along the geomagnetic meridian, accumulating at the edge of

the water droplet. Their direction of swimming can be readily changed by approaching the droplet with a bar magnet. Bacteria collected in the Northern Hemisphere swim consistently to the geomagnetic north whereas those collected in the Southern Hemisphere swim to the south (Blakemore & Frankel 1981). This difference in behaviour of northern and southern hemisphere bacteria was predicted on the supposition that a likely biological survival advantage of magnetotaxis was that bacteria would use the inclination of the geomagnetic field in order to direct themselves downwards so keeping themselves near the substrate, well away from surface waters.

Magnetotactic bacteria are anaerobic or micro-aerophilic, occurring in concentrations of around 100 or 1000 per millilitre in a wide range of environments (Moench & Konetzka 1978). They are typically about 3–5 μm in length and have been found in salt and freshwater peat bogs, marshes, sewage treatment plants, and lake, estuarine and marine surface sediments. They appear to be ubiquitously distributed in aquatic sediments with pH values in the range 6–8, although only rarely occurring in the overlying water columns. Moreover, the relative importance of their contribution to soil and to lake sediment magnetism remains unevaluated (Chs 8 & 10).

Transmission electron micrographs of magneto-tactic bacteria reveal striking chains or clusters of opaque iron-rich particles in the cytoplasm of the cells. Mossbauer absorption spectra of freeze-dried cells demonstrate that the iron is primarily in the form of magnetite. Some other compounds, presumably precursors of magnetite in its biochemical pathway, are also found in the precipitates or magnetosomes. Magnetosomes are produced in bacteria cultured in media with soluble iron concentrations in excess of 1 mg l^{-1}. They are not produced if the iron concentrations fall much below this level. The magnetite particles have lengths of around 0.05–0.1 μm, axial ratios (width/length) ranging from 0.6 to 0.9 and shapes varying from cubic or octahedral to sharply pointed forms. Clumps of magnetite particles are found outside the bacteria cells suggesting that the particles are released after death and can accumulate in the substrate. The size of the particles produced by the magnetotactic bacteria falls neatly in the range 0.04–0.2 μm which ensures that they are single domain (Kirschvink & Gould 1981). This condition assures efficient use of the magnetite in its rôle as a compass. Larger multidomain particles would have a lower specific remanent magnetisation and smaller superparamagnetic particles would have a lower fluctuating moment, which again would result in a much reduced effective magnetic moment. The configuration of the magnetic particles in chains is an enterprising biochemical arrangement designed to exploit magnetic interactions which serve to align the magnetic moments of all the particles along the length of the chain and hence to form a most efficient magnetic disposition.

Of course the magnetic particles do not cause a magnetic force which can pull the bacteria along the geomagnetic field lines. The torque produced by the Earth's magnetic field acting on the chain of magnetised magnetite particles only serves to point the bacteria: they have to swim along the field lines using their flagellae. This orientational rôle is clearly demonstrated by the observation that dead bacteria do not move along magnetic field lines, but only align with the field direction. The usual number of a few tens of magnetite particles in natural magnetotactic bacteria seems to be controlled by the magnetic moment needed to produce a useful torque.

A novel experiment (Kalmijn & Blakemore 1978) can be performed in order to (a) demonstrate that the bacteria do indeed possess a magnetic remanence which is used to sense the geomagnetic field and (b) at the same time to determine the coercivity of the biomagnetic remanence. The procedure of this neat experiment is first to align the bacteria in a weak magnetic field and then to hit them with a short one-shot magnetic field pulse antiparallel to the weak aligning field. Production of a sufficiently strong magnetic field pulse reverses the biomagnetic remanence and instantly causes the bacteria to turn around and swim in the opposite direction. The strength of the field required to produce this change in behaviour reflects the coercivity of remanence of the magnetite particles. It varies from species to species ranging from at least 35 to 55 mT. Exposure of a population of magnetotactic bacteria to an alternating field demagnetising coil also produces a change in behaviour. After magnetic cleaning half of the population swims to the north and half to the south. This result is that expected for a chain of magnetite particles as their magnetic moments cannot be destroyed, they can only be switched from one easy direction to another. Natural populations of mixtures of both north- and south-seeking bacteria are found near the magnetic equator. Magnetotaxis is presumably useful there in preventing harmful upward excursions.

The evolution of bacterial communities when subjected to changing magnetic field conditions has also been studied (Blakemore 1982). Such investigations permit speculation about shifts in magnetotactic bacterial population densities and their magnetic characteristics during geomagnetic field changes (Kirschvink 1982a). On culturing a natural population of normal polarity bacteria in a reversed field a significant change in bacterial polarity is found after only a few days and almost complete population reversal takes place in a month. These types of response to magnetic field change clearly must occupy a few generations of bacteria. The polarity of a bacterium with a chain of magnetic particles is conserved in its daughters by the act of partitioning of the magnetic chain during cell division. Any particles later synthesised by the daughters become magnetised in the chain polarity, i.e. the polarity of the parent. Although such an ability to synthesise magnetic particles can be genetically encoded, the polarity of the particles cannot be encoded. So if a bacterium without magnetosomes begins to manufacture them it can be either north or south seeking. In every generation a few bacteria can end up with the wrong polarity. These few can proliferate under a change of field conditions.

A decrease in geomagnetic field intensity can be expected to lead to an increase in the number of magnetosomes, in order to produce an equivalent magnetic torque to that of modern day natural bacteria. A fall in field intensity below a critical limit however is likely to halt the production of magnetosomes altogether, as the synthesis of additional magnetosomes becomes too much of a burden. If magnetotactic bacteria were making a significant contribution to the total magnetic mineral content of a substrate then geomagnetic field intensity fluctuations could be imagined as being reflected in mineral magnetic concentration variations.

In summary, magnetotactic bacteria are bottom-dwelling, swimming organisms that are passively steered by the torque of the Earth's magnetic field on their biomagnetic compass. Being anaerobic or micro-aerophilic they swim downwards in order to avoid the toxic effect of greater oxygen concentrations in surface waters. They are able to control the size of their precipitated magnetite crystals, synthesising chains of single-domain particles from soluble iron in their environment. Concentrations of around 1000 magnetotactic bacteria per millilitre in many surface sediments are quite sufficient to indicate that magnetosomes might be an important part of the mineral magnetic component of some sediment and soil samples.

A SENSE OF MAGNETISM

The most convincing evidence that certain higher animals can use the geomagnetic field as a navigational aid is provided by studies of homing pigeons (Keeton 1971). Pigeons need both a map and compass-sense to be able to return home after being released at an unfamiliar site. A number of controlled experiments on pigeon homing have indicated that magnetic field information is used to aid orientation. For example, pigeons have been found to navigate poorly on cloudy days if small magnets or current-carrying coils are affixed near their heads (Keeton 1971, Walcott & Green 1974). In older experienced pigeons the sun is the preferred compass. The use of the sun as a compass to aid navigation on sunny days is well established since clock-shift experiments (in which the internal clocks of homing pigeons have been phase shifted by several hours) deflect departure bearings of pigeons from their release locations.

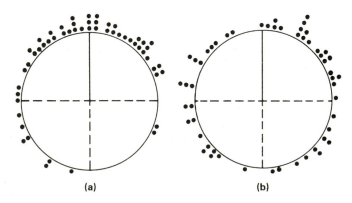

Figure 15.1 The influence of magnetic fields on the departure bearings of pigeons. Clock-shifted pigeons (i.e. ones that had never seen the sun before noon) neverthe-less were homeward (solid line) orientated on morning release (a) but were disorientated (b) when carrying magnets. Dots mark the vanishing bearings of individual birds. Data from Wiltschko *et al.* (1981).

(a)   (b)

However, on cloudy days released, clock-shifted pigeons depart directly homeward, presumably navigating by the geomagnetic field. The predominant use of the geomagnetic field compass even on sunny days is illustrated in Figure 15.1 for young pigeons which have no previous use of the sun as a compass. Figure 15.1 plots out the well orientated homeward departure direction of control pigeons and the random departure directions of birds carrying small bar magnets.

The ability of pigeons to sense magnetic fields is presumably related to the magnetite particles which have been found between their brain and skull in a small (0.5 ml) region of tissue which also contains nerve fibres (Walcott *et al.* 1979). The remanent magnetic moments of pigeons range from 10^{-4} to 10^{-3} A m^2. Magnetite is thought to be the dominant magnetic mineral in pigeons' heads because the magnetic particles have yielded Curie temperatures of 575 °C. The magnetite is thought to be largely single domain, but the domains of individual particles are not well aligned as an isothermal remanence about ten times the magnitude of the natural remanence can be produced by a 0.2 T field.

Bees also apparently orientate in the geomagnetic field (Kirschvink 1982b). Their dances, used to communicate the direction of a food source (with respect to the sun), make small regular errors. Cancellation of the ambient magnetic field causes these errors to disappear. Another indication that bees have a sense of magnetism is that the orientation of combs constructed by swarms otherwise deprived of orientation cues are reported to be in the same magnetic direction as that of the parent hive. Many bees have been found to possess a magnetic remanence of about 10^{-3} A m^2. The remanence tends to lie in their horizontal plane and to have its source almost exclusively in the anterior part of the abdomen. Again magnetite is indicated to be the main magnetic mineral by a 580 °C Curie temperature. Cooling experiments, however, point to the presence of a large proportion of superparamagnetic magnetite particles. In addition to bees, five other invertebrate species have been found to have magnetic moments approaching 10^{-3} A m^2.

Magnetic minerals have even been isolated from the heads of dolphins (Zoeger *et al.* 1981). These magnetic minerals, which are in part magnetite, have a low natural magnetic moment of 2×10^{-8} A m^2. The weakness of their natural remanence is mainly due to an exceedingly low stability as revealed by a median

destructive field of 2 mT in alternating field demagnetisation. The magnetic tissue is located near the roof of the skull of the dolphin, apparently associated with nerve fibres. If magnetite is used by dolphins as part of a biomagnetic geomagnetic field detection system, then a torque on the induced moment of the multidomain magnetite is a likely receptor mechanism.

The ability of humans to navigate by means of sensing a magnetic field has been examined, but the experimental trials have yielded conflicting results (Gould & Able 1981). Some navigation and orientation tests have led to different aptitudes being recorded for subjects wearing magnets and subjects wearing equivalent dummy weights. These experimental results suggested that man may unknowingly use a magnetic sense. However, attempts to replicate the experiments have failed. So it now appears that any human ability to sense the Earth's magnetic field is at most marginal and is unlikely to match that of our conventional senses.

Biogenic magnetite is clearly produced by many organisms and can be used as a magnetic compass. An interesting question concerns the time in the geological past when biogenic minerals were first produced and whether biomagnetic precipitates have been preserved in the geological record. At present the magnetite of chitons' teeth (used as a strengthening material) probably makes the largest contribution to the biomagnetic mineral content of any sediments (Kirschvink & Lowenstam 1979). A rough calculation of the chiton teeth magnetite input to continental shelf sediments, assuming that all the biomagnetite is preserved, shows that it could account for average magnetite concentrationssuming that all the biomagnetite is preserved, shows that it could account for average magnetite concentrations of up to one part in 10^5 (Kirschvink & Lowenstam 1979), which is comparable with typical shallow marine total magnetite concentrations. Bacteria with natural population densities of 1000 cells per millilitre would only account for magnetite concentrations of one part in 10^{10}, but this amount could be increased to perhaps one part in 10^5 through the accumulation of successive generations of dead bacteria. Distinguishing between biogenic, volcanic and pedogenic magnetite is not an easy task as the size ranges, shapes and magnetic properties of all these magnetite types overlap. The possibilities of bacterial biomagnetic mineral concentration changes reflecting geomagnetic field intensity fluctuations merit careful study, although

181

recent sediment mineralogy records do not obviously track or mirror the well established Holocene geomagnetic 6500 year BP intensity minimum and the 2500 year BP maximum of Figure 5.9.

15.3 Pneumomagnetism

One of the first practical medical applications of biomagnetism concerned the measurement of the remanent magnetic field of magnetic contaminants in the lung (Cohen 1973). Cohen showed that it was possible to estimate the amount of magnetic dust (cf. Ch. 11) in a person's lungs by magnetising the dust. and then by measuring the newly created magnetic field at the subject's chest.

Such pneumomagnetic techniques have found application in both research and occupational health areas. A particularly helpful aspect of magnetic techniques in studying the hazards from respired particles is that they are non-invasive. Magnetic methods are also potentially very useful for studying contaminant clearance rates. For example they have suggested that smokers may have distinctly lower clearance rates than non-smokers (Cohen *et al.* 1979). Occupational exposure to airborne particles can lead to large numbers of particles being inhaled. In certain cases an appreciable fraction of the particles is magnetic and the smaller particles (>2 μm) will be carried into the lung and may be deposited on the large alveolar surface. This is an unhealthy situation as it is known that iron oxides retained in the lungs are responsible for the development of pulmonary sideosis.

The non-invasive pneumomagnetic method basic-ally consists of magnetising the contaminant particles by standing the subject for about 5 minutes between a pair of Helmholtz coils giving a magnetising field of some 0.02 T and then of mapping the remanent field parallel to the applied field (i.e. normal to the skin). Measuring times using a superconducting magneto-meter capable of sensing fields as low as 10^{-5} nT are around 5 minutes. The average error, as deduced from measurements on control groups, works out to be 0.02 nT with the largest error signals from the abdominal area – presumably related to ingested food (Kalliomaki *et al.* 1976, 1981). Some of the strongest signals have come from the right lungs of welders (Fig. 15.2). Their magnetic contaminants produced remanent fields of several hundred nanotesla, several orders of magnitude above those of the control subjects. The magnetic technique turns out in many instances to be more sensitive than others, such as radiography, in assessing lung contamination. The threshold of radiographs corresponds to a remanent field of around 1 nT which in turn is equivalent to about 100 mg of dust. At this and higher con-centrations good correlations have been found between magnetic measurements and radiographic findings.

After magnetisation the remanent field of the lung shows characteristic decay with time (Cohen 1973) dropping to a small fraction of the original in around one hour. These changes in remanence are thought to relate to movement of the magnetised particles in the lung tissue rather than to viscous mineral magnetic behaviour properties of the contaminants. Variations in relaxation rate may be related to pulmonary diseases which can change the mechanical properties of pulmonary tissue.

Magnetopneumography has thus been successfully applied to meet the needs of many workers including arc welders, coal miners and shipyard and foundry workers by assessing particle deposition in the lung (Kalliomaki *et al.* 1981). Being a non-invasive method of characterising particulate burden it can play an important rôle in both research and occupational

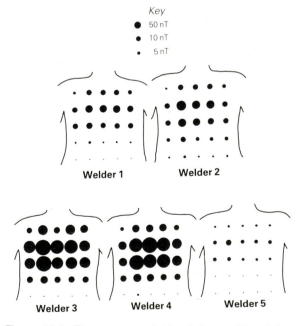

Key
● 50 nT
● 10 nT
· 5 nT

Welder 1   Welder 2

Welder 3   Welder 4   Welder 5

Figure 15.2 The remanent fields of five welders (after Kalliomaki *et al.* 1976).

health. A weakness of the magnetic technique is that high remanent fields might derive from a relatively small number of dust particles with strong individual magnetisations rather than from many particles caught on the lungs as a result of a long exposure to dust that is magnetically weaker. However, such disadvantages are countered by a number of extra research potentials such as the possibility of guiding particles towards regions of the lung of special interest through the use of magnetic field gradients.

15.4 Cardiomagnetism

One of the most studied and indeed one of the strongest biomagnetic phenomena is the magneto-cardiogram (Williamson & Kaufman 1981). Magnetic fields of 50 pT corresponding to magnetic moments of 0.1 µA m² are produced by the electric currents of the cardiac cycle. A normal magnetocardiogram is shown in Figure 15.3. Clinical applications of the magneto-cardiogram include (a) a convenient, rapid screening method for large numbers of people, (b) a complementary research approach to the electro-cardiogram in investigations of the nature of electrical

activity in the heart and (c) a method of characterising abnormal electrocardiograms. Most of the current interest in cardiomagnetism is involved with its diagnostic applications as a supplement to the electrocardiogram.

Early measurements of the heart's magnetic field were made by induction coils (Baule & McFee 1963) but superconducting magnetometers now provide better spatial resolution, lower noise levels and a better frequency range. In practice it is useful to demagnetise any magnetic contaminant particles found lodged in the body by the application of a magnetic recording tape eraser before measuring the magnetocardiogram. Contributions from the susceptibility of the changing volume of blood in the heart need to be taken into account in detailed research studies but are unnecessary for clinical applications. Magnetocardiograms can be displayed in a number of ways many of which closely resemble methods used in palaeomagnetic and geomagnetic studies. They may be represented as a set of time series; alternatively they may be modelled as a dipole heart vector and the locus of the vector traced on orthogonal plots. In another approach higher order heart terms are modelled using spherical harmonic analyses, and the variations of the multipole coefficients are plotted against time.

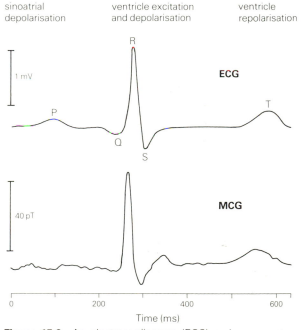

sinoatrial depolarisation   ventricle excitation and depolarisation   ventricle repolarisation

Figure 15.3 An electrocardiogram (ECG) and a magneto-cardiogram (MCG) recorded in a hospital (after Williamson & Kaufman 1981).

15.5 Neuromagnetism

Magnetic signals from the nerves have been studied in attempts to bridge the gap between psychological and psychophysical aspects of behaviour and perception, and the functioning of individual neurons (Williamson & Kaufman 1981). A possible advantage of studying magnetic signals over electric signals is that they are less affected by dispersion through intervening tissue. This observation suggests that a magnetoencephalogram recorded outside the head should be able to resolve the site of neural activity more accurately than can an electroencephalogram. The two main lines of neuromagnetic research which have been pursued are first, analysis of spontaneous brain activity, such as the alpha rhythm, and secondly, investigation of activity evoked by sensory stimuli. Neuromagnetic signals can be extremely weak ranging from amplitudes of a few picotesla down to a few femtotesla. Well balanced SQUID gradiometers (Fig. 6.4a) or particularly well shielded rooms are needed for these remarkably sensitive measurements.

The most common spontaneous neuromagnetic activity is the theta brain rhythm with a frequency of around 6 Hz. Magnetic and electrical activity of the brain at this frequency are often similar although correlation is lost during sleep. Magnetic activity as a response to sensory stimuli includes (a) visual cortex fields of a few 100 fT evoked by light flashes, (b) somatically evoked field responses such as from stimuli to the wrist and finally, (c) fields near the auditory cortex produced by click stimuli or prolonged sounds. The sources of some of these evoked fields have been located within the head to an accuracy of around 1 cm. Signal processing has been used to extract somatically evoked magnetic activity of below 50 fT in strength (Okada *et al.* 1982). This type of neuromagnetic experimentation represents one of the finest, most sensitive magnetic detection systems yet realised.

15.6 Summary

The long-held belief that magnetic fields can influence the behaviour of certain animals and organisms has been strikingly confirmed in recent years through the detection of magnetised iron oxides in bacteria, pigeons, dolphins and a host of other animals. Iron oxides, particularly magnetite, are now recognised as being produced biochemically by many phyla.

The recognition of biochemical magnetite in sediments and soils is hampered by its similarity to authigenic or volcanic magnetite. Consequently, the importance of biochemical magnetite in environmental studies remains enigmatic. The profusion of magnetotactic bacteria in many present day aquatic environments suggests that the possible presence of biochemical magnetite crystals must be carefully considered in interpreting the magnetic properties of environmental samples, particularly those with low concentrations.

Airborne iron oxide dusts can be trapped in our lungs; thus non-invasive magnetic methods can be used to investigate the health hazards to people such as welders, shipyard workers and miners who are particularly prone to pulmonary problems. Sensitive cryogenic magnetometers are also contributing to cardiological and neurological work by measuring the very weak but diagnostic magnetic signals emanating from the heart and the nervous system.

[16]

The Rhode River, Chesapeake Bay, an integrated catchment study

Let the downpour roil and toil!
The worst it can do to me
Is carry some garden soil
A little nearer the sea.

Robert Frost
In time of cloudburst

16.1 Physical setting

The Rhode River is a tidal estuary on the western shore of Chesapeake Bay (Fig. 16.1) some 10 km south of Annapolis, Maryland. The surface waters of the estuary comprise some 485 ha and the total catchment is 3332 ha. The tidal range is low (<1 m) and the estuary is shallow with no strongly developed deep channels and a maximum water depth of about 4 m at its mouth. The open water is bordered for the most part by gentle wooded slopes though there are extensive salt marshes along the southern shoreline and, more locally, actively eroding cliff sections exposed to wave action at the present day. In particular, the islands in the middle of the 'River' are scarred by conspicuous cliffs. Relief overall is low with no part of the catchment above 180 m. The underlying bedrock comprises a variety of sedimentary types. Very restricted exposure of Marlboro Clay are overlaid by the sandy, glauconitic, Eocene, Nanjemoy formation, the sandy and diatomaceous sediments of the Miocene Calvert formation and the alluvium of the

Talbot formation. The Nanjemoy and Talbot formations are the only important lithologies in the lower parts of the catchment and around the water's edge. Of the actively eroding cliffs, only one is in the Talbot alluvium, the rest are in Nanjemoy sands. Most of the higher ground in the catchment is developed on Calvert sediments or the locally overlying Sunderland terrace deposits. The well drained parts of the catchment are deeply weathered and soils are mostly eluviated sandy loams. More locally, gleyed soils occur in streamside locations, and the main drainage system is flanked in its lower reaches by an extensive swamp.

The climate is humid through most of the year with a mean annual average precipitation of 1120 mm (Brush *et al.* 1980). There is a tendency to sudden heavy rain storms in spring and summer as well as occasional severe hurricanes. No significant transport of material coarser than 3 mm occurs in the catchment, most of which lacks particles larger than this. Only very locally are stream channels incised and active channel erosion is very limited. Character-

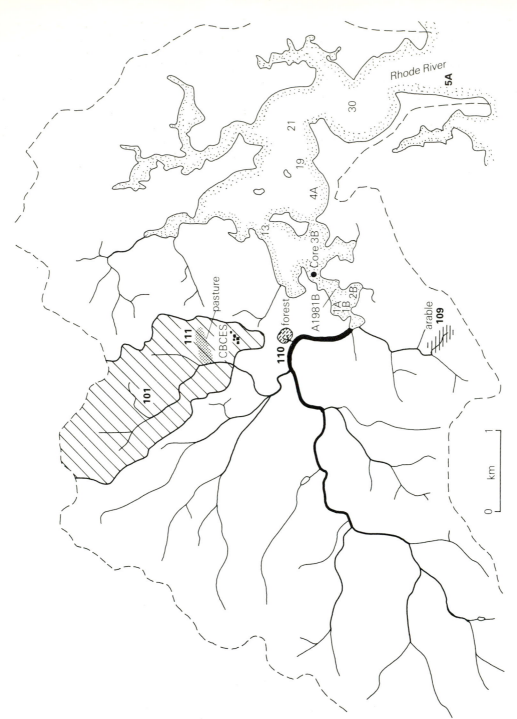

Figure 16.1 Map of the Rhode River catchment. Subcatchments 101 (large, mixed land use). 109 (arable) 110 (forest) and 111 (pasture) are located, as are the main coring sites in the estuary. Two of the islands shown are the sites of the magnetic reference profiles on eroding cliff sections (Fig. 16.4).

istically, at low flow, streams dwindle to a narrow thread of water within a residual sandy-bedded channel between banks of graded and subhorizontally bedded sands, silts and clays laid down during the receding levels of preceding floods. It is apparent from the channel morphology that large volumes of fine sediment are both stored and moved within the system.

As part of the Rhode River ecosystem monitoring programme operated by the Smithsonian Institution, several stream gauging stations are maintained, four of which are located on Figure 16.1. Three of these are at the outfall of small predominantly single land-use

catchments – number 109 (cultivated crops – corn and tobacco), 110 (mixed hardwood forest) and 111 (pasture); the fourth 101, is a large mixed land-use catchment which includes 111. Further details of geology and land use and of the hydrological/sedimentological monitoring programme are given in Correll (1977). The main overall aim of the monitoring programme is to assess the contribution of non-point pollution sources to the waters of the Bay by making a detailed study of the relationship between land use and water and sediment quality in a small rural watershed.

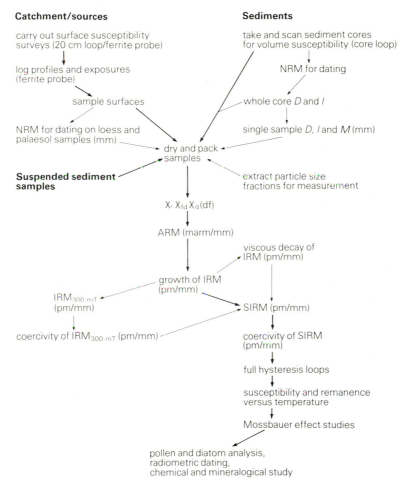

Figure 16.2 A flow diagram of measurements used to characterise the magnetic mineralogy of the Rhode River sediments and sources, to measure the record of palaeomagnetic secular variations in the sediments and to identify samples for subsequent more detailed study. pm = pulse magnetiser; mm = magnetometer; em = electromagnet; D.F. = dual frequency magnetic susceptibility sensor.

16.2 Sediment sources

At the outset there are three potential major sediment sources for the estuary:

(a) The conspicuous eroding cliffs provide large volumes of predominantly Nanjemoy bedrock and associated deeply weathered subsoil at least close to the cliffs themselves. This is the local expression of a much more widespread coastal erosion problem round Chesapeake Bay (Slaughter *et al.* 1976).

(b) The terrestrial surfaces of the catchment are extensively cultivated. Pierce and Dulong (1977)

using data from the suspended sediment sampling programme for the mixed land-use catchment 101, calculate a loss of 511 kg ha^{-1} a^{-1} for 1975. This implies that erosion rates for the catchment may range from 5 to 16 cm/1000 years depending on the extent to which the total output is ascribed to the cultivated parts of the catchment.

(c) It is quite possible that there is a significant flux of sediment into the Rhode River from the open bay. Donoghue (1981) proposes that this may contribute as much as 17% of the current sediment budget. Clearly all these sources and others less significant at present, will have

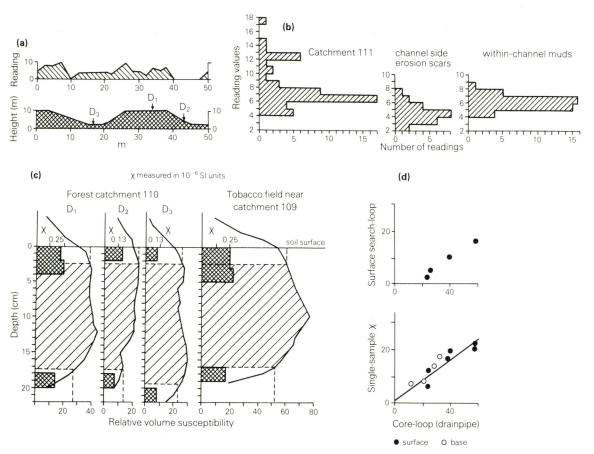

Figure 16.3 Magnetic survey and initial characterisation of the Rhode River catchment: (a) surface 20 cm search-loop readings in relation to topography in the forest catchment (110); (b) frequency plots of surface 20 cm search-loop readings for the pasture catchment (111) (note the bimodal distribution resulting from high values on Calvert soils in the upper part of the catchment), stream channels side minor erosion features (note the low values) and within-channel sediments; (c) drainpipe cores of surface soils providing core-loop scans of volume susceptibility, and single samples from top and base for individual measurements; (d) plots of the drainpipe core-loop readings at the ends of the scans (see c) versus surface search-loop readings and single-sample measurements.

responded to both physiographic and anthropogenic changes over the past few centuries. Notable among these are the isostatic sea-level rise estimated by Donoghue at over 2.7 mm per year for the last millenium, the land-use changes that have occurred subsequent to the first colonial settlements in the area some 300 years ago, and the recent dramatic increase in power-boating and associated developments along the shores of the 'river'.

16.3 Study aims

The Rhode River magnetic study was started in 1980 with a number of methodological and substantive aims. In terms of methodology, the application of the newly developed environmental magnetic techniques

to an estuarine context was a significant extension from previous experience based on lacustrine systems which are for the most part more 'closed', subject to lower energy levels and sedimentologically less complex. Moreover, unlike the previous study localities used in developing and evaluating mineral magnetic techniques, the Rhode River area was chosen with no prior knowledge of its lithology and hence in ignorance of its primary magnetic mineralogy. Within this new environment two further methodological challenges were identified as especially important. The first of these was to develop and evaluate a comprehensive magnetic approach to catchment studies through every stage from field survey to detailed source and sediment characterisation. The second was to overcome the possibility of coincidental or invalid sediment source identification, by attempting a detailed magnetic characterisation of

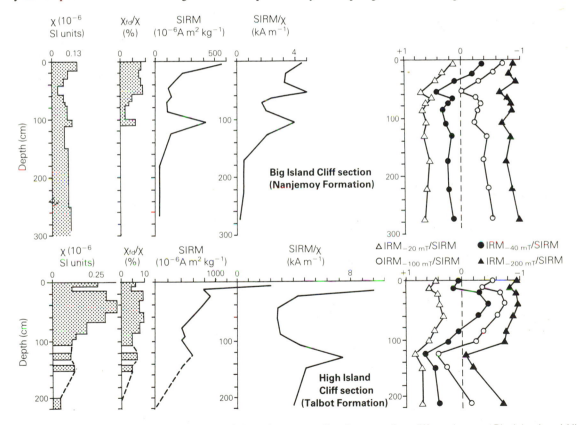

Figure 16.4 Mineral magnetic characterisation of the reference profiles from eroding cliff sections on Big Island and High Island (Fig. 16.1). The two parameters which most consistently identify the upper part of the profiles are χ_{fd}/χ (%) and SIRM. In both profiles there are lower peaks of χ_{fd}/χ and SIRM as well as higher ('harder') IRM/SIRM in the strongly illuviated zone. Note also the very low SIRM values in the basal samples. In the Nanjemoy section which is representative of most of those in the estuary, this SIRM minimum is associated with a steep viscous loss of SIRM and very low SIRM/χ ratios.

material in all phases of the system, using, in so far as possible, mutually independent magnetic parameters measured on a particle size related basis (Olfdield *et al.* 1985c).

The substantive aims were:

(a) to reconstruct spatial and temporal variations in sedimentation since pre-colonial times in terms of rates and sources;

(b) to characterise magnetically particulate flux within the system at the present day;

(c) to establish the implications of (a) and (b) in terms of erosion rates and of their relation to land-use changes and other human activities.

16.4 Methods

As a result of experience gained in the first two field seasons (1980 and 1981) and of improved instrument design taking place *pari passu*, the methodology outlined in Figure 16.2 was gradually developed and applied. It can now be advanced as a general scheme for subsequent catchment-based magnetic studies. The initial survey stage in the catchment involved extensive surface measurements using both the 20 cm 'search loop' and the ferrite probe attachments of the portable Bartington susceptibility equipment. Figure 16.3 shows plots of some of these results. Soil 'drainpipe scans' and single-sample measurements at selected sites followed, and from the extensive field survey stage involving many hundreds of readings, representative sites were chosen for more detailed study.

At the beginning of the project, estuarine cores from 1 to 2.5 m in length were already available for study (Donoghue 1981) and others were taken subsequently. Samples from several cores had already been used for ^{14}C, ^{210}Pb and ^{137}Cs assay. All cores were scanned for volume susceptibility variations using successive versions of the Bartington whole core measuring equipment. Examples of whole core traces are shown in Oldfield (1983a). Subsequently cores were subsampled for single-sample measurement.

Pilot sediment samples from cores taken in 1981 confirmed that at least for parts of the record, the NRM signal was strong, stable, repeatable and compatible with recent geomagnetic inclination values. Subsequent NRM measurements were carried

out on subsamples from two cores at the mouth of the estuary using the methods outlined in Chapter 14.

During the course of the study, samples were taken for mineral magnetic characterisation from deep reference soil profiles in Nanjemoy, Calvert and Talbot formation exposures; a variety of field surface flow features in the aftermath of heavy storms; stream channels and the banks of fine contemporary alluvial sediment on either side; areas of water ponding behind the weirs built at each gauging station; the suspended sediment samples taken at each gauging station during flood events as part of the routine monitoring programme; the suspended sediments of the open estuary; and surface sediments within the Rhode River and beyond its mouth. Particle size measurements were carried out on a variety of soil and sediment samples within the system. Standard magnetic characterisation for virtually all samples comprised measurement of χ, χ_{fd}, SIRM, $IRM_{-20\,mT}$/SIRM, $IRM_{-40\,mT}$/SIRM, $IRM_{-100\,mt}$/SIRM and $IRM_{-300\,mT}$/

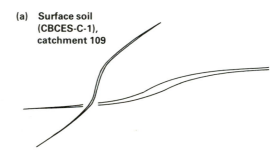

(a) **Surface soil (CBCES-C-1), catchment 109**

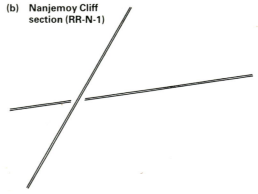

(b) **Nanjemoy Cliff section (RR-N-1)**

Figure 16.5 Hysteresis loop plots for a representative surface soil and an unweathered Nanjemoy cliff section sample. Full plots up to ±00 T are superimposed on expanded plots of the central part of each loop. The basal Nanjemoy sample is almost exclusively paramagnetic. The surface soil has both a paramagnetic and a ferrimagnetic component identifiable in the plot.

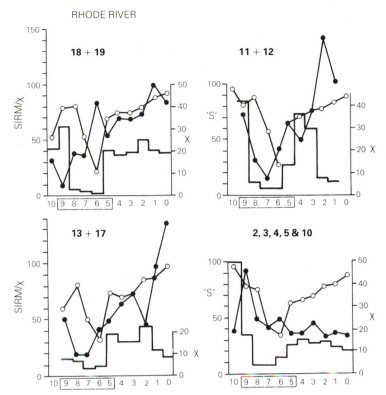

RHODE RIVER

Key

●────●SIRM/χ
○────○S
S = IRM$_{-100\,mT}$/SIRM × 100

Figure 16.6 χ, SIRM, SIRM/χ and IRM$_{-100\,mT}$/SIRM measured on particle size fractions for a range of catchment samples. The 'surface samples' are taken from residual 'sediments' retained on the arable catchment (109) after heavy rainfall. Each graph plots the mean values of between 2 and 4 samples. Each sample was divided into size fractions down to 4 phi by dry sieving then further subdivided by the pipette method. Thus each of the phi sizes 5 to 9 include the finer sizes as well. IRM$_{-100\,mT}$/SIRM values show a similar trend in all samples with a peak of 'hardness' in the fine sands. SIRM/χ increases towards the fines except in the case of the channel muds (see text).

SIRM. Full coercivity of SIRM profiles were produced for 24 samples using 12 to 14 reverse fields. For selected samples, ARM was measured, hysteresis loops were plotted and Mössbauer spectra determined at 300 K, 77 K and 4.2 K. Pollen diagrams were constructed for Cores 3B and 1982A.

16.5 The magnetic mineralogy of the Rhode River catchment

In attempting to characterise the magnetic mineralogy of the potential sources in the catchment, particular attention has been paid to Nanjemoy bedrock exposures, to reference soil profiles (Fig. 16.4) in both Nanjemoy and Talbot formations and to the surface soils developed on all the catchment lithologies. This reflects the need to characterise especially material derived from cliff recession and from surface soil erosion.

As a result of the range of measurements summarised above, the following four distinctive magnetic components can be identified in the system:

(a) A *primary ferrimagnetic component* present in the unweathered Nanjemoy sands below the conspicuous zones of ferrugination found at all well

Table 16.1 Range of mineral magnetic parameters for potential sediment sources from the Rhode River catchment.

Potential sources	χ (10^{-6} m^{3}kg^{-1})	χ_{fd}/χ (%)	SIRM (10^{-6} Am^{2}kg^{-1})	SIRM/χ (kAm^{-1})	$\dfrac{IRM_{-100\,mT}}{SIRM}$
surface soils	0.1–0.8	6–15	500–6000	3–8	−0.30 to −0.95
weathered/illuviated substrates	<0.1	0–3	<700		+0.50 to −0.40
unweathered substrates	0.08–0.2	0–1	<150	<0.8	−0.10 to −0.85

drained sites. Very low or zero χ_{fd} values (Fig. 16.4) coupled with low $(B_0)_{CR}$ and rapid viscous loss of isothermal suggest that this material is predominantly multidomain.

(b) A *secondary antiferromagnetic component* especially significant in iron-enriched illuviated subsoil horizons (Fig. 16.4 & 6). In extreme cases up to 95% of the SIRM remains unsaturated in a reverse field of 0.3 T and $(B_0)_{CR}$ exceeds 0.2 T. This component could be either haematite, goethite, or both.

(c) A *secondary ferrimagnetic component* present in all non-gleyed surface soils. This gives rise to near-surface peaks in χ, χ_{fd}/χ, SIRM and SIRM/χ (Fig. 16.4). All the mineral magnetic character-istics indicate a stable single-domain fine viscous and superparamagnetic assemblage, typical of surface enhancement whether by fire or 'fermentation'. Magnetite, maghaemite or both may be represented.

(d) A high *paramagnetic component* abundant in all

but the eluviated A horizons of freely drained soils. Mössbauer spectra and hysteresis loop plots (Fig. 16.5) suggest that paramagnetic forms of iron are probably dominant in all phases of the system. In the Nanjemoy sections this gives rise to very low SIRM/χ values.

Components (a) and (d) dominate the magnetic characteristics of bulk samples in the extensive unweathered Nanjemoy cliff sections. At higher levels in the regolith components (b) and (d) dominate, although residual primary ferrimagnetic crystals are still present. In surface soils component (c) dominates the magnetic characteristics though, especially where soils are mixed by ploughing, all four components will be present. Measurements of particle size splits of these mixed soils shows that component (c) dominates in the finest clay fractions, component (b) and (d), the medium to fine sands, and component (a) the coarser sands (Fig. 16.6). Components (a), (b) and (c) may be regarded as conservative components on the time-scales of interest to us in the present account (cf. Ch.

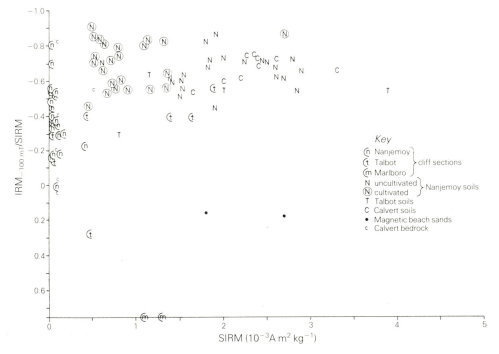

Figure 16.7 SIRM versus IRM$_{-100mT}$/SIRM for potential sediment source types within the Rhode River catchment. 90% of the cliff section and bedrock samples have SIRM values less than 500 × 10^{-6}Am^{2}kg^{-1}, though the IRM$_{-100mT}$/SIRM values range from −0.8 to +0.3 The values for the soil samples as a whole are much more varied in terms of SIRM (~500 to 4000 × 10^{-6}Am^{2}kg^{-1}) but the IRM$_{-100mT}$/SIRM range is narrower (−0.3 to −0.9). Note the contrast in SIRM between cultivated and uncultivated Nanjemoy soil samples.

8). The paramagnetic component includes readily soluble iron which is likely to change phase during erosion, transport and subsequent deposition. The change in SIRM/χ versus particle size between soils and channel sediments (Fig. 16.6) confirms visual impressions that during removal from the field surfaces and temporary deposition in the channel zone the soluble iron is dissociated from the sand fraction to become associated with finer clay and silt-sized particles. For this reason SIRM/χ has not been used here as a key parameter in source characterisation.

Table 16.1 summarises the magnetic characterisation of the potential sediment sources within the Rhode River catchment and identifies those used in subsequent comparisons.

Figure 16.7 plots SIRM versus IRM$_{-100\,mT}$/SIRM for all the catchment samples measured. Plotting these parameters confirms that there is very little overlap between the values for surface soils and those for the range of underlying parent materials. On a particle size basis (Fig. 16.8a) the sand and clay components of cultivated soils can also be clearly

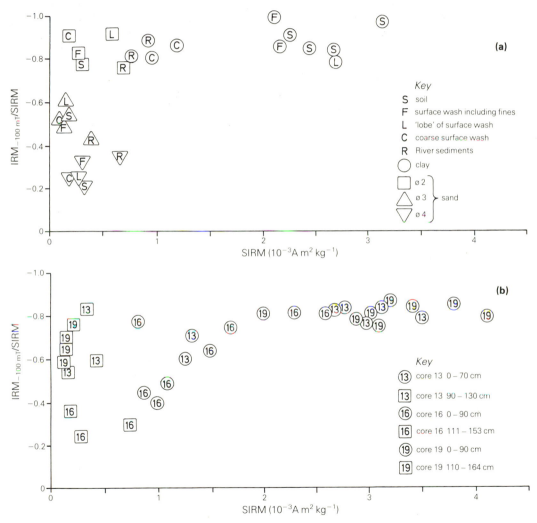

Figure 16.8   SIRM versus IRM$_{-100\,mT}$/SIRM for (a) sand and clay fractions from catchment surface samples and (b) for clay fractions from estuarine sediments. The catchment surface clay fractions and the clay fractions from the upper parts of each core are comparable with bulk surface soils (Fig. 16.7). The clays from the lower part of each core have low SIRMs and are more comparable with surface *sands* and with bulk parent material samples (see text).

differentiated, the former overlapping with the parent material envelope, the latter corresponding more closely with the surface soils.

16.6 Suspended sediment samples

Figure 16.9a summarises the range of mineral magnetic values obtained from the 88 suspended sediment samples measured to date. All three estuarine samples and all but one of the 85 catchment samples have the high SIRM and low $IRM_{-100\,mT}/$ SIRM values characteristic of surface soils. The extremely high SIRM of the estuarine sample from farthest down 'river' suggests some selective transport of relatively more ferrimagnetic fines into the open estuary. As would be expected from previously published data (Pierce & Dulong 1977), the bulk of the suspended stream sediments abstracted from the gauging stations fall within the envelope of values for *cultivated* rather than uncultivated soils in Figure 16.7.

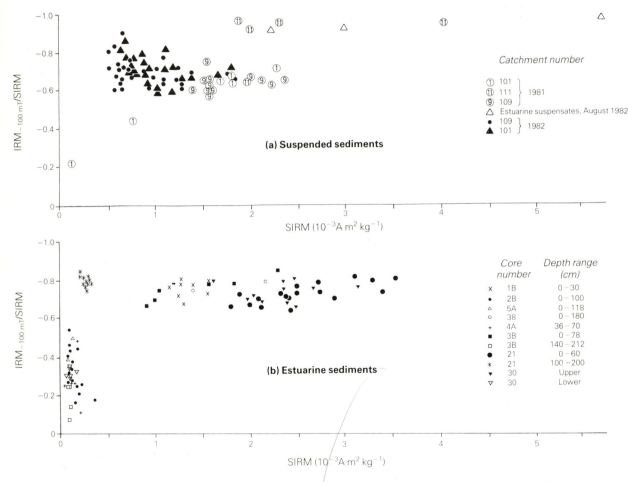

Figure 16.9 SIRM versus $IRM_{-100\,mT}/$SIRM for bulk suspended sediments and estuarine sediment. (a) The suspended sediments were taken during spring and early-summer floods by means of an automatic integrated water sampler installed by the weir at the outfall of each catchment. (b) The estuarine sediment samples all come from cores obtained by J. Donoghue. In the case of marginal cores (2B, 4A and 5A) the samples are not separately identified by depth range. The samples from the 'central' cores are grouped into upper and lower sets separated by the horizon of increasing χ and SIRM values (see text and Fig. 16.11).

16.7 Estuarine sediment cores: mineral magnetic characteristics

The sediment cores subsampled and measured so far fall into two types (Fig. 16.9b). Those close to eroding shoreline sites (e.g. 4A and 5A) are dominated by magnetic mineral assemblages comparable to those of the nearby eroding cliff sections (cf. Fig. 16.7) with low SIRM, SIRM/χ and variable IRM$_{-100\,mT}$/SIRM. At these sites, save for occasional horizons of higher SIRM there seems to have been little variation in sediment source for the past few centuries at least, and sedimentation appears to have been largely dominated by the coastal exposures. Sediments in these cores tend to be relatively coarse.

Sediment cores from the central parts of the estuary, whether close to the head of any of the tributary creeks or in the open waters of the middle and lower reaches show a different mineral magnetic pattern. Although whole core scans do not correlate in detail over more than a few metres, the vast majority of cores record an increase in volume susceptibility in the upper levels (Oldfield 1983a). Single-sample measurements make it possible to subdivide the central cores into an upper part characterised by high SIRM, SIRM/χ and χ_{fd} and a lower part characterised by low SIRM, SIRM/χ and χ_{fd}. Samples from the upper part of each core have bulk magnetic parameters which correspond with those of the surface soils of the catchment and with the contemporary suspended sediments. Lower samples correspond much more closely with material derived from much deeper in the regolith. Every 'central' core from the head of the estuary to its mouth records this shift (Fig. 16.9b) and in each case it is a dramatic and irreversible feature.

On a particle size specific basis (Figs 16.8b & 16.10) we see that above the shift clays dominate the bulk magnetic characteristics and have a range of values directly comparable both with surface soil and with the clay fraction therein. The small sand fractions above the shift may be comparable to either weathered or unweathered parent material as is the case in the sand fraction of contemporary soils. The magnetic mineralogy confirms a surface soil source for most of this material. Below the shift, both the sands and clays have low SIRM values suggesting that the latter are derived from comminution of 'primary' parent material rather than from magnetically enhanced surface soil.

In summary all the evidence obtained so far is interpretable in terms of within-catchment shifts of magnetic mineral assemblages and hence, of sediment sources. The central cores record a major shift in sediment source from bedrock and subsoil-derived material to surface soil-derived material. However, the characteristics of material derived from the open bay remain unresolved at present, as does the possibility that external sources are significant in the lower reaches of the 'river'. These topics form the main focus of continuing study.

16.8 Chronology and links with land-use change

Depending on the model of ^{210}Pb dating used (Oldfield & Appleby 1984) the date at which the shift in

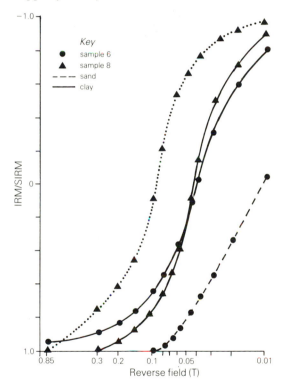

Figure 16.10 Coercivity curves for sand and clay size fractions from two sediment samples. Core 21 is located on Figure 16.1. The samples plotted here lie above 60 cm (see Fig. 16.9b). Values for the clay size samples compare with those for the soil clay fractions and the bulk surface soil samples. The sand size samples can have either very low or relatively high coercivities. They may thus be comparable with sands from either the weathered and illuviated or the unweathered sands from eroding cliff sections.

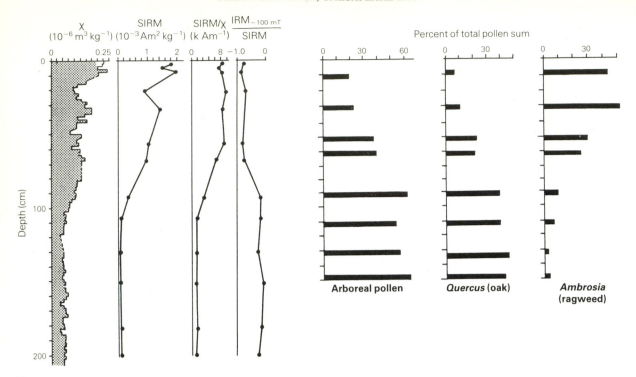

Figure 16.11 Rhode River estuarine core 3B (see Fig. 16.1). The horizon at which χ, SIRM, SIRM/χ and IRM$_{-100\,mT}$/SIRM all change from values typical of eroding parent material to those characteristic of soils corresponds with a major change in pollen assemblage summarised here in the total arboreal pollen, *Ambrosia* and *Quercus* values (cf. Brush *et al.* 1982).

sedimentation takes place ranges from *c.*1840 to 1880 AD in cores 3B and 30. Associated pollen analysis in cores 3B and 1982B links the sedimentation shift to a dramatic change in the *Quercus* (oak) to *Ambrosia* (ragweed) ratio, a development indicative of more extensive forest clearance and cultivation (Fig. 16.11). This pollen assemblage change is dated by Brush *et al.* (1982) to *c.*1780 along the north shores of the bay and to the first half of the 19th century further south. The pollen-analytical change is interpreted as a reflection of land-use intensification and a greater concentration on corn as against tobacco cultivation. Additional chronological insights are provided by the NRM record in Cores 1982A and B from the mouth of the estuary. In both cores, the directional measurements vary widely from sample to sample above 30 cm. Below this, the intensity of NRM and the repeatability of the directional measurements are closely correlated with SIRM and hence the changes in magnetic mineralogy associated with the shift in sedimentation found in these cores around 150 cm. Above the shift, NRM intensities are relatively high

and directional measurements are repeatable, below it, intensities are too close to instrumental noise level for the directional measurements to be reliable. Within the zone of repeatable measurements the main distinctive feature replicated in both cores is the increase in inclination around 100–120 cm (cf. Figs 5.5, 14.2 and 14.3 and Table 14.1). This would indicate a date of *c.*1850 AD for the 120 cm level, some 30 cm above the beginning of the main rise in χ and SIRM.

Thus from all the lines of evidence available so far it would appear that the shift in magnetic mineralogy from parent material to surface soil characteristics took place some time around 1800 AD. Since then, soil sources have dominated sedimentation through to the present day.

16.9 Summary and implications

As a result of the work completed so far, it is possible to attempt a brief appraisal of the value and limitations of the magnetic approach to the Rhode

196

River catchment study in the light of the methodological and substantive aims declared at the outset.

The procedures outlined in Figure 16.2 have provided a basis for within-catchment source characterisation right through from the initial extensive survey stage to the point of detailed diagnostic measurements using SIRM, $IRM_{-100\,mT}$/ SIRM and χ_{fd} for differentiation. With the advent of the portable pulse magnetiser (6.6.2), not available until the last stage of the work, it would now be possible to complete all these stages on site with remarkable speed and economy. Sediment characterisation can be quickly completed in the same way beginning with whole core scans, each done in a matter of minutes, and continuing through to single-sample measurements, again easily carried out on site. The magnetic methods are suitable for measuring material in all phases of the system, though input from the open bay still remains to be characterised. Source–sediment linkages have emerged and can be confirmed not only by largely independent measurements on bulk samples (e.g. SIRM and χ_{fd}) but also by measurements on particle size fractions. Moreover the record of palaeomagnetic secular variation in the recent sediments provides an additional source of evidence for sedimentation rates. At the end of this full range of magnetic measurements the samples remain available for palaeoecological study, radiometric dating or mineralogical analysis.

The substantive results point to a pattern of sediment flux within the system over the past 100–150 years dominated by the export of soil-derived particulates from the land surfaces of the catchment. Results from the Potomac Estuary (Oldfield & Maher 1984) confirm that this pattern also prevails in most of the cores taken between Washington DC and the mouth of the estuary. The high ARM, SIRM and χ_{fd}/χ values associated with this material in both the Rhode River and the Potomac are found in the clay-sized particles, and the magnetic parameters here can be confidently regarded as indicators of the loss of fines from the cultivated soils. In the Rhode River, the contemporary suspended sediment record confirms that this process is continuing and the magnetic traces from many of the cores suggest that the material now coming in is itself rather more depleted in the finest fractions than was the sediment deposited during the early stages after the shift to soil sources. The evidence points to selective loss of fines over a long period leading to their progressive depletion and eventual relative paucity in contemporary soils. This process has serious implications for soil moisture retention, resistance to future erosion and the maintenance of both soil structure and fertility. The mineral magnetic approach can thus highlight not only quantitative but also important qualitative aspects of erosion for both the present day and the past.

[17]
Prospects

The history of science, like the history of all human ideas, is a history of irresponsible dreams, of obstinacy and of error. But science is one of the very few human activities – perhaps the only one – in which errors are systematically criticized and fairly often, in time, corrected.

Karl R. Popper
Conjectures and refutations

Clearly most of the chapters from 7 onwards are concerned with relatively new applications of magnetic measurements to environmental systems, and in virtually every case, there is considerable scope for further development. Future prospects range from the further extension and refinement of established approaches (for example secular variation studies of lake sediments), through the transfer of newly developed techniques to different types of environment (for example magnetic tagging and tracing of beach sands), to the possibility of identifying a range of entirely new opportunities in fields such as medical and forensic science. This final chapter is concerned with summarising some of these prospects as they appear at the present time.

17.1 Palaeomagnetism of recent sediments

Chapters 13 and 14 deal with palaeomagnetic aspects of investigations of recent sediments and rocks. As the laboratory instrumentation and magnetic techniques used in these palaeomagnetic studies were largely developed in the 1950s and refined in the 1960s, progress in secular variation magnetostratigraphy is perhaps now more likely to arise from improved dating techniques, or through the development of a practical

sediment palaeointensity method, than through further advances in instrumentation dealing with palaeomagnetic directions.

A palaeointensity method which can yield full vector information for recent sediments is sorely needed. Geomagnetic interpretations of palaeomagnetic directional data are seriously hampered by lack of intensity control. Present attempts at extracting palaeointensities from recent sediments have not yet been demonstrated to be capable of yielding repeatable between-lake intensity *and* direction logs. Palaeointensity investigations which provide only intensity data without accompanying directional data, or which yield only information on selected sections of sediment core are likely to be viewed with suspicion.

Substantial efforts have been invested in tens of thousands of NRM and partial demagnetisation measurements and in extensive geomagnetic modelling computations in palaeomagnetic investigations of Holocene lake sediments, while little attention has been directed by palaeomagnetists to the topic of dating. This seems a surprising situation considering all the possible sources of error associated with lake sediment chronologies. There are two immediate prospects for improving the dating of the palaeo-limnomagnetic logs.

First, there is the recent development of the accelerator method of the detection of radiocarbon which allows minute samples to be dated. Sections of core, 20 cm long, need no longer be the prime source of radiocarbon dates. Small fractions of sediment, such as seeds or chemical extracts, can now be analysed and so provide a practical possibility for the reduction of natural contamination errors.

Secondly, annually laminated lake sediments are being discovered in increasing abundance. These potentially accurate chronological successions are being unearthed largely on account of the development of the freezer corer technique (Swain 1973, Saarnisto 1975, Huttunen & Merilainen 1978). O'Sullivan (1983) comprehensively reviews the distribution and formation of laminated lake sediments. He describes four main types of annual laminations, namely biogenic, calcareous, ferrogenic and clastic. Of these four, calcareous laminations possibly hold the greatest prospect for palaeomagnetic studies. Calcareous laminations are not suppressed by allochthonous clastic input, so the laminated sediment can hold a stable (post-) depositional remanence (Thompson & Kelts 1974). Deep flat-bottomed lakes with semi-permanent stratification and with a scarcity of oxygen in the bottom waters (which discourages mixing and inhibits benthic activity) are prime candidates for annual laminations (O'Sullivan 1983). Calcareous laminations occur in lakes with high internal carbonate loading and are formed in the summer months, during the time of high water temperatures, by precipitated $CaCO_3$. O'Sullivan (1983) concludes that the present scarcity of annually laminated lake sediments is more apparent than real, and that a search for laminated sediments, not just in lakes of the North Temperate zone, but in many parts of the Earth should prove fruitful. It has been found that certain laminations are much more clearly seen in material collected by freezing and that some biogenic laminations only become apparent after long exposure to air (Renberg 1982). This suggests that it will probably be advantageous in palaeomagnetic studies to collect freezer cores for detailed lamination counts, especially those of the surface sediments, and to take parallel piston cores for magnetic work.

Palaeomagnetic dating and correlation has relied on pattern matching by eye and subjective assessment of goodness of fit. Clark (pers. comm.) has adapted Gordon's (1973) sequence slotting method to allow it to accept data on the unit sphere. This development will permit formal dating of palaeomagnetic secular variation direction records by matchings with previously established secular variation time series using sequence slotting algorithms. Clark's approach will also allow a formal statistical testing of the discordance between the palaeomagnetic record and the master time series. Current developments of sequence slotting programmes may permit matching of several sequences simultaneously. Applications of such advances would include across lake correlations in multi-core mineral magnetic studies and between site stacking of secular variation records.

17.2 The mineral magnetic approach

The present account barely serves to introduce the full range of possible applications in soil and weathering studies. Systematic application of the techniques on appropriate spatial scales is required to evaluate the potential rôle of mineral magnetic parameters in soil characterisation. Initial results suggest that they may be valuable on all spatial scales from local survey to world-wide classification. The value of mineral magnetic parameters derives both from the close links apparent between magnetic mineral assemblages and soil forming processes, and from the relative ease with which these assemblages can be characterised and distinguished even in very low concentrations well beyond the reach of other techniques. Further exploration of the relationship between magnetic minerals and soil forming processes, using both natural and synthetic materials, will establish a rôle for magnetic measurements in, for example, both descriptive and experimental studies of rates of weathering and soil formation, of processes such as gleying, lessivage and ferrugination, and of the rôle of organic matter in the soil environment. There is also considerable scope for exploiting more fully the sensitivity of magnetic minerals to the effects of fire in the soil as well as in peats where there is every prospect of being able to use them to reconstruct site-specific fire histories. Although some indication of the potential of mineral magnetism in archaeological studies has begun to emerge not only from the studies outlined in Section 8.10, but also from recent studies of obsidian (McDougall *et al.* 1983) and iron ore artifacts (Pires-Ferreira 1976), the potential in palaeosol studies is virtually unevaluated. Mineral magnetic aspects have also, for the most part, been neglected in published magnetic studies of loess deposits.

Within the broad area of particulate flux between terrestrial and fresh-water systems the scope for extending the mineral magnetic approach on a much wider range of temporal and spatial scales is very attractive. Most of the sediment-based studies reported so far are concerned with timescales ranging from 10^2 to 10^4 years and the scope for capitalising on the conservation of magnetic properties, in order to link these to contemporary process studies through integrated long-term lake-watershed ecosystem-based projects, has not been fully realised. Nor have magnetic measurements been applied to the reconstruction of environmental variables at the level of detailed temporal resolution made possible by laminated sediments (cf. Rummery 1983). At the other extreme, mineral magnetic studies of longer-term deposition sequences in low latitude lakes will offer new insights into climatic variation and its bearing on erosion and sedimentation. In fact there are few aspects of Quaternary studies that cannot benefit from mineral magnetic measurements, and where, as in lake or in rapidly accumulating marine sediments, there is often an opportunity to obtain geochronological information from the palaeomagnetic record at the same time, the present repertoire of techniques provides a valuable new methodology.

The approach to sediment source identification and to tagging and tracing experiments, though initially developed in small river basins on carefully selected lithologies, is capable of extension not only to larger and more complex contexts (cf. Oldfield & Maher 1984) but to new environments – for example in studies of coastal sediment movement, erosion, accretion and shoreline protection. Moreover, the link between magnetic measurements and erosive processes, which was one of the first areas of possible application to emerge, is now seen to be potentially of qualitative as well as quantitative value in view of the diagnostic nature of the parameters in terms of both source type and particle size.

Mineral magnetic characterisation has not hitherto been one of the recognised research tools in sedimentology or petrology, nor has it been used as a logging technique in hard-rock exploration. Given the range of studies now completed on contemporary and recent sediments and on soft rocks, the time is ripe to extend the techniques to older materials. Results reported from Pleistocene marine sediments point to the sensitivity of the magnetic mineral assemblages to diagenetic changes. Moreover, initial trials using rapid magnetic susceptibility probe and loop readings on core samples from a deep borehole in the Trias (Robinson, pers. comm.) confirm their probable geotechnical value.

Chapter 12 outlines the challenge of reconstructing the flux of magnetic minerals to the oceans and confirms its palaeoclimatic significance. Like so many aspects of environmental science, it calls on the researcher to unite insights from contemporary process monitoring and long-term sediment-based reconstruction, and it requires a comprehensive appraisal of the linkage between environmental systems, in this case atmospheric and oceanographic. The scale of study implied clearly requires both the continued international co-ordination and collaboration that has been characteristic of much research in this field in recent years, and wider recognition of the virtues of mineral magnetic characterisation in terms of its intrinsic value and compatibility with other lines of study.

One of the threads running through several of the later chapters is the apparent link between mineral magnetic properties and particulate pollution as a result of fossil-fuel combustion and other industrial processes such as smelting and the manufacture of iron and steel. Although the main emphasis so far has been on the use of magnetic measurements in monitoring heavy metal deposition this is not the only possible application. The association of long-distance sulphur dispersal with the particulate component of atmospheric aerosols to which the soluble phases become temporarily adsorbed may indicate a rôle for magnetic measurements in studies of acid rain sources and trajectories. Equally, the recent interest in the use of fly-ash as a source of magnetite (Chaddha & Seehra 1982) makes magnetic characterisation especially important since we may expect a quantifiable relationship between susceptibility or SIRM and crystalline iron concentration in any given fly-ash. At present, the studies completed fall somewhat uneasily between reinforcing the view that magnetic/heavy metal ratios are sometimes sufficiently constant to encourage the use of magnetic measurements as a surrogate monitoring technique, and suggesting that variations in these same ratios and in mineral magnetic parameters indicative of changes in magnetic assemblages will be of value in identifying aerosol and sediment sources. To a large extent these apparently contradictory indications must be a function of spatial scale and hence of proximity to distinctive sources.

This points the way to one urgent area of future study.

Just as the ease with which magnetic minerals can be characterised and identified makes them attractive as tracers in aquatic systems, it also makes them potentially valuable in atmospheric transport and deposition experiments. This is especially so in contexts where because of the complexities of surface roughness, neither mathematical models nor empirical observations have been completely successful. For example, experiments using magnetic powders can complement and extend into much finer particle size ranges, e.g. the type of study carried out by Chamberlain (1966) using radioactively tagged *Lycopodium* spores. In addition, some empirical appraisal of the relative effectiveness of trees as aerosol filters should be possible using magnetic measurements.

In the realm of historical reconstruction, the possible use of magnetic mineral assemblages in reconstructing atmospheric deposition has barely been extended beyond the beginning of the Industrial Revolution. On longer timescales using both ombrotrophic peat and ice core samples, one may expect to see evidence emerge of significance in the reconstruction of past atmospheric circulation patterns, local and global dust-veil variations and sediment supply to the world's oceans. All these will have direct or indirect palaeoclimatic implications.

Looking beyond the scope of the present themes, further applications in medical and forensic science are not difficult to envisage. Identification of inhaled particulate types and sources should be possible from lung tissue measurements as well as characterisation of work environments in terms of aerosol loadings and types.

In relation to all the above, one of the notable limitations of the mineral magnetic approach, as outlined in the foregoing chapters, is its failure to provide a basis for expressing results in quantitative terms with regard to mineral composition and domain size assemblages.

One quantitative approach currently being pursued is the development of a procedure which compares mineral magnetic measurements on natural samples with equivalent measurements on synthetic specimens. In this approach, measurements of susceptibility and of isothermal remanences and coercivities are matched with synthetic sample results so that the magnetic characteristics of natural materials can be described as being equivalent to the properties of various concentrations of synthetic

minerals. One method of applying such a procedure is to determine the magnetic properties of particular combinations of synthetic minerals by consulting tables of results from a great range of synthetic minerals and well characterised mixtures. Such a task can be greatly simplified and undertaken in a relatively straightforward manner through a minimisation algorithm.

The procedure currently under development involves a simplex minimisation technique (Nelder & Mead 1965). In the procedure the magnetic properties of synthetic minerals are expressed as continuous functions. For example, isothermal properties are represented by smooth bicubic spline surfaces (Hayes & Halliday 1974) which relate back fields and remanences. By using this type of functional approach possible mathematical problems associated with local minima can be largely avoided and rapid searches of the mineral magnetic parameter space can be performed. The simplex minimisation procedure is arranged to hunt for a best fit to the mineral magnetic measurements by varying the concentrations and grain sizes of the constituent minerals and the concentration of a (super)paramagnetic component. While it is recognised that such a formalised approach cannot cover all possible natural magnetic mixtures, because the range of natural magnetic minerals and their properties is just so large and the range of magnetic mixtures is so vast and complex, it does appear that the simplex minimisation formulation is able to handle the majority of environmental samples that we have encountered.

Problems that arose during the development of the procedure have highlighted two particular areas in which further research is needed. First, many natural samples appear to contain intermediate coercivity minerals, with remanence coercivities of around 0.1 T, which are difficult to explain. Secondly, distinguishing between paramagnetic and superparamagnetic components has not proved to be simple and additional measurements appear desirable to aid the distinction. The uses of frequency-dependent or quadrature susceptibility, low temperature susceptibility variations and saturation magnetisation measurements are being investigated in connection with this second problem, while further high-field measurements on synthetic and well characterised imperfect antiferromagnets are being carried out in relation to the first problem.

A further development being pursued is a method of calculating the errors associated with concentration and grain-size determinations. Such estimates of the ranges of likely interpretations of mineral magnetic measurements appear to be very desirable. Divergences of opinion in the interpretation of mineral magnetic data between various research groups, particularly on the importance of superparamagnetic grains, can lead to very different environmental models being proposed. Such disparities should be clarified by a quantitative approach which is capable of handling magnetic mixtures and of assessing possible plausible ranges. An important test of the minimisation formulation will be to compare mineral magnetic estimates of iron oxide concentrations with chemical, X-ray and Mössbauer determinations.

Not only is progress urgently needed in the direction of quantifying mineral magnetic properties in terms of magnetic domain states and mineralogy, but, given the size of data sets readily obtainable by mineral magnetic measurements, the time is now ripe for a much more rigorous statistical approach. This will be increasingly important in the more 'open' atmospheric (cf. Hunt 1986) and marine systems where linkages are more extended and complex and much less direct than in the more materially bounded soil and watershed systems considered in most of the studies completed so far.

Measurement of both major and minor hysteresis loop characteristics coupled with initial and anhysteretic magnetisation properties, using a slow field cycling technique combining a bipolar supplied electromagnet and a new software controlled electronic integrator, has allowed Jiles and Atherton (1984) to test a simple theory of hysteresis based on mean field approximation. Trial experiments (Jiles pers. comm.) on a weakly magnetic granite cylinder and red sandstone sample suggest that the equipment, with some small modifications, could be used to analyse the great majority of 'environmental' samples. More complete characterisation of magnetic properties using such equipment would yield many advantages, over current methods, especially in the quantification of the mineral magnets approach.

Perhaps a final and more general qualification of the mineral magnetic studies summarised in the foregoing chapters is in order. The approach has been almost exclusively empirical and observational rather than theoretical and experimental. This reflects in large part a reluctance to make heavy investments of time and resources in more closely controlled studies until apparently significant (though often initially only circumstantially supported) empirical relationships and patterns can be demonstrated. Such an approach carries with it inevitable dangers – of exaggerating the potential significance of possibly coincidental relationships, of failing to perceive and give adequate consideration to exceptions, and of avoiding the critical evaluation of assumptions for as long as favoured explanatory paradigms can be sustained. In consequence, several of the major assumptions, which have proved practical and consistent with all the reported observations so far, now require closer scrutiny under conditions which will allow them to be more securely defined and qualified. Specifically, there is a need to establish the quantitative significance of bacterial magnetite in depositional environments, to specify more fully the conditions under which authigenic and diagenetic processes are magnetically significant and to evaluate questions of magnetic mineral survival in peats and sediments where extremely acid and/or reducing conditions prevail.

Glossary of magnetic terms

anhysteretic remanent magnetisation (ARM) The remanence produced during the smooth decay of a strong alternating **field** in the presence of a weak steady **field**.

anisotropy of susceptibility The variation of magnetic **susceptibility** with direction. Only found in recent sediments if a fabric is present. Dominated by the shape of **ferrimagnetic** grains, e.g. alignment of elongated magnetite grains in tills.

antiferromagnetism A type of magnetic behaviour arising from crystals having lattices in which adjacent atoms have antiparallel spins. The **susceptibility** is very low and the **remanence** zero. Exhibited by MnO and FeO, Antiferromagnetic material becomes **paramagnetic** above the Néel temperature.

blocking temperature The critical temperature at which thermal energy is just sufficient to cause spontaneous reversals of **magnetic moment**. On cooling, a magnetic grain acquires a spontaneous magnetisation at the **Curie temperature**. However, at such elevated temperatures, thermal energy is generally able to reorientate the **magnetic moment** and the moment is not stabilised until the grain is cooled below the blocking temperature. **Superparamagnetic** grains, on account of their small volumes, have blocking temperatures below room temperature.

blocking volume The threshold volume which separates **superparamagnetic** and **stable single-domain** behaviour. Grains smaller than the blocking volume have lower magnetic energies than thermal energies and their **magnetisation** consequently fluctuates in the same way as in a **paramagnetic** gas. Chemical growth through the blocking volume locks in the average **magnetisation** of an assemblage of grains as a stable **remanence**.

canted antiferromagnetism An imperfect form of **antiferromagnetism** in which the spins are not quite antiparallel. Can result in weak but very stable **remanences**, e.g. haematite.

chemical remanent magnetisation (CRM) The remanence acquired when a magnetic material is chemically formed or crystallised in a magnetic field.

coercivity or coercive force $(B_0)_C$ The reverse field required to reduce the **magnetisation** to zero from **saturation**.

coercivity of remanence $(B_0)_{CR}$ The reverse **field** required to reduce the **remanent magnetisation** to zero after **saturation**.

Curie temperature or **Curie point** (T_C) The temperature above which a **ferromagnetic** or **ferrimagnetic** substance becomes **paramagnetic**.

declination (D) The angle between geographical (or true) north and a magnetic **remanence** or magnetic **field**. A westerly variation is recorded as a positive declination, an easterly variation as negative.

demagnetisation The process of depriving a specimen of its **magnetisation**.

(a) *Alternating field*. Demagnetisation is achieved by subjecting a specimen to an alternating magnetic **field** which smoothly decreases to zero.

(b) *Thermal*. Demagnetisation is achieved by cooling a specimen from above its **Curie temperature** in zero magnetic **field**.

detrital remanent magnetisation (DRM) The **remanence** found in sediments in which the magnetic particles have inherited a **remanence** from the material from which they have been eroded and have tended to be aligned in the **geomagnetic field** as the sediment formed.

diamagnetism (cross-magnetism) A phenomenon occurring in all substances, due to a change in orbital motion of electrons about the nucleus in an applied field. Causes a rod of material, when suspended between the poles of a magnet, to arrange itself across the line joining the poles. Results in a weak negative **susceptibility** which is often masked by the much greater effect of **paramagnetism** or **ferromagnetism**. Exhibited by quartz, feldspars, calcite.

dipole field   The magnetic **field** pattern associated with, for example, a small bar magnet. On the surface of the Earth (or any sphere of radius a) surrounding a **magnetic dipole** (of moment m):

(a)   the total **field intensity** (F) is related to the distance ($\pi/2 - \lambda$) from the **pole position** by the equation $F = m\,(1 + 3\sin^2\lambda)^{\frac{1}{2}}/a^3$;

(b)   the **inclination** (I) of the field is given by the equation $\tan I = 2\tan\lambda$;

(c)   the **declination** (D) of the **field** is $0°$ by definition.

domain   A region of parallel atomic **magnetic moments** in a crystal, which behaves as a unit during change in **magnetisation**. Domains form because they minimise the potential energy associated with a **magnetised** sample.

excursion   A change in direction of the **geomagnetic field** in which the virtual **geomagnetic pole** migrates through more than $45°$ for some 10^2 to 10^4 years and then returns to its original **polarity**. Such changes have not yet been demonstrated to be worldwide phenomena.

ferrimagnetism   A phenomenon similar to **ferromagnetism** but where the exchange interactions favour both parallel and antiparallel alignment of the **magnetic moments** of groups of atoms. A net **magnetisation** is observed. Exhibited by spinel-structured minerals, e.g. magnetite, maghaemite.

ferromagnetism   A phenomenon of some crystalline substances due to unbalanced electron spins combined with an ionic spacing such that very large forces, called exchange interactions, cause coupling and alignment of all the individual **magnetic moments** of millions of atoms to give highly magnetic domains. Results in positive and relatively large **susceptibilities** and large **remanence** and **hysteresis**. Exhibited by iron and some other metals.

field (B_0)   The magnetic field of force resulting from the presence of either a permanent magnet or an electric current in the neighbourhood. The free space **induction**, measured in tesla. $B_0 = \mu_0 H$.

frequency-dependent susceptibility (χ_{fd})   The time delay between the application of a **field** and the **magnetisation** response causes **susceptibility** to vary with frequency. At vanishingly low frequencies (in static measurements) **magnetisation** and **field** remain in phase. At higher frequencies relaxation phenomena result in a decrease in **susceptibility** and in associated energy losses which appear as heat in the sample.

geocentric axial dipole field   The **field** due to a **dipole** situated at the centre of the Earth and aligned along the Earth's axis of rotation. This is a good approximation to the time average of the **geomagnetic field** over a period of about 10^6 years.

geomagnetic field   The magnetic **field** associated with the Earth. A good approximation of the present geomagnetic field is that of a **dipole**, situated at the centre of the Earth, inclined at $11\frac{1}{2}°$ to the Earth's rotation axis.

hysteresis   A physical phenomenon where the **magnetisation** produced by an applied **field** lags behind the **field**, with a consequent energy loss.

inclination (I)   The angle between the horizontal plane and a **magnetic remanence** or magnetic **field**. A downwards dip of the north-seeking pole is recorded as a positive inclination and an upwards orientation as negative.

induction (B)   The induction of magnetism in a medium by an external magnetic **field**.

intensity   The magnitude of a **magnetic remanence** (M) or the magnitude of a magnetic **field** (B).

intensity or excitation (H)   The strength of a **magnetic field** measured in ampere per metre. Defined through Ampere's law. Rarely used alone in the SI system.

isothermal remanent magnetisation (IRM)   The **remanence** grown by the application and subsequent removal of a magnetic **field**. (The ordinary sense of magnetisation.)

magnetic moment (m)   The couple exerted on a magnet placed at right angles to a uniform magnetic **field** with unit flux density. Measured in ampere metre^2.

magnetisation (M)   The **magnetic moment** per unit volume of a magnetised body. Measured in ampere per metre. Generally made up of two components:

(a)   the **remanent magnetisation** (which remains after removal of the external **field**),

(b)   the **induced magnetisation** (which disappears after removal of the **field**).

multidomain   A magnetic grain which contains more than one **domain**. Large grains divide into more than one **domain** in order to reduce their total energy. Multidomain grains have much lower **remanences** and **coercivities** than **stable single-domain grains**.

natural remanent magnetisation (NRM)   The residual **magnetisation** possessed by rocks and other *in situ* materials.

Néel temperature   The critical temperature below which the atomic moments of an **antiferromagnet** or **ferrimagnet** are arranged alternately parallel and antiparallel.

paramagnetism (parallel-magnetism)   A phenomenon occurring in substances with unpaired electrons. Causes a rod of the substance in a magnetic **field** to arrange itself parallel to the magnetic **field**. A small positive **susceptibility** arises from the alignment of the **magnetic moment** of individual atoms of the substance in an applied **field**. Displayed for example by substances with rare earth or transition series members, e.g. clays, pyroxenes, amphiboles.

permeability of a vacuum (free space) (μ_0)   The ratio of the magnetic flux density, B_0, in a vacuum to the external **field intensity**, H. i.e. $\mu_0 = B_0/H$. The exact value of μ_0 is $4\pi \times 10^{-7}$ henry per metre.

polarity reversal   A **geomagnetic field** change in which both the **declination** and **inclination** move through $180°$ and then remain stable. Polarity reversals are worldwide phenomena.

pole position   The point of intersection of the extension of a dipole axis and the Earth's surface. Two pole positions result, named the north and south poles.

    (a) **geomagnetic pole position** Associated with the Earth's present **geocentric dipole field**. The north geomagnetic pole lies in north-west Greenland at $78\frac{1}{2}°$N, $70°$W.

remanent magnetisation or remanence   The **magnetisation** remaining in the absence of an external magnetic **field**.

saturation isothermal remanent magnetisation (M_{RS} or σ_{RS} or SIRM)   The maximum **remanence** attainable. It is produced by the application and removal of a powerful magnetic **field**. We use the symbol SIRM for remanent magnetisations produced in the strongest readily available laboratory field (often 1 tesla). We recognise that our SIRM will fall short of the true saturation remanence (σ_{RS}) of the physicist when high **coercivity**, **antiferromagnetic** minerals are incorporated in our samples.

saturation magnetisation (M_S or σ_S)   The strongest possible **magnetisation** which can be produced in a specimen by applying a powerful **field**.

secular variation   **Geomagnetic field** fluctuations of smaller angles than are involved in excursions but of a similar timescale of $10^1–10^4$ years. Such **field** changes are recognisable over regions of 'continental extent'.

specific magnetisation (σ)   The **magnetic moment** per unit mass of a magnetised body. Measured in A m^2 kg^{-1}.

specific susceptibility (χ)   Magnetic **susceptibility** expressed in terms of unit mass. We measure specific susceptibility in units of m^3 kg^{-1}.

spinel   An oxide mineral with cubic symmetry with the general formula $R^{2+}O.R_2^{3+}O_3$ where R^{2+} is a divalent metal and R^{3+} a trivalent metal. Magnetite ($Fe^{2+}O.Fe_2^{3+}O_3$) forms a continuous series with ulvospinel ($Fe_2^{2+}Ti^{4+}O_4$).

stable single-domain grain   A magnetic grain with just one **domain** which at normal temperatures is capable of retaining its **remanent magnetisation** unaltered for millions of years.

superparamagnetism   The phenomenon of the rapid decay in **remanence** of magnetic grains. Superparamagnetic behaviour occurs in **ferro-** and **ferrimagnetic** grains below a critical size. For spherical magnetite grains the critical diameter is about 10^{-8} m. Superparamagnetic grains are also characterised by a noticeably high **susceptibility** and by zero **coercivity**.

susceptibility (κ)   A measure of the degree to which a substance can be magnetised. The ratio of the **magnetisation** (M) produced in a substance to the **intensity** of the magnetic **field** (H) to which it is subject. $\kappa = M/H$. With this convention, susceptibility defined using unit volume is dimensionless.

thermoremanent magnetisation (TRM)   The **remanence** a sample gains by cooling from above the **Curie temperature** in a magnetic field.

viscosity   The change of **magnetisation** with time. In natural materials viscous **magnetisation** changes are generally produced by thermal effects which cause changes in **domain** wall positions or **domain** alignment or else cause reversal in **magnetisation** direction of **single-domain grains** close to the critical **superparamagnetic** grain-size boundary. Viscous **magnetisation** and **remanence** changes are normally proportional to the logarithm of time and to the **intensity** of the ambient **field**.

viscous remanent magnetisation (VRM)   The **remanence** produced by a weak magnetic **field** applied over a long period of time.

References

Aaby, B., J. Jacobsen and O. S. Jacobsen 1978. Pb-210 dating and lead deposition in the ombrotrophic peat bog Draved Mose, Denmark. *Danm. Geol. Unders. Arbog.* 45–68.

Abrahamsen, N. and K. L. Knudsen 1979. Indication of a geomagnetic low-inclination excursion in supposed middle Weichselian interstadial marine clay at Rubjerg, Denmark. *Phys. Earth Planet. Interiors* 18, 238–46.

Abrahamsen, N. and P. W. Readman 1980. Geomagnetic variations recorded in older (>2300 BP) and younger Yoldia clay (~14 000 BP) at Nore Lyngby, Denmark. *Geophys. J. R. Astr. Soc.* 62, 329–44.

Abrahamsen, N., G. Schoenharting and M. Heinesen 1984. Palaeomagnetism of the Vestmanna core and magnetic age and evolution of the Faeroe Islands. In *The deep drilling project 1980–1981 in the Faeroe Islands*. 93–108 Ann. Soc: et. Sci. Faeorensis Suppl. IX 159. O. Berthelsen, A. Noe-Nygaard and J. Rasmussen (eds).

Ade-Hall, J. M. 1964. The magnetic properties of some submarine oceanic lavas. *Geophys. J.* 9, 85–92.

Ade-Hall, J. M., H. C. Palmer and T. P. Hubbard 1971. The magnetic and opaque petrological response of basalts to regional hydrothermal alternation. *Geophys. J. R. Astr. Soc.* 24, 137–74.

Aitken, M. J. 1974. *Physics and archaeology*. Oxford: Clarendon Press.

Aitken, M. J. 1978. Archaeological involvements of Physics. *Physics Reports* (Section C of *Physics Letters*) 40C, 5, 277–351.

Aksenov, V. V. and S. S. Lapin 1967. Theory and equipment of the induction method of measuring the magnetic susceptibility of rock specimens. *Isv. Earth Physics* 10, 106–12. English translation 1968. *Physics of the Solid Earth*, 697–701.

Allan, T. D. 1969. A review of marine geomagnetism. *Earth Sci. Rev.* 5, 217–54.

Anderson, T. W., R. J. Richardson and J. H. Foster 1976. *Late Quaternary palaeomagnetic stratigraphy from east-central Lake Ontario*. Rep. of Activities, Part C: Geol. Surv. Can. Paper 76–1C.

Appleby, P. G., J. A. Dearing and F. Oldfield 1985. Magnetic studies of erosion in a Scottish lake-catchment. I. Core chronology and correlation. *Limnol. Oceanogr.* 30, 144–153.

Arkell, B. 1984. *Magnetic tracing of river bedload*. Unpublished PhD thesis. University of Liverpool.

Arkell, B., G. Leeks, M. Newson and F. Oldfield 1982.

Trapping and tracing: some recent observations of supply and transport of coarse sediment from upland Wales. *Spec. Publ. Int. Assoc. Sediment.* 6, 117–29.

As, J. A. 1967. The a.c. demagnetization technique, in *Methods in palaeomagnetism*, D. W. Collinson, K. M. Creer and S. K. Runcorn (eds). Amsterdam: Elsevier.

Bailey, M. E. and D. J. Dunlop 1983. Alternating field characteristics of pseudo-single-domain (2–14μm) and multidomain magnetite. *Earth Planet. Sci. Letters* 63, 335–52.

Banerjee, S. K., S. P. Lund and S. Levi 1979. Geomagnetic record in Minnesota lake sediments – absence of the Gothenburg and Erieu excursions. *Geology* 7, 588–91.

Banerjee, S. K., J. King and J. Marvin 1981. A rapid method for magnetic granulometry with applications to environmental studies. *Geophys. Res. Letters* 8, 333–6.

Banerjee, S. K. 1981. Experimental methods of rock magnetism and palaeomagnetism. *Adv. Geophys.* 23, 25–99.

Barber, K. E. 1976. History of vegetation. In S. B. Chapman (ed.). *Methods in plant ecology* 5–83. Oxford: Blackwell.

Barraclough, D. R. 1974. Spherical harmonic analyses of the geomagnetic field for eight epochs between 1600 and 1910. *Geophys. J. R. Astr. Soc.* 36, 497–513.

Barton, C. E. 1978. *Magnetic studies of some Australian lake sediments*. Unpubl. PhD thesis, Australian National University, Canberra.

Barton, C. E. and M. W. McElhinny 1979. Detrital remanent magnetisation in five slowly redeposited long cores of sediment. *Geophys. Res. Letters* 6, 229–32.

Barton, C. E. and M. W. McElhinny 1981. A 10 000 yr geomagnetic secular variation record from three Australian maars. *Geophys. J. R. Astr. Soc.* 67, 465–85.

Bates, L. F. 1961. *Modern magnetism*, 4th edn. London: Cambridge University Press.

Battarbee, R. W. 1978. Observations on the recent history of Lough Neagh and its drainage basin. *Phil. Trans. R. Soc. Lond.* B 281, 303–45.

Bauer, L. A. 1896. On the secular motion of a free magnetic needle, II. *Phys. Rev.* 3, 34–48.

Baule, G. M. and R. McFee 1963. Detection of the magnetic field of the heart. *Am. Heart J.* 66, 95–6.

Bean, C. P. and J. D. Livingston 1959. Superparamagnetism. *J. Appl. Phys.* 30, 1205–95.

Beckwith, P. R., J. B. Ellis, D. M. Revitt and F. Oldfield

1984. Identification of Pollution Sources in Urban Drainage Systems Using Magnetic Methods. In *Urban Storm Drainage*, P. Balmer (ed.), Göteborg.

Beckwith, P. R., J. B. Ellis, D. M. Revitt and F. Oldfield 1986. Heavy metal and magnetic relationships for urban source sediments. *Phys. Earth Plant Int.* **42**, 67–75.

Bengtsson, L. and T. Persson 1978. Sediment changes in a lake used for sewage reception. *Polskie Archivum Hydrobiologii* **25**, 17–33.

Bhandari, N., J. R. Arnold and D. Parkin 1968. Cosmic dust in the stratosphere. *J. Geophys. Res.* **73**, 1837–45.

Bhathal, R. S. and F. D. Stacey 1969. Frequency independence of low-field susceptibility of rocks. *J. Geophys. Res.* **74**, 2025–7.

Bjorck, S., J. A. Dearing and A. Jonsson 1982. Magnetic susceptibility of late Weichselian deposits in S. E. Sweden. *Boreas* **11**, 99–111.

Blackett, P. M. S. 1952. A negative experiment relating to magnetism and the Earth's rotation. *Phil. Trans. R. Soc. London* **A245**, 309–70.

Blakemore, R. P. 1975. Magnetotactic bacteria. *Science* **190**, 377–9.

Blakemore, R. P., and R. B. Frankel 1981. Magnetic navigation in bacteria, *Scient. Am.* **245**, 6, 58–65.

Blakemore, R. P. 1900. Magnetotactic bacteria. *Ann. Rev. Microbiol.* **36**, 217–38.

Bloch, F. 1930. Zur theorie des ferromagnetismus, *Zeit. Phys.* **61**, 206–19.

Bloemendal, J. 1980. Paleoenvironmental implications of the magnetic characteristics of sediments from deep sea drilling project site 514 southeast Argentine basin. *Initial Reports of the D.S.D.P. Washington* **LXXI**, 1097–8.

Bloemendal, J., F. Oldfield and R. Thompson 1979. Magnetic measurements used to assess sediment influx at Llyn Goddionduon. *Nature* **280**, 5717, 50–3.

Bloemendal, J. 1982. *The quantification of rates of total sediment influx to Llyn Goddionduon, Gwynedd.* Unpubl. PhD thesis. University of Liverpool.

Bonhommet, N. and J. Babkine 1967. Sur la presence d'aimantations inversees dans la chaines des puys. *C. R. Acad. Sci. Paris* **264**, 92–4.

Borman, F. H. and G. E. Likens 1969. The watershed–ecosystem concept and studies of nutrient cycles. *The ecosystem concept in natural resource management.* G. M. Van Dyne (ed.).

Bradshaw, R. H. W. and R. Thompson 1985. The use of magnetic measurements to investigate the mineralogy of some Icelandic lake sediments and to study catchment processes. *Boreas.* **14**, 203–15.

Braginsky, S. I. 1963. Structure of the F layer and reasons for convection in the Earth's core. *Dokl. Akad. Nauk SSR* **149**, 1311–14.

Brownlee, D. E. Extraterrestrial Components. In *The sea*, vol. 7. *The oceanic lithosphere*, C. Emiliani (ed.). New York: Wiley.

Bruckshaw, J. M. and E. I. Robertson 1948. The measure-ment of the magnetic properties of rocks. *J. Sci. Instrum.*, **25**, 444–6.

Bruckshaw, J. M., and B. S. Rao 1950. Magnetic hysteresis of igneous rocks. *Proc. Phys. Soc. London* **63**, 931–8.

Brush, G. S., C. Leak and J. Smith 1980. The natural forests of Maryland: an explanation of the vegetation map of Maryland. *Ecol. Monogr.* **50**, 77–92.

Brush, G. L., E. A. Martin, R. S. De Fries and C. A. Rice 1982. Comparison of ^{210}Pb and pollen methods for determining rates of estuarine sediment accumulation. *Quat. Res.* **18**, 196–217.

Bucha, V. 1973. The continuous pattern of variations of the geomagnetic field in the Quaternary and their causes. *Stud. Geophys. Geod.* **17**, 218–31.

Bullard, E. C. 1948. The secular change in the Earth's magnetic field. *Mon. Not. Roy. Astr. Soc. Geophys. Suppl.* **5**, 248–57.

Burns, R. G. and V. M. Burns 1981. Authigenic oxides. In *The sea* vol. 7. *The oceanic lithosphere*, C. Emiliani (ed.), 875–914.

Butler, R. F. and S. K. Banerjee 1975. Theoretical single-domain grain size range in magnetite and titano-magnetite, *J. Geophys. Res.* **80**, 4049–58.

Campbell, B. L., R. J. Loughran and G. L. Elliott 1982. Caesium–137 as an indicator of geomorphic processes in a drainage basin system. *Aust. Geog.* **20**, 49–64.

Carling, P. A. and N. A. Reader 1982. Structure, composition and bulk properties of upland stream gravels. *Earth Surfaces Processes and Landforms* **7**, 349–65.

Chaddha, G. and M. S. Seehra 1983. Magnetic components and particle size distribution of coal fly ash. *J. Phys. D: Appl. Phys.* **16**, 1767–76.

Chamberlain, A. C. 1966. Transport of *Lycopodium* spores and other small particles to rough surfaces. *Proc. R. Soc. Lond. A* **296**, 45–70.

Channel, J. E. T., J. G. Ogg and W. Lowrie 1982. Geomagnetic polarity in the early Cretaceous and Jurassic. *Phil. Trans. R. Soc. Lond. A* **306**, 137–46.

Chester, R. and L. R. Johnson 1971. Atmospheric dusts collected off the Atlantic coasts of north Africa and the Iberian Peninsula. *Marine Geol.* **11**, 251–60.

Chester, R. 1972. Geological, geochemical and environmental implications of the marine dust veil. In *The changing chemistry of the oceans* **291**, 20th Nobel Symposium Volume. D. Dyssen and P. Jaguer (eds).

Chester, R., E. J. Sharples, G. Sanders, F. Oldfield and A. C. Saydam 1984. The distribution of natural and non-crustal ferrimagnetic minerals in soil-sized particulates from the Mediterranean atmosphere. *Water, Air and Soil Pollut.* **23**, 25–35.

Chevallier, R. and S. Mattieu 1943. Propietes magnetique des poudres d'hematites – influence des dimensions des grains. *Ann. Phys.* **18**, 258–88.

Chikazumi, S. 1964. *Physics of magnetism.* New York: Wiley.

Chorlton, J. 1981. *Source identification of suspended sediment by magnetic measurements.* Unpubl. BSc dissertation. Department of Geography, University of Liverpool.

Cisowski, S. W. 1980. The relationship between the magnetic properties of terrestrial igneous rocks and the composition and internal structure of their component Fe-Oxide grains. *Geophys. J. R. Astr. Soc.* **60**, 107–22.

Clark, A. J. 1980. Magnetic dating. *Sussex Arch. Collection* **18**, 7–12.

Clark, D. A. 1983. Comments on magnetic petrophysics. *Bull. Aust. Soc. Explor. Geophys.* **14**, 49–62.

Clark, D. A. 1984. Hysteresis properties of sized dispersed monoclinic pyrrhotite grains. *Geophys. Res. Letters*, **11**, 173–6.

Clark, H. C. and J. P. Kennett 1973. Palaeomagnetic excursion recorded in latest Pleistocene deep-sea sediment. Gulf of Mexico. *Earth Planet. Sci. Letters* **19**, 267–74.

Clark, R. M. and R. Thompson 1978. An objective method for smoothing palaeomagnetic data. *Geophys. J. R. Astr. Soc.* **52**, 205–13.

Clark, R. M. and R. Thompson 1979. A new approach to the alignment of time series. *Geophys. J. R. Astr. Soc.*, **58**, 593–607.

Clement, B. M. and D. V. Kent 1984. Latitudinal dependency of geomagnetic polarity transition durations. *Nature* **310**, 488–91.

Clough, W. S. 1973. Transport of particles to surfaces. *Aerosol Science* **4**, 227–34.

Coe, R. S., M. Prevot, E. A. Mankinen and C. S. Gromme 1983. Behaviour of the complete field vector during the Steens mountain reversal. *IAGA Bulletin* **48**, 179–80.

Cohen, I. D. 1967. A shielded facility for low-level magnetic measurements. *J. Appl. Phys.* **38**, 1295–6.

Cohen, D. 1970. Large-volume conventional magnetic shields. *Rev. Phys. Appl.* **5**, 53–8.

Cohen, D. 1973. Ferromagnetic contamination in the lungs and other organs of the human body. *Science* **180**, 745–8.

Cohen, D., S. R. Arai and J. D. Brain 1979. Smoking impairs long-term dust clearance from the lung. *Science* **204**, 514–17.

Colani, C. and M. J. Aitken 1966. A new type of locating device. *Archaeometry* **9**, 9–19.

Colley, S., J. Thompson, T. R. S. Wilson and N. C. Higgs 1984. Post-depositional migration of elements during diagenesis in brown clay and turbidite sequence in the north east Atlantic. *Geochim. Cosniochim. Acta.* **48**, 1223–1236.

Collinson, D. W. 1975. Instruments and techniques in palaeomagnetism and rock magnetism. *Rev. Geophys. Space. Phys.* **13**, 659–86.

Collinson, D. W. 1983. *Methods in rock magnetism and palaeomagnetism – techniques and instrumentation.* London: Chapman and Hall.

Collinson, D. W., K. M. Creer and S. K. Runcorn (eds) 1967. *Methods in palaeomagnetism.* Amsterdam: Elsevier.

Correll, D. L. 1977. An overview of the Rhode river watershed program. In *Watershed research in eastern North America.* D. L. Correll (ed.). Smithsonian Institution, Washington. 105–20.

Courtillot, V., J. Ducruix and J.-L. le Mouël 1978. Sur un accélération récente de la variation séculaire du champ magnétique terrestre. *C.R. Hebd. Seances Acad. Sci. Ser. D.*, **287**, 1095–8.

Cox, A. 1969. Geomagnetic reversals. *Science* **163**, 237–44.

Cox, A. 1973. *Plate tectonics and geomagnetic reversals.* New York: W. H. Freeman.

Cox, A., R. R. Doell and G. B. Dalrymple 1963. Geomagnetic polarity epochs and Pleistocene geo-chronometry. *Nature* **198**, 1049–51.

Craik, D. J. 1971. *Structure and properties of magnetic materials.* London: Pion.

Crangle, J. 1975. SI Units in magnetism. *Physics Bulletin* **26**, 539.

Crangle, J., 1977. *The magnetic properties of solids.* London: Edward Arnold.

Creer, K. M. 1959. A.C. demagnetization of unstable Triassic Keuper Marls from S. W. England. *Geophys. J. R. Astr. Soc.* **2**, 261–75.

Creer, K. M., T. W. Anderson and C. F. M. Lewis 1976. Late Quaternary geomagnetic stratigraphy recorded in Lake Erie sediments. *Earth Planet. Sci. Letters.* **31**, 37–47.

Creer, K. M., R. Thompson, L. Molyneux and F. J. H. Mackereth 1972. Geomagnetic secular variation recorded in the stable magnetic remanence of recent sediments. *Earth Planet. Sci. Letters* **14**, 115–27.

Curie, P. 1895. Propriétés magnetiques des corps à diverses temperatures. *Ann. de Chim. et Phys.* **5**, 289.

Currie, R. G. and B. D. Bornhold 1983. The magnetic susceptibility of continental-shelf sediments, west coast Vancouver Island, Canada. *Marine Geology* **51**, 115–27.

Dalrymple, G. B. 1972. Potassium–argon dating of geomagnetic reversals and North American glaciations. In *Calibration of hominid evolution: recent advances in isotopic and other dating methods applicable to the origin of man.* W. W. Bishop and J. A. Miller (eds). New York: Scottish Academic Press.

Dankers, P. H. 1978. *Magnetic properties of dispersed natural iron oxides of known grain size.* PhD thesis. University of Utrecht.

Davidson, R. L., D. F. S. Natusch, J. R. Wallace and C. A. Evans Jr. 1974. Trace elements in fly ash: dependence of concentration on particle size. *Environ. Sci. Technol.* **8**, 1107–13.

Davis, M. B. 1976. Erosion rates and land use history in southern Michigan. *Environ. Conserv.* **3**, 139–48.

Davis, M. B. and M. S. Ford 1982. Sediment focusing in

Mirror Lake, New Hampshire. *Limnol. Oceanogr.* **27**, 137–50.

Davis, P. M. and M. E. Evans 1976. Interacting single-domain properties of magnetite intergrowths. *J. Geophys. Res.* **81**, 989–94.

Davis, R. B. and R. W. Doyle 1969. A piston corer for upper sediment in lakes. *Limnol. Oceanogr.* **14**, 643–8.

Day, R., M. Fuller and V. A. Schmidt 1977. Hysteresis properties of titanomagnetites: grain-size and compositional dependence. *Phys. Earth Planet. Interiors* **13**, 260–7.

Dearing, J. A. 1979. *The applications of magnetic measurements to studies of particulate flux in lake–watershed ecosystems.* Unpubl. PhD thesis. University of Liverpool.

Dearing, J. 1983. Changing patterns of sediment accumulation in a small lake in Scania, southern Sweden. *Hydrobiologia* **103**, 59–64.

Dearing, J. A., J. K. Elner, and C. M. Happey-Wood 1981. Recent sediment influx and erosional processes in a Welsh upland lake–catchment based on magnetic susceptibility measurements. *Quat. Res.* **16**, 356–72.

Dearing, J. A., and R. J. Flower 1982. The magnetic susceptibility of sedimenting material trapped in Lough Neagh, Northern Ireland and its erosional significance. *Limnol. Oceanogr.* **17**, 969–75.

Dearing, J. A., B. A. Maher and F. Oldfield 1985. Geomorphological linkages between soils and sediments: the role of magnetic measurements. In *Geomorphology and soils*, K. Richards (ed.). London: George Allen & Unwin.

Dearing, J. A., R. I. Morton, T. W. Price and I. D. L. Foster 1986. Tracing movements of topsoil by magnetic measurements: two case studies. *Phys. Earth Planet Int.* **42**, 93–104.

Dell, C. I. 1972. An occurrence of greigite in Lake Superior sediments. *Am. Mineralogist* **57**, 1303–4.

Denham, C. R. 1974. Counter clockwise motion of palaeomagnetic directions 24 000 years ago at Mono lake, California. *J. Geomag. Geoelectr.* **26**, 487–98.

Denham, C. R. 1975. Spectral analysis of palaeomagnetic time series. *J. Geophys. Res.* **80**, 1897–901.

Denham, C. R. 1976. Blake polarity episode in two cores from the greater Antilles outer ridge. *Earth Planet Sci. Letters* **9**, 422–34.

Denham, C. R. 1981. Numerical correlation of recent paleomagnetic records in two Lake Tahoe cores. *Earth Planet. Sci. Letters* **54**, 48–52.

Denham, C. R. and A. Cox 1971. Evidence that the Laschamp polarity event did not occur 13 300–30 400 years ago. *Earth Planet. Sci. Letters* **13**, 181–90.

Digerfeldt, G. 1978. *A simple corer for sediment sampling in deep water.* Department of Quaternary Geology Report, University of Lund.

Doake, S. M. 1977. A possible effect of ice ages on the earth's magnetic field. *Nature* **267**, 415–17.

Dodson, M. A. and E. A. McClelland-Brown 1980. Magnetic blocking temperatures of single-domain grains during slow cooling. *J. Geophys. Res.* **85**, 2625–37.

Doe, S.-J. and W. K. Steel 1983. The late Pleistocene geomagnetic field as recorded by sediments from Fargher Lake, Washington, U.S.A. *Earth Planet. Sci. Letters*, **63**, 385–98.

Donoghue, J. F. 1981. *Estuarine sediment transport and Holocene depositional history, Upper Chesapeake Bay.* Unpubl. PhD thesis. University of Southern California.

Doyle, J. L., T. L. Hopkins and P. R. Betzer 1976. Black magnetic spherule fallout in the eastern Gulf of Mexico. *Science*, 1157–9.

Dunlop, D. J. 1972. Magnetic mineralogy of unheated and heated red sediments by coercivity spectrum analysis. *Geophys. J. R. Astr. Soc.*, **27**, 37–55.

Dunlop, D. J. 1973a. Superparamagnetic and single-domain threshold sizes in magnetite. *J. Geophys. Res.* **78**, 1780–93.

Dunlop, D. J. 1973b. Theory of magnetic viscosity in lunar and terrestrial rocks. *Rev. Geophys. Space Phys.* **11**, 855–901.

Dunlop, D. J. 1981. The rock magnetism of fine particles. *Phys. Earth Planet Interiors* **26**, 1–26.

Dunlop, D. J. 1983a. Determination of domain structure in igneous rocks by alternating field and other methods. *Earth Planet. Sci. Letters* **63**, 353–67.

Dunlop, D. J. 1983b. Viscous magnetization of 0.04–100µm magnetites. *Geophys. J. R. Astr. Soc.* **74**, 667–87.

Dunlop, D. J. and M. Prevot 1982. Magnetic properties and opaque mineralogy of drilled submarine intrusive rocks. *Geophys. J. R. Astr. Soc.* **69**, 763–802.

Dzyaloshinsky, I. E. 1958. A thermodynamic theory of 'weak' ferromagnetism of antiferromagnetics. *J. Phys. Chem. Solids* **4**, 241–55.

Edgington, D. N. and J. A. Robbins 1975. Records of lead deposition in Lake Michigan sediments since 1800. *Environ. Sci. Technol.* **10**, 266–73.

Edwards, K. J. 1978. *Palaeoenvironmental and archaeological investigations in the Howe of Cromar, Grampian region, Scotland.* Unpubl. PhD thesis. University of Aberdeen.

Edwards, K. J. and K. M. Rowntree 1980. Radiocarbon and palaeoenvironmental evidence for changing rates of erosion at a Flandrian stage site in Scotland in *Timescales in geomorphology*, R. A. Cullingford, D. A. Davison and J. Lewis (eds).

Ellis, J. B. 1979. The nature and sources of urban sediments and their relation to water quality. In *Man's impact on the hydrological cycle in the UK*, G. E. Hollis (ed.), 199–216.

Ellsasser, H. W. 1975. The upward trend in airborne particulates that isn't. In *The changing global environment*. S. F. Singer (ed.), 235–69. Dordrecht: D. Reidel.

Ellsasser, W. M. 1946. Induction effects in terrestrial magnetism. *Phys. Rev.* **69**, 106–16.

Ellwood, B. B. 1980. Induced and remanent magnetic

properties of marine sediments as indicators of depositional processes. *Marine Geology* 38, 233–44.

Epp, R. J., J. W. Tukey and G. S. Watson 1971. Testing unit vectors for correlation. *J. Geophys. Res.* 76, 8480–3.

Ewing, J. A. 1900. *Magnetic induction in iron and other metals.* The Electrician Publishing Company.

Fisher, R. 1953. Dispersion on a sphere. *Proc. R. Soc. London A* 217, 295–305.

Foex, G. and R. Forrer 1926. Sur un appareil sensible pour la mesure precise des coefficients d'aimantation à diverses temperatures. *J. Phys. et Radium* 7, 180–7.

Foner, S. 1959. Versatile and sensitive vibration magnetometer. *Rev. Sci. Instrum.* 30, 548–57.

Freed, W. K. and N. Healy 1974. Excursions of the Pleistocene geomagnetic field recorded in Gulf of Mexico sediments. *Earth Planet. Sci. Letters* 24, 99–104.

Fuller, M. 1963. Magnetic anistropy and palaeomagnetism. *J. Geophys. Res.* 68, 293–309.

Fuller, M. 1974. Lunar magnetism. *Rev. Geophys. and Space Phys.* 12, 23–70.

Fuller, M., I. Williams and K. A. Hoffman 1979. Paleomagnetic records of geomagnetic field reversals and the morphology of the transitional fields. *Rev. Geophys. and Space Phys.* 17, 179–203.

Furr, A. K., T. F. Parkinson, R. A. Hinrichs, D. R. Van Campen, C. A. Bache, W. H. Gutermann, L. E. St. John, I. S. Pakkala and D. J. Lisk 1977. National survey of elements and radioactivity in fly ashes. Absorption of elements by cabbage grown in fly ash–soil mixture. *Environ. Sci. Technol.* 11, 1194–201.

Galt, J. K. 1952. Motion of a ferromagnetic domain wall in Fe_3O_4. *Phys. Rev.* 85, 664–9.

Gauss, C. F. 1833. Intensities vis magneticae terrestris ad mensuram absolutam revocata. *Gottingen Comment* 8, 3–44.

Gauss, C. F. 1839. Allgemeine Theorie des Erdmagnetismus (general theory of terrestrial magnetism). In *Scientific memoirs selected from the transactions of foreign academies and learned societies and from foreign journals*, 2, 184–251 (1841).

Gellibrand, H. 1635. *A discourse mathematical on the variation of the magnetical needle.* London: William Jones.

Gillingham, D. E. W. and F. D. Stacey 1971. Anhysteretic remanent magnetization (A.R.M.) in magnetite grains. *Pure Appl. Geophys.*, 91, 160–5.

Goldberg, E. D. 1975. Man's role in the major sedimentary cycle. In *The changing global environment*, S. F. Singer (ed.), 275–94. Dordrecht: D. Reidel.

Gordon, A. D. 1973. A sequence–comparison statistic and algorithm, *Biometrika* 60, 197–200.

Gordon, A. D. 1982. An investigation of two sequence–comparison statistics. *Aust. J. Stat.* 24, 332–42.

Goree, W. S. and M. Fuller 1976. Magnetometers using RF-driven squids and their applications in rock magnetism and palaeomagnetism. *Rev. Geophys. and Space Phys.* 14, 591–608.

Goudie, A. S. (ed.) 1981. *Geomorphological techniques.* London: George Allen & Unwin.

Gould, J. L. and K. P. Albe 1981. Human homing an elusive phenomenon. *Science* 212, 1061–3.

Gouy, M. 1889. Sur l'energie potentielle magnetique et la mesure des coefficients d'aimantation. *Compt. Rend. Acad. Sci. Paris* 109, 935–7.

Graham, I. 1976. The investigation of the magnetic properties of archaeological sediments. In *Geoarchaeology*, D. A. Davidson and M. L. Shackley (eds).

Granar, L. 1958. Magnetic measurements on Swedish varved sediments. *Arkiv forr Geofysik* 3, 1–40.

Gregory, K. J. and D. E. Walling 1973. *Drainage basin form and process: a geomorphological approach.* London: Edward Arnold.

Griffiths, D. H., R. F. King, A. I. Rees and A. E. Wright 1960. Remanent magnetism of some recent varved sediments. *Proc. R. Soc. London.* A 256, 359–83.

Haggerty, S. E. 1970. Magnetic minerals in pelagic sediments. *Ann. Rept. Geophys. Lab. Carnegie Institute Year Books*, 68. Washington 1560. 332–6.

Haggerty, S. E. 1976. Opaque mineral oxides in terrestral igneous rocks. In *Oxide minerals* 3. D. Rumble III (ed.), Mineralogical Society of America. Blacksburg: Southern Printing Co.

Hakanson, L. and M. Jansson 1983. *Principles of lake sedimentology.* Berlin: Springer

Hallam, J. S., J. N. Edwards, B. Barnes and K. J. Stuart 1973. A late glacial Elk with associated barbed points from High Furlong, Lancashire. *Proc. Prehist. Soc.* 39, 100–28.

Halley, E. 1692. An account of the cause of the change of the variation of the magnetical needle; with an hypothesis of the structure of the internal part of the Earth. *Phil. Trans. R. Soc. London* 17, 563–78.

Hamilton, A. C., W. Magowan and D. Taylor 1986. Use of the Bartington Meter to determine the magnetic susceptibility of organic-ride sediments from western Uganda. *Phys. Earth Plant Int.* 42, 5–9.

Hamilton, N. and A. I. Rees 1970. The use of magnetic fabric in paleocurrent estimation. In *Paleogeophysics*, S. K. Runcorn (ed.), 445–64. London: Academic Press.

Hammond, S. R., F. Theyer and G. H. Sutton 1974. Palaeomagnetic evidence of the northward movement of the Pacific plate in deep-sea cores from the central Pacific basin. *Earth Planet. Sci. Letters* 22, 22–8.

Hansen, L. D., D. Silberman and G. L. Fisher 1981. Crystalline components of stack-collected, size-fractionated coal fly ash. *Environ. Sci. Technol.* 15, 1057–62.

Harland, W. B., A. V. Cox, P. G. Llewellyn, C. A. G. Pickton, A. G. Smith and R. Walters 1982. *A geologic time scale.* Cambridge: Cambridge University Press.

Harrison, C. G. A. and J. M. Prospero 1977. Reversals of the earth's magnetic field and climatic changes. *Nature* **250**, 563–4.

Harvey, A. M., A. F. Baron, F. Oldfield and G. W. Pearson 1981. Dating of post-glacial landforms in the central Howgills. *Earth Surface Processes* **6**, 401–12.

Hayes, J. G. and J. Halliday 1974. The least-squares fitting of cubic spline surfaces to general data sets. *J. Inst. Maths Applics* **14**, 89–103.

Heirtzler, J. R., G. O. Dickson, E. M. Herron, W. C. Pitman and X. Le Pichon 1968. Marine magnetic anomalies, geomagnetic field reversals and motions of the ocean floor and continents. *J. Geophys. Res.* **73**, 2119–35.

Heller, F. 1980. Self-reversal of natural remanent magnetisation in the Olby–Laschamp lavas. *Nature* **284**, 334–5.

Heller, F. and T. S. Liu 1982. Magnetostratigraphic dating of loess deposits in China. *Nature* **300**, 431–3.

Henry, W. M. and K. T. Knapp 1980. Compound forms of fossil fuel fly ash emissions. *Environ. Sci. Technol.* **14**, 450–6.

Henshaw, P. C. and R. T. Merrill 1979. Characteristics of drying remanent magnetization in sediments. *Earth Planet. Sci. Letters* **43**, 315–20.

Henshaw, P. C. and R. T. Merrill 1980. Magnetic and chemical changes in marine sediments. *Rev. Geophys. Space Phys.* **18**, 483–504.

Hide, R. and P. H. Roberts 1956. The origin of the main geomagnetic field. *Physics and Chemistry of the Earth* **4**, 35–98. Oxford: Pergamon.

Hidy, G. M. and J. R. Brock 1971. An assessment of the global sources of tropospheric aerosols. In *Proceedings of the 2nd international clean air congress*, H. M. Englund and W. T. Beery (eds), 1088–97. New York: Academic Press.

Higgitt, S. E. 1985. *The palaeocology of the Lac d'Annecy and its drainage basin.* Unpubl. PhD thesis. University of Liverpool.

Hilton, J. and J. P. Lishman 1985. The effect of redox change on the magnetic susceptibility of sediments from a seasonally anoxic lake. *Limnol. Oceanogr.*, **30**, 907–909.

Hirons, K. R. 1983. *Paleoenvironmental investigations in east Co. Tyrone, Northern Ireland.* Unpubl. PhD thesis. Queens University, Belfast.

Honda, K. and T. Sone 1914. Uber die magnetische untersuchung der Strukturanderungen in Eisen-und Chromverbindungen bei hohoheren Temperaturen. *Sci. Rept. Tohuku Imperial University* **3**, 224–34.

Horie, S., K. Yaskawa, A. Yamanoto, T. Yokoyama and M. Hyodo 1980. Paleolimnology of Lake Kizaki. *Arch. Hydrobiol.* **89**, 407–15.

Hulett, L. D., A. J. Weinberger, M. Ferguson, K. J. Northcutt, and W. S. Lyon 1981. *Trace element and phase relations in fly ash.* DE-81028555 EPR1-EA-1822.

Hughes, S. J. 1978. *The recent paleoecology and magnetic stratigraphy of Ringinglow Bog, near Sheffield.* Unpubl. BSc dissertation. University of Liverpool.

Hunt, A. 1986. The application of mineral magnetic methods to atmospheric aerosol discrimination. *Phys. Earth Planet Int.* **42**, 10–21.

Hunt, A., J. Jones and F. Oldfield 1984. Magnetic measurements and heavy metals in atmospheric particulates of anthropogenic origin. *The Science of the Total Environment*, **33**, 129–39.

Huttunen, P. and J. Merilainen 1978. New freezing device providing large, unmixed sediment samples from lakes. *Ann. Bot. Fenn.* **15**, 128–30.

Huttunen, P. and J. Stober 1980. Dating of palaeomagnetic records from Finnish lake sediment cores using pollen analysis. *Boreas* **9**, 193–202.

Hyodo, M. 1984. Possibility of reconstruction of past geomagnetic field from homogeneous sediments. *J. Geomag. Geoelectr.* **36**, 45–62.

Irving, E. 1964. *Paleomagnetism and its application to geological and geophysical problems.* New York: Wiley.

Irving, E. 1977. Drift of major continental blocks since the Devonian. *Nature* **270**, 304–9.

Irving, E. and A. Major 1964. Post-depositional detrital remanent magnetization in a synthetic sediment. *Sedimentology* **3**, 135–43.

Irving, E., L. Molyneux and S. K. Runcorn 1966. The analysis of remanent intensities and susceptibilities of rocks. *Geophys. J. R. Astr. Soc.* **10**, 451–64.

Irving, E. and G. Pullaiah 1976. Reversals of the geomagnetic field, magnetostratigraphy, and relative magnitude of palaeosecular variation in the phanerozoic. *Earth. Sci. Rev.* **12**, 35–64.

Isbell, R. F., P. J. Stephenson, G. G. Murtha and G. P. Gillman 1976. *Red basaltic soils in North Queensland.* Div. Soils. Tech. paper no. 28. Australia: CSIRO.

Ising, G. 1943. On the magnetic properties of varved clay. *Ark. Mat. Astr. Fys.* **29**, no. 5, 1–37.

Jacobs, J. A. 1975. *The Earth's core.* London: Academic Press.

Jaep, W. F. 1971. Role of interactions in magnetic tapes. *J. Appl. Phys.* **42**, 2790–4.

Jiles, D. C. and D. L. Atherton 1984. Theory of ferromagnetic hysteresis. *J. Appl. Phys.* **55**, 2115–20.

Johnston, M. J. S., B. E. Smith and R. Mueller 1976. Tectonomagnetic experiments and observations in western U.S.A. *J. Geomag. Geoelec.*, **28**, 85–97.

Johnston, M. J. S., and F. D. Stacey 1969. Transient magnetic anomalies accompanying volcanic eruptions in New Zealand. *Nature* **224**, 1289–90.

Johnson, E. A. and A. G. McNish 1938. An alternating current apparatus for measuring small magnetic moments. *Terr. Mag. Atmos. Electr.*, **43**, 393–9.

Johnson, E. A., T. Murphy and O. W. Torreson 1948. Prehistory of the Earth's magnetic field. *Terr. Mag.* **53**, 349–72.

Johnson, H. P., H. Kinoshita, and R. T. Merrill 1975. Rock

magnetism and paleomagnetism of some North Pacific deep sea sediments. *Geol. Soc. Am. Bull.* 86, 412–20.

Jones, B. F. and C. J. Bowser 1978. The mineralogy and related chemistry of lake sediments. A. Lerman (ed.). *Lakes: Chemistry, Geology, Physics*. New York: Springer.

Joseph, J. H., A. Manes and D. Ashbel 1973. Desert aerosols transported by Khamsinic depressions and their climatic effect. *J. Appl. Meteorol.*, 12, 792–7.

Judson, S. 1968. Erosion of the land, or what's happening to our continents? *American Scientist*, 56, 356–74.

Kalliomaki, K., K. Aittoniemi, P. L. Kalliomaki and M. Moilanen 1981. Measurement of lung-retained contaminants in vivo among workers exposed to metal aerosols. *Am. Ind. Hyg. Assoc.* J (42), 234–8.

Kalliomaki, P. L., P. J. Karp, T. Katila, P. Makipaa, P. Saar and A. Tossavainen 1976. Magnetic measurements of pulmonary contamination. *Scand. J. Work Environ. Health* 4, 232–9.

Kalmijn, A. J. and R. P. Blakemore 1978. The magnetic behaviour of mud bacteria: in *Animal migration, navigation and homing*. K. Schmidt-Koenig and W. T. Keeton (eds). Berlin: Springer.

Karlin, R. and S. Levi 1983. Diagenesis of magnetic minerals in Recent hemipelagic sediments. *Nature* 303, 327–30.

Keeton, W. T. 1971. Magnets interfere with pigeon homing. *Proc. Natl. Acad. Sci. USA* 68, 102–6.

Kennett, J. P. 1981. Marine tephrochronology. In *The sea*, vol. 7. 1373–1436. *The oceanic lithosphere*. C. Emiliani (ed.). New York: J. Wiley and Sons.

Kennett, J. P. and R. C. Thunell 1975. Global increase in Quaternary explosive volcanism. *Science* 187, 497–503.

Kent, D. V. 1973. Post-depositional remanent magnetisation in deep sea sediment. *Nature* 246, 32–4.

Kent, D. V., B. M. Honnorez, N. D. Opdyke and P. J. Fox 1978. Magnetic properties of dredged oceanic gabbros and the source of marine magnetic anomalies. *Geophys. J. R. Astr. Soc.* 55, 513–37.

Kent, D. V. 1982. Apparent correlation of palaeomagnetic intensity and climatic records in deep-sea sediments. *Nature* 299, 538–9.

Kent, D. V. and W. Lowrie 1974. Origin of magnetic instability in sediment cores from the central north Pacific. *J. Geophys. Res.* 79, 2987–3000.

Kent, D. V. and N. D. Opdyke 1977. Paleomagnetic field intensity variation recorded in a Brunhes epoch deep-sea sediment core. *Nature* 266, 5598, 156–9.

Kershaw, A. P. 1978. Record of the last integlacial–glacial cycle from northeastern Queensland. *Nature* 272, 159–61.

Keyser, T. R., D. F. S. Natusch, C. A. Evans Jr. and R. W. Linton 1978. Characterizing the surfaces of environmental particles. *Environ. Sci. Technol.* 12, 768–73.

Kidson, C. 1982. Sea level changes in the Holocene. *Quat. Sci. Rev.* 1, 121–51.

King, J., S. Banerjee, J. Marvin and O. Ozdemir 1982. A comparison of different magnetic methods for determining the relative grain size of magnetite in natural materials: some results from lake sediments. *Earth Planet. Sci. Letters* 59, 404–19.

King, R. F. 1955. Remanent magnetism of artificially deposited sediments. *Mon. Not. R. Soc. Geophys. Suppl.* 7, 115–34.

King, R. F. and A. I. Rees 1962. The measurement of the anisotropy of magnetic susceptibility of rocks by the torque method. *J. Geophys. Res.* 67, 1565–72.

King, R. F. and A. I. Rees 1966. Detrital magnetism in sediments: an examination of some theoretical models. *J. Geophys. Res.* 71, 561–71.

Kirschvink, J. L. 1982a. Palaeomagnetic evidence for fossil biogenic magnetite in western Crete. *Earth Planet Sci. Lett.* 59, 388–92.

Kirschvink, J. L. 1982b. Birds, bees and magnetism. *Trends in Neurosciences* 5, 5, 160–7.

Kirschvink, J. L. and J. L. Gould 1981. Biogenic magnetite as a basis for magnetic field detection in animals. *Biosystems* 13, 181–201.

Kirschvink, J. L. and S.-B. R. Chang 1984. Ultrafine-grained magnetite in deep-sea sediments: Possible bacterial magnetofossils. *Geology*, 12, 559–62.

Kirschvink, J. L. and H. A. Lowenstam 1979. Mineralization and magnetization of chiton teeth: palaeomagnetic, sedimentologic, and biologic implications of organic magnetite. *Earth Planet. Sci Letters* 44, 193–204.

Kittel, C. 1949. Physical theory of ferromagnetic domains. *Rev. Mod. Phys.* 21, 541–83.

Kleinman, M. T., B. S. Pasternack, M. Eisenbad and T. J. Knelp 1980. Identifying and estimating the relative importance of sources of airborne particulates. *Environ. Sci. Technol.* 14, 62–5.

Kneller, E. 1980. Static and anhysteretic magnetic properties of tapes. IEEE MAG-16, 36–41.

Kneller, E. F. and F. E. Luborsky 1963. Particle size dependence of coercivity and remanence of single-domain particles. *J. Appl. Phys.* 134, 656–8.

Kobayashi, K. and M. D. Fuller 1967. Vibration magnetometer. In *Methods in paleomagnetism*. D. W. Collinson, K. M. Creer and S. K. Runcorn (eds.). Amsterdam: Elsevier.

Kobayashi, K., K. Kitazawa, T. Kanaya and T. Sakai 1971. Magnetic and micropaleontological study of deep-sea sediments from the west-central Pacific. *Deep-Sea Research*, 18, 1045–62.

Kobayashi, K. and M. Nomura 1972. Iron sulphides in the sediment cores from the Sea of Japan and their geophysical implications. *Earth Planet. Sci. Letters* 16, 200–8.

Kobayashi, K. and M. Nomura 1974. Ferromagnetic minerals in the sediment cores collected from the Pacific basin. *J. Geophys.* 40, 501–12.

Kodoma, P. and A. Cox 1978. The effects of a constant volume deformation on the magnetization of an artificial sediment. *Earth Planet. Sci. Letters* 38, 436–42.

Koenigsberger, J. G.   1938. Natural residual magnetism of eruptive rocks, parts I and II. *Terr. Magn. Atmos. Elec.* **43**, 119–27; 299–320.

Kopper, J. S. and K. M. Creer   1976. Palaeomagnetic dating and stratigraphic interpretation in Archaeology. *MASCA newsletter*, **12**. No. 1 1–4. Applied Science Center for Archaeology. The University Museum, Pennsylvania.

Kovacheva, M.   1982. Archaeomagnetic investigations of geomagnetic secular variations. *Phil. Trans. R. Soc. London.*, *A***306**, 79–86.

Krawiecki, A.   1982. *The burning of the hillfort at Maiden Castle, Bickerton Hill, Cheshire.* Unpubl. BSc thesis. Department of Geography, University of Liverpool.

Kristjansson, L. and A. Gudmundsson   1980. Geomagnetic excursion in late-Glacial basalt outcrops in South-Western Iceland. *Geophys. Res. Letts.* **7**, 337–40.

LaBrecque, J. L., D. V. Kent and S. C. Cande   1977. Revised magnetic polarity time scale for late Cretaceous and Cenozoic time. *Geology* **5**, 330–5.

Lamb, H. H.   1970. Volcanic dust in the atmosphere: with a chronology and assessment of its meteorological significance. *Phil. Trans. R. Soc. Lond.* **266**, 425–533.

Lancaster, D. E.   1966. Electronic metal detection. *Electronics World*, (Dec.), 39–62.

Larson, R. L. and W. C. Pitman III   1972. World-wide correlation of Mesozoic magnetic anomalies, and its implications. *Geol. Soc. Am. Bull.* **83**, 3645–62.

Lauf, R. J., L. A. Harris and S. S. Rawiston   1982. Pyrite framboids as the source of magnetite spheres in fly ash. *Environ. Sci. Technol.* **16**, 218–20.

Le Borgne, E.   1955. Susceptibilité magnétique anormale du sol superficiel. *Ann. Geophys.* **11**, 399–419.

Le Borgne, E.   1960. Influence du feu sur les propriétés magnétiques du sol et du granite. *Ann. Geophys.* **16**, 159–95.

Lehman, J. T.   1975. Reconstructing the rate of accumulation of lake sediments: the effect of sediment focusing. *Quat. Res.* **5**, 541–50.

Liddicoat, J. C. and R. S. Coe   1979. Mono lake geomagnetic excursion. *J. Geophys. Res.* **84**, 261–71.

Likhite, S. D., C. Radhakrishnamurty and P. W. Sahasrabudhe, 1965. Alternating current electromagnet-type hysteresis loop tracer for minerals and rocks. *Rev. Sci. Instrum.* **36**, 1558–64.

Lins de Barros, D. N. S., J. Esquivel. J. Danon and L. P. H. de Oliviera   1981. Magnetotactic algae. *Acad. Bras. Notas. Fis.* CBPF-NF-48.

Linton, R. W., D. F. S. Natusch, R. L. Soloman and C. A. Evans Jr.   1980. Physiochemical characterization of lead in urban dusts. A microanalytical approach to lead tracing. *Environ. Sci. Technol.* **14**, 159–64.

Longworth, G. and M. S. Tite   1977. Mossbauer and magnetic susceptibility studies of iron oxides in soils from archaeological sites. *Archaeometry* **19**, 3–14.

Longworth, G., L. W. Becker, R. Thompson, F. Oldfield, J. A. Dearing and T. A. Rummery   1979. Mossbauer and magnetic studies of secondary iron oxides in soils. *J. Soil. Sci.* **30**, 93–110.

Lorrain, P. and D. R. Corson   1978. *Electromagnetism: principles and application.* New York: W. H. Freeman.

Lovlie, R.   1974. Post-depositional remanent magnetization in a re-deposited deep-sea sediment. *Earth Planet. Sci. Letters* **21**, 315–20.

Lowes, F. J.   1984. The geomagnetic dynamo-elementary energetics and thermodynamics. *Geophysical Surveys* **7**, 91–105.

Lowrie, W.   1979. *Geomagnetic reversals and ocean crust magnetization in deep drilling results in the Atlantic Ocean: Ocean crust*, M. Talwami, C. G. A. Harrison and D. E. Hayes (eds). Washington: American Geophysical Union.

Lowrie, W. and W. Alvarez   1981. One hundred million years of geomagnetic polarity history. *Geology*, **9**, 392–7.

Lowenstam, H. A.   1981. Magnetite biomineralization by organisms (Abs.), *EOS* **62**, 849.

Lukshin, A. A., T. I. Rumyantseva, and V. P. Kovrigo   1968. Magnetic susceptibility of the principal soil types in the Udmuit Asociation. *Soviet Soil Sci.* **3**, 88–93.

Mackereth, F. J. H.   1958. A portable core sampler for lake deposits. *Limnol. Oceanogr.* **3**, 181–91.

Mackereth, F. J. H.   1965. Chemical investigation of lake sediments and their interpretation. *Proc. R. Soc.* **161**, 295–309.

Mackereth, F. J. H.   1966. Some chemical observations on post-glacial lake sediments. *Phil. Trans. R. Soc. Lond. B* **250**, 165–213.

Mackereth, F. J. H.   1969. A short core-sampler for sub-aqueous deposits. *Limnol. Oceanogr.* **14**, 145–51.

Mackereth, F. J. H.   1971. On the variation in direction of the horizontal component of remanent magnetisation in lake sediments. *Earth Planet. Sci. Letters* **12**, 332–8.

Maher, B. A.   1981. *The effects of gleying on magnetic minerals in the soil.* Unpubl. BSc dissertation. Department of Geography, University of Liverpool.

Maher, B. A.   1984. Origins and transformations of magnetic minerals in soils, unpublished Ph.D Thesis, University of Liverpool.

Maher, B. A.   1986. Characterization of soil by mineral magnetic measurement. *Phys. Earth Planet Int.* **42**, 76–92.

Malin, S. R. C. and E. C. Bullard   1981. The direction of the Earth's magnetic field at London. 1570–1975. *Phil. Trans. R. Soc. Lond. A* **299**, 357–423.

Malin, S. R. C. and B. M. Hodder   1982. Was the 1970 geomagnetic jerk of internal or external origin? *Nature* **296**, 726–8.

Mardia, K. V.   1972. *Statistics of directional data.* London: Academic Press.

Maxted, R.   1983. *The measurement of the atmospheric heavy metal pollution on leaf surfaces using magnetic analysis techniques.* Unpubl. BSc dissertation. Department of Geography, University of Liverpool.

McCaig, M. 1977. *Permanent magnets in theory and practice.* London: Pentech Press.

McDougall, I. 1979. The present status of the geomagnetic polarity time scale. In *The Earth: its origin, structure and evolution*, M. W. McElhinny (ed.). London: Academic Press.

McDougall, I., K. Saemundsson, H. Johannesson, N. D. Watkins and L. Kristjansson 1977. Extension of the geomagnetic polarity time scale to 6.5 m.y: K–Ar dating, geological and palaeomagnetic study of a 3,500-m lava succession in western Iceland. *Geol. Soc. Am. Bull.* **88**, 1–15.

McElhinny, M. W. 1973. *Palaeomagnetism and plate tectonics.* London: Cambridge University Press.

McElhinny, M. W. and W. E. Senanayake 1982. Variations in the geomagnetic dipole 1: the past 50 000 years. *J. Geomag. Geoelectr.* **34**, 39–51.

McNish, A. G. and E. A. Johnson 1938. Magnetization of unmetamorphosed varves and marine sediments. *Terr. Mag.* **43**, 401–7.

McWilliams, M. O., R. T. Holcolm and D. E. Champion 1982. Geomagnetic secular variation from ^{14}C dated lava flows on Hawaii and the question of the Pacific non-dipole low. *Phil. Trans. R. Soc. Lond. A* **306**, 211–21.

Merrill, R. T. and M. W. McElhinny 1983. *The Earth's magnetic field.* London: Academic Press.

Mitchell, A. Crichton 1939. Chapters in the history of terrestrial magnetism – II. The discovery of the magnetic declination. *Terr. Mag.* **37**, 105–46.

Mitchell, J. M. 1970. A preliminary evaluation of atmospheric pollution as a cause of the global temperature fluctuation of the last century. In *Global effects of environmental pollution.* S. F. Singer (ed), 139–55. New York: Springer.

Miura, N., G. Kido, M. Akihiro and S. Chikazumi 1979. Production and usage of megagauss fields for solid state physics. *J. Magnetism Magnetic Materials* **11**, 275–83.

Moench, T. T. and W. A. Konetzka 1978. A novel method for the isolation and study of a magnetotactic bacterium. *Arch. Microbiol.* **119**, 203–12.

Molyneux, L. and R. Thompson 1973. Rapid measurement of the magnetic susceptibility of long cores of sediment. *Geophys. J. R. Astr. Soc.* **32**, 479–81.

Molyneux, L., R. Thompson, F. Oldfield and M. E. McCallan 1972. Rapid measurement of the remanent magnetization of long cores of sediment. *Nature* **237**, 42–3.

Mooney, H. M. 1952. Magnetic susceptibility measurements in Minnesota. Part I: technique of measurement. *Geophysics* **XVII**, 531–43.

Mooney, H. M. and R. Bleifuss 1953. Magnetic susceptibility measurements in Minnesota. Part II: analysis of field results. *Geophysics* **XVIII**, 383–93.

Morgan, W. J. 1968. Rises, trenches, great faults and crustal blocks. *J. Geophys. Res.* **73**, 1959–82.

Morner, N.-A., J. P. Lanser and J. Hospers 1971. Late Weichselian paleomagnetic reversal. *Nature Phys. Sci.* **234**, 173–4.

Mortimer, C. H. 1942. The exchange of dissolved substances between mud and water in lakes III and IV. *J. Ecol.* **30**, 147–201.

Mosley, P. 1980. Mapping sediment source in a New Zealand mountain watershed. *Environ. Geol.* **3**, 85–95.

Mosteller, F. and J. W. Tukey 1977. *Data analysis and regression.* Reading: Addison-Wesley.

Mullen, R. E., D. A. Darby and D. L. Clark 1972. Significance of atmospheric dust and ice rafting for Arctic Ocean sediment. *Geol. Soc. Am. Bull.* **83**, 205–12.

Mullins, C. E. 1974. The magnetic properties of the soil and their application to archaeological prospecting. In *Technische und Naturwinsenschaftliche Beitrag zur Feldarchaologie.* Klohn: Rheinland-Verlag.

Mullins, C. E. 1977. Magnetic susceptibility of the soil and its significance in Soil Science: a review. *J. Soil Sci.* **28**, 223–46.

Mullins, C. E. and M. S. Tite 1973. Magnetic viscosity, quadrature susceptibility and frequency dependence of susceptibility in single-domain assemblages of magnetite and maghaemite. *J. Geophys. Res.* **78**, 804–9.

Murray, J. 1876. On the distribution of volcanic debris over the floor of the ocean – its character, source and some of the products of its disintegration and decomposition. *Proc. R. Soc. Edinb.* **9**, 247–61.

Nagata, T. 1953. *Rock magnetism.* Tokyo: Maruzen.

Nagata, T., T. Rikitake and K. Akasi 1943. The natural remanent magnetism of sedimentary rocks. *Bull. Earthquake Res. Inst. Tokyo Univ.* **31**, 276–96.

Nagata, T., S. Uyeda and S. J. Akimoto 1952. Self-reversal of thermoremanent magnetism of igneous rocks. *Geomag. Geoelectr.* **4**, 22–38.

Nagata, T. 1976. Principles of the ballistic magnetometer for the measurement of remanence. In *Methods in palaeomagnetism*, D. W. Collinson, K. M. Creer and S. K. Runcorn (eds). Amsterdam: Elsevier.

Nakajima, T., K. Yaskawa, N. Natsuhara, N. Kawai and S. Horie 1973. Very short period geomagnetic excursion 18 000 yr BP. *Nature Phys. Sci.* **244**, 8–10.

National Academy of Sciences 1979. *Airborne particles.* New York.

Neel, L. 1948. Proprietes magnetiques des ferrites; ferrimagnetism et antiferromagnetism. *Ann. Phys.* **3**, 137–98.

Neel, L. 1955. Some theoretical aspects of rock-magnetism. *Adv. Phys.* **4**, 191–243.

Negrini, R. M., J. O. Davis and K. L. Verosub 1984. Mono Lake geomagnetic excursion found at Summer Lake, Oregon. *Geology* **12**, 643–6.

Nelder, J. A. and R. Mead 1965. A simplex method for function minimization. *Computer J.* **7**, 308–13.

Nelmes, R. J. 1984. *Palaeolimnological studies of Rostherne*

Mere (Cheshire) and Ellesmere (Shropshire). Unpubl. PhD thesis. Liverpool Polytechnic.

Newson, M. D. 1980. The erosion of drainage ditches and its effect on bed load yield in mid–Wales. Reconnaissance case studies. *Earth Surface Processes* **5**, 190–275.

Noel, M. and D. H. Tarling 1975. The Laschamp geomagnetic 'event'. *Nature* **253**, 705–6.

Noltimier, H. C. and P. A. Colinvaux 1976. Geomagnetic excursion from Imuruk Lake, Alaska. *Nature* **259**, 197–200.

Norman, R. 1581. *The newe attractive*. London: John Kyngston for Richard Ballard.

NRC 1978. *The tropospheric transport of pollutants and other substances to the oceans*. Washington, DC: National Reseach Council.

Oades, J. M. and W. N. Townsend 1963. The detection of ferromagnetic minerals in soils and clays. *J. Soil Sci.* **14**, 179–87.

Okada, Y. C., L. Kaufman, D. Brenner and S. J. Williamson 1982. Modulation transfer functions of the human visual system revealed by magnetic field measurements. *Vision Res.* **22**, 319–33.

Oldfield, F. 1977. Lakes and their drainage basins as units of sediment-based ecological study. *Prog. Phys. Geog.* **3**, 460–504.

Oldfield, F. 1981. Peat and lake sediments: formation, stratigraphy, description and nomenclature. 306–26. In *Geomorphological techniques*, A. Goudie (ed.). London: George Allen & Unwin.

Oldfield, F. 1983a. The role of magnetic studies in palaeo-hydrology. In *Background to palaeohydrology*. K. J. Gregory (ed.), 141–65. Chichester: Wiley.

Oldfield, F. 1983b. Man's impact on environment. *Geography* **68**, 245–56.

Oldfield, F. unpubl. *Mineral magnetic measurements on particulates from a Greenland Ice Core – a preliminary report*. Available from the author.

Oldfield, F. and P. G. Appleby 1984. Empirical testing of ^{210}Pb dating models for lake sediments. In *Lake sediments and environmental history*, E. Y. Haworth and J. W. G. Lund (eds), 93–114. Leicester University Press.

Oldfield, F. and B. Maher 1984. A mineral magnetic approach to erosion studies. In report of conference on *Drainage Basin Erosion and Sedimentation*, Newcastle, N.S.W., Australia, May 1984.

Oldfield, F. and A. Mannion (in prep.). *Magnetic measurements of ice and snow samples from the Okstinden area of Norway*.

Oldfield, F. and S. G. Robinson 1985. Geomagnetism and palaeoclimate. In *The climatic scene*, M. J. Tooley and G. Sheil (eds). London: George Allen & Unwin.

Oldfield, F., J. A. Dearing, R. Thompson and S. E. Garret-Jones 1978. Some magnetic properties of lake sediments and their possible links with erosion rates. *Polskie Archive. Hydrobiologia* **25**, 321–31.

Oldfield, F., P. G. Appleby, R. W. Cambray, J. D. Eakins, K. E. Barber, R. W. Battarbee, G. W. Pearson and J. W. Williams 1979a. Lead-210, caesium-137 and plutonium-239 profiles in ombrotrophic peat. *Oikos* **33**, 40–5.

Oldfield, F., A. Brown and R. Thompson 1979b. The effect of microtopography and vegetation on the catchment of airborne particles measured by remanent magnetism. *Quat. Res.* **12**, 326–32.

Oldfield, F., T. A. Rummery, R. Thompson and D. E. Walling 1979c. Identification of suspended sediment sources by means of magnetic measurements: some preliminary results. *Water Resources Res.* **15**, 211–18.

Oldfield, F., P. G. Appleby and R. Thompson 1980. Palaeoecological studies of three lakes in the Highlands of Papua New Guinea. 1. The chronology of sedimentation. *J. Ecol.* **68**, 457–77.

Oldfield, F., R. Thompson and D. P. E. Dickson 1981a. Artificial enhancement of stream bedload: a hydrological application of superparamagnetism. *Phys. Earth Planet. Int.* **26**, 107–24.

Oldfield, F., K. Tolonen and R. Thompson 1981b. History of particulate atmospheric pollution from magnetic measurements in dated Finnish peat profiles. *Ambio* **10**, 185–8.

Oldfield, F., C. Barnosky, E. B. Leopold and J. P. Smith 1983a. Mineral magnetic studies of lake sediments: a brief review. In *Proceedings of the 3rd International Symposium on Palaeolimnology. Hydrobiol.* **103**, 37–44.

Oldfield, F., J. A. Dearing and R. W. Battarbee 1983b. New approaches to recent environmental change. *Geog. J.* **149**, 167–81.

Oldfield, F., A. Krawiecki, B. Maher, J. T. Taylor and S. Twigger 1984. The role of mineral magnetic measurements in archaeology. In *Proceedings of the Association for Environmental Archaeology Conference*. Sheffield, 1983.

Oldfield, F., P. G. Appleby and A. T. Worsley (nee O'Garra) 1985a. Evidence from lake sediments for recent erosion rates in the Highlands of Papua New Guinea. In *Environmental change and tropical geomorphology*, I. Douglas and T. Spencer (eds). London: George Allen & Unwin.

Oldfield, F., A. Hunt, M. D. H. Jones, R. Chester, J. A. Dearing, L. Olsson and J. M. Prospero 1985b. Magnetic differentiation of atmospheric dusts. *Nature* **317**, 516–518.

Oldfield, F., B. A. Maher, J. Donaghue and J. Pierce 1985c. Particle-size related magnetic source-sediment linkages in the Rhode River catchment, Maryland, U.S.A. *J. Geol. Soc. London*, **142**, 1035–1046.

Olson, K. W. and R. K. Skogerboe 1975. Identification of soil lead compounds from automotive sources. *Environ. Sci. Technol.* **9**, 227–30.

Olsson, I. U. 1974. Some problems in connection with the evaluation of C^{14} dates. *Geol. Foren. Stockh. Forh.* **96**, 311–20.

Ondov, J. M., R. C. Ragaini and A. H. Biermann 1979. Emission and particle-size distribution of minor and trace elements at two western coal-fired power plants equipped with cold-side electrostatic precipitators. *Environ. Sci. Technol.* **13**, 946–53.

Opdyke, N. D. 1972. Palaeomagnetism of deep-sea cores. *Rev. Geophys. and Space Phys.* **10**, 213–49.

Opdyke, N. D., L. H. Burckle and A. Todd 1974. The extension of the magnetic time scale in sediments of the central Pacific Ocean. *Earth Planet. Sci. Letters* **22**, 300–6.

Opdyke, N. D., D. Ninkovich, W. Lowrie and J. D. Hayes 1972. The palaeomagnetism of two Aegean deep-sea cores. *Earth Planet. Sci. Letters* **14**, 145–9.

O'Reilly, W. 1976. Magnetic minerals in the crust of the Earth. *Rep. Prog. Phys.* **39**, 857–908.

O'Reilly, W. 1984. *Rock and mineral magnetism.* Glasgow: Blackie.

O'Sullivan, P. E. 1979. The Ecosystem–watershed concept in the environmental sciences – a review. *J. Environ. Sci.* **13**, 273–81.

O'Sullivan, P. E. 1983. Annually-laminated lake sediments and the study of Quaternary environmental changes – a review. *Quat. Sci. Rev.* **1**, 245–313.

O'Sullivan, P. E., F. Oldfield and R. W. Battarbee 1972. Preliminary studies of Lough Neagh sediments I. Stratigraphy, chronology & pollen analysis. in H. J. B. Birks and R. G. West, *Quaternary plant ecology.* Oxford: Blackwells.

Ottow, J. C. G. and H. Glathe 1971. Isolation and identification of iron reducing bacteria from gley soils. *Soil. Biol. Biochem.* **3**, 43–55.

Ozdemir. O. and S. K. Banerjee 1982. A preliminary magnetic study of soil samples from west-central Minnesota. *Earth Planet. Sci. Lett.* **59**, 393–403.

Ozima, M. and M. Ozima 1971. Characteristic thero-magnetic curves in submarine basalts. *J. Geophys. Rev.*, **76**, 2051–6.

Parry, L. G. 1965. Magnetic properties of dispersed magnetite powders. *Phil. Mag.* **11**, 303–12.

Patton, B. J. and J. L. Fitch 1962. Anhysteretic remanent magnetization in small steady field. *J. Geophys. Res.*, **67**, 307–11.

Payne, M. A. and K. L. Verosub 1982. The acquisition of post-depositional detrital remanent magnetization in a variety of natural sediments. *Geophys. J. R. Astr. Soc.* **68**, 625–42.

Peddie, N. W. 1982. International geomagnetic reference field: the third generation. *J. Geomag. Geoelectr.* **34**, 309–26.

Peirce, J. W. and M. J. Clark 1978. Evidence from Iceland on geomagnetic reversal during the Wisconsinan Ice Age. *Nature* **273**, 456–8.

Pennington, W., E. Y. Haworth, A. P. Bonny and J. P. Lishman 1972. Lake sediments in northern Scotland. *Phil. Trans. R. Soc. B* **264**, 191–294.

Peterson, J. T. and C. G. Junge 1971. Sources of particulate matter in the atmosphere. In *Man's impact on the climate.* E. H. Matthews, W. W. Kellogg and G. D. Robinson (eds), 310–20. Cambridge, Mass: MIT Press.

Phillips, J. D. and D. Forsyth 1972. Plate tectonics, palaeomagnetism and the opening of the Atlantic. *Geol. Soc. Am. Bull.* **83**, 1579–1600.

Pierce, J. W. and F. T. Dulong 1977. Discharge of suspended particulates from Rhode River subwatersheds. In *Watershed research in eastern North America.* D. L. Correll (ed.), 531–53. Washington: Smithsonian Institution.

Pouillard, E. 1950. Sur le comportement de l'alumine et de l'oxyde de titane vis-a-vis des oxydes de fer. *Ann. Chim.* **5**, 214–164.

Poutiers, J. 1975. *Sur les propriétés magnetiques de certains sediments continentaux et marins; applications.* Thèse de doctorat. Université de Bordeaux.

Prasad, B. and B. P. Ghildyal 1975. Magnetic suscepti-bility of lateritic soils and clays. *Soil Sciences* **120**, 219–29.

Press, F. and R. Siever 1974. *Earth.* New York: W. H. Freeman.

Prospero, J. M. 1968. Atmospheric Dust studies on Barbados. *Bull. Am. Meteorol. Soc.* **49**, 645.

Prospero, J. M. 1981. Eolian transport to the world's oceans. In *The sea*, vol. 7, *The oceanic lithosphere.* C. Emiliani (ed.), 875–914. New York: Wiley

Prospero, J. M., R. A. Glaccum and R. T. Nees 1981. Atmospheric transport of Soil Dust from Africa to South America. *Nature* **289**, 570–7.

Puffer, J. H., E. W. B. Russell and M. R. Rampino 1980. Distribution and origin of magnetite spherules in air, water and sediments of the greater New York city area and the north Atlantic Ocean. *J. Sed. Petrol.* **50**, 247–56.

Puranen, R. 1977. Magnetic susceptibility and its anisotropy in the study of glacial transport in northern Finland. In *Prospecting in areas of glaciated terrain*, L. K. Lawrence (ed.), 111–19. London: Institution of Mining and Metallurgy.

Radhakrishnamurty, C., S. D. Likhite, B. S. Amin and B. L. K. Somayajulu 1968. Magnetic susceptibility stratigraphy in ocean sediment cores. *Earth Planet. Sci. Letters* **4**, 464–8.

Radhakrishnamurty, C., S. D. Lidhite, E. R. Deutch and G. S. Murthy 1978. Nature of magnetic grains in basalts and implications for palaeomagnetism. *Proc. Indian Acad. Sci.* **87**, 235–43.

Rees, A. I. 1961. The effect of water currents on the magnetic remanence and anisotropy of susceptibility of some sediments. *Geophys. J. R. Astr. Soc.* **5**, 235–51.

Renberg, I. 1982. Varved lake sediments – a geochrono-logical record of the Holocene. *Geol. Foren. i Stockh. Forh.* **104**, 85–90.

Revitt, D. M., J. Bryan Ellis and F. Oldfield 1981. Variations in heavy metals of stormwater sediments in a separate sewer system. In *Urban storm drainage*, B. C. Yen (ed.), 49–58. New York: Pentech Press.

Richardson, N. 1986. The mineral magnetic record in recent ombrotrophic peat synchronised by fine resolution pollen analysis. *Phys. Earth Planet Int.* 42, 48–56.

Robinson, S. G. 1982. Two applications of mineral–magnetic techniques to deep-sea sediment studies. *Geophys. J. R. Astr. Soc.* 69, 294.

Robinson, S. G. 1986. The late Pleistocene palaeoclimatic record of North Atlantic deep-sea sediments revealed by mineral-magnetic measurements. *Phys. Earth Planet Int.*

Rosen, J. M. 1969. Stratospheric dust and its relationship to the meteoric influx. *Space Science Reviews* 9, 58–89.

Rubin, L. G. and P. A. Wolff 1984. High magnetic fields for Physics. *Physics Today* (Aug), 24–33.

Ruddiman, W. F. 1971. Pleistocene sedimentation in the equatorial Atlantic stratigraphy and faunal palaeoclimatology. *Geol. Soc. Am. Bull.* 82, 283–302.

Ruddiman, W. F., G. A. Jones, T. H. Peng, L. K. Glover, B. P. Glass and P. J. Liebertz 1980. Tests for size and shape dependency in deep-sea mixing. *Sed. Geol.* 25, 257–76.

Rudman, A. J. and R. F. Blakely 1976. *Fortran program for correlation of stratigraphic time series.* Dept. Nat. Resources. Geol. Survey Occasional Paper. Bloomington, Indiana.

Rumble, III, D., 1976a. Oxide minerals in metamorphic rocks. In *Oxide minerals*, D. Rumble III (ed.), R1–20, Mineralogical Society of America. Blacksburg: Southern Printing Co.

Rumble, III, D. (ed.) 1976b. *Oxide minerals.* Mineralogy Society of America Short Course Notes Vol. 3. Blacksburg: Southern Printing Co.

Rummery, T. A., J. Bloemendal, J. Dearing, F. Oldfield and R. Thompson 1979. The persistence of fire-induced magnetic oxides in soils and lake sediments. *Ann. Geophys.* 35, 103–7.

Rummery, T. A. 1981. *The effects of fire on soil and sediment magnetism.* Unpubl. PhD thesis. University of Liverpool.

Rummery, T. A. 1983. The use of magnetic measurements in interpreting the fire histories of lake drainage basins. *Hydrobiologia* 103, 53–8.

Runcorn, S. K. 1955. On the theory of the geomagnetic secular variation. *Ann. Geophys.* 15, 87–92.

Runcorn, S. K. 1956. Palaeomagnetic comparisons between Europe and North America. *Proc. Canad. Geol. Assoc.* 8, 77–85.

Saarnisto, M. 1975. Pehmeiden jarvisedimenttien naytteenottoon soveltuva jaadytysmenetelma. *Geologi* 26, 37–9.

Sandlands, Judith 1983. *Magnetic measurements applied to a study of river channel erosion on the river Severn.* Unpublished BSc dissertation. University of Liverpool.

Sarajma, S., S. Nishimura and K. Hirooka 1984. The Blake geomagnetic event as inferred from late Brunhes ignimbrites in Southwest Japan and West Indonesia. *J. Geomag. Geoelectr.* 36, 203–14.

Schwertmann, U. and R. M. Taylor 1977. Iron oxides. In *Minerals in soil environments.* J. B. Dixon (ed.), *Soil. Sci. Soc. Am.* 145–80.

Sclater, J. G. and B. Parsons 1981. Oceans and continents: Similarities and differences in the mechanism of heat loss. *J. Geophys. Res.* 86, 11535–52.

Scollar, I. 1965. A contribution to magnetic prospecting in archaeology. *Archaeo-Physicka* 1, 21–92.

Scoullos, M., F. Oldfield and R. Thompson 1979. Magnetic monitoring of marine particulate pollution in the Elefsis Gulf, Greece. *Marine Pollut. Bull.* 10, 287–91.

Shackleton, N. J. 1977. The oxygen isotope stratigraphic record of the late Pleistocene. *Phil. Trans. R. Soc. B* 280, 169–80.

Singh, G., A. P. Kershaw and R. Clark 1979. Quaternary vegetation and fire history in Australia. In *Fire and Australian biota.* A. M. Gill, R. A. Groves and I. R. Noble (eds). Canberra: Australian Academy of Science.

Slaughter, T. H., R. T. Kershin, B. G. McMullan, G. Cocoros, and D. Vanko 1976. *Shoreline conditions.* Annotated map published by the State of Maryland, Department of Natural Resources and the Maryland Geological Survey.

SMIC 1971. *Inadvertent climate modification.* Report of the study of man's impact on climate. Cambridge, Mass: MIT Press.

Smit, J. and H. P. J. Wijn 1954. Physical properties of ferrites. *Adv. Electron. Electron. Phys.* 6, 69–136.

Smit, J. and H. P. J. Wijn 1959. *Ferrites.* Eindhoven: Philips Tehnical Library.

Snoek, J. L. 1948. Dispersion and absorption in magnetic ferrites at frequencies above one megacycle. *Physica* 14, 207–17.

Stacey, F. D. 1963. The physical theory of rock magnetism. *Adv. Phys.* 12, 45–133.

Stevenson, A. 1971. Single domain grain distributions. *Phys. Earth Planet. Interiors,* 4, 353–60, 361–9.

Stober, J. C. 1978. *Palaeomagnetic secular variation studies on Holocene lake sediments.* Unpubl. PhD thesis. University of Edinburgh.

Stober, J. C. and R. Thompson 1977. Palaeomagnetic secular variation studies of Finnish lake sediment and the carriers of remanence. *Earth Planet. Sci. Letters* 37, 139–49.

Stober, J. C. and R. Thompson 1979. Magnetic remanence acquisition in Finnish lake sediments. *Geophys. J. R. Astr. Soc.* 57, 727–39.

Stoker, M. S., A. C. Skinner, J. A. Fyfe and D. Long 1983. Palaeomagnetic evidence for early Pleistocene in the central and northern North Sea. *Nature* 304, 332–4.

Stoner, E. C. and E. P. Wohlfarth 1948. A mechanism of magnetic hysteresis in heterogeneous alloys. *Phil. Trans. R. Soc. Lond. A* 240, 599–642.

Stott, A. P. 1986. Sediment tracing in a reservoir catchment system using a magnetic mixing model. *Phys. Earth Plant Int.* 42, 105–14.

Strangway, D. W. 1970. *History of the Earth's magnetic field.* New York: McGraw-Hill.

Stuiver, M. 1972. On climate changes. *Quat. Res.* **2**, 409–11.

Stupavsky, M., C. P. Gravenor and D. T. A. Symons 1979. Palaeomagnetic stratigraphy of the medowcliffe till, Scarborough bluffs, Ontario; a late Pleistocene excursion? *Geophys. Res. Lett.* **6** (4), 269–72.

Sugden, D. E. and C. M. Clapperton 1980. West Antarctic ice sheet fluctuations in the Antarctic Penninsula. *Nature* **286**, 378–81.

Sutton, M. 1982. *Magnetic catenas in the Cotswolds.* Unpubl. BSc dissertation, Department of Geography, University of Liverpool.

Swain, A. 1973. A history of fire and vegetation as recorded in lake sediments. *Quat. Res.* **3**, 383–96.

Tarling, D. H. 1983. *Palaeomagnetism.* London: Chapman and Hall.

Taylor, R. M. and U. Schwertmann 1974. Maghemite in soils and its origin. *Clay Minerals* **10**, 289–310.

Theis, T. L. and J. L. Wirth 1977. Sorptive behaviour of trace metals on fly ash in aqueous systems. *Environ. Sci. Technol.* **11**, 1095–100.

Thellier, E. 1933. Magnetometre insensible aux champs magnetiques troubles des grandes villes, *C.R. Acad. Sci. Paris* **197**, 224–34.

Thellier, R. 1938. Sur l'aimantation des terres cuites et ses applications geophysiques. *Ann. Inst. Physique du Globe. Univ. Paris* **16**, 157–302.

Thellier, E. 1946. Sur la thermoremanence et la theorie du metamagnetisme. *C.R. Acad. Sci. Paris* **233**, 319–21.

Thellier, E. 1966. Methods of alternating current and thermal demagnetization, in *Methods and techniques in Geophysics*, vol. 2. S. K. Runcorn (ed.). London: Interscience.

Thellier, E. 1981. Sur la direction du champ magnetique terrestre, en France, durant les deux derniers millenaires. *Phys. Earth Planet. Interiors* **24**, 89–132.

Thellier, E. and O. Thellier 1959. Sur l'intensite du champ magnetique terrestre dans le passe historique et geologique. *Ann. Geophys.* **15**, 285–376.

Thomas, A. K. 1968. Magnetic shielded enclosure design in the d.c. and VLF region. *IEEE Trans. Electromagn. Compat.* **10**, 142–52.

Thompson, R. 1973. Palaeolimnology and palaeomagnetism. *Nature* **242**, 182–4.

Thompson, R. 1982. A comparison of geomagnetic secular variation as recorded by historical, archaeomagnetic and palaeomagnetic measurements. *Phil. Trans. R. Soc. Lond. A* **306**, 103–12.

Thompson, R. 1983. ^{14}C dating and magnetostratigraphy. *Radiocarbon* **25**, 229–38.

Thompson, R. 1984. Geomagnetic evolution: 400 years of change on planet Earth. *Phys. Earth Planet. Interiors* **36**, 61–77.

Thompson, R., M. J. Aitken, P. L. Gibbard and J. J. Wymer 1974. Palaeomagnetic study of Hoxnian lacustrine sediments. *Archaeometry* **16**, 233–45.

Thompson, R. and D. R. Barraclough 1982. Geomagnetic secular variation based on spherical harmonic and cross validation analyses of historical and archaeomagnetic data. *J. Geomag. Geoelectr.* **34**, 245–63.

Thompson, R., R. W. Battarbee, P. E. O'Sullivan and F. Oldfield 1975. Magnetic susceptibility of lake sediments. *Limnol. Oceanogr.* **20**, 687–98.

Thompson, R. and B. Berglund 1976. Late Weichselian geomagnetic 'reversal' as a possible example of the reinforcement syndrome. *Nature* **263**, 5577, 490–1.

Thompson, R. and R. M. Clark 1981. Fitting polar wander paths. *Phys. Earth Planet Interiors* **27**, 1–7.

Thompson, R. and K. Kelts 1974. Holocene sediments and magnetic stratigraphy from Lakes Zug and Zurich, Switzerland. *Sedimentology* **21**, 577–96.

Thompson, R. and D. J. Morton 1979. Magnetic susceptibility and particle-size distribution in recent sediments of the Loch Lomond drainage basin, Scotland. *J. Sed. Petrol.* **49**, 3, 801–12.

Thompson, R., G. M. Turner, M. Stiller and A. Kaufman 1985. Near East palaeomagnetic secular variation recorded in sediments from the sea of Galilee (Lake Kinneret). *Quat. Res.* **23**, 175–88.

Tite, M. S. and R. E. Linington 1975. Effect of climate on the magnetic susceptibility of soils. *Nature* **265**, 565–6.

Tolonen, K., A. Siiriainen and R. Thompson 1975. Prehistoric field erosion sediment in Lake Lovojarvi, S. Finland and its palaeomagnetic dating. *Ann. Bot. Fenn.* **12**, 161–4.

Tolonen, K. and F. Oldfield 1986. The record of the magnetic-mineral and heavy metal deposition at Regent Street Bog, Fredicton, New Brunswick, Canada. *Phys. Earth Planet Int.* **42**, 57–66.

Tucker, P. 1980. Stirred remanent magnetization: a laboratory analogue of post-depositional realignment. *J. Geophys.* **48**, 153–7.

Turner, G. M. 1979. *Geomagnetic investigation of some recent British sediments.* Unpubl. PhD thesis. University of Edinburgh.

Turner, G. M. and R. Thompson 1979. Behaviour of the Earth's magnetic field as recorded in the sediments of Loch Lomond. *Earth Planet. Sci. Letters* **42**, 412–26.

Turner, G. M. and R. Thompson 1981. Lake sediment record of the geomagnetic secular variation in Britain during Holocene times. *Geophys. J. R. Astr. Soc.* **65**, 703–25.

Undzendor, B. A. and V. A. Shapiro 1967. Seismomagnetic effect in a magnetite deposit. *Izv. Earth Physics* **1**, 121–6. English translation, 1968, *Physics of the solid Earth.* 69–72.

Vadyunina, A. F. and V. F. Babanin 1972. Magnetic

susceptibility of some soils in the U.S.S.R. *Soviet Soil. Sci.* 6, 106–10.

Veinberg, B. P. and V. P. Shibaev 1969. *Catalogue. The results of magnetic determination at equidistant points and epochs, 1500–1940* (ed.-in-Chief A. N. Pushkov). Moscow: IZMIRAN. Translation no. 0031 by Canadian Department of the Secretary of State, Translation Bureau, 1970.

Verhoogen, J. 1973. Thermal regime of the Earth's core. *Phys. Earth Planet Int.* 7, 47–58.

Verosub, K. L. 1977. Depositional and post-depositional processes in the magnetization of sediments. *Rev. Geophys. Space Phys.* 15 (2), 129–43.

Verosub, K. L. 1982. Geomagnetic excursions: a critical assessment of the evidence as recorded in sediments of the Brunhes Epoch. *Phil. Trans. R. Soc. Lond.* A 306, 161–8.

Vevosub, K. L., J. O. Davis and S. Valastro 1980. A palaeomagnetic record from Pyramid Lake, Nevada, and its implications for proposed geomagnetic excursions. *Earth Planet. Sci. Letters* 49, 141–8.

Verosub, K. L., R. A. Ensley and J. S. Ulrich 1979. The role of water content in the magnetization of sediments. *Geophys. Res. Letters* 6, 226–8.

Verwey, E. J. W. and P. W. Haayman 1941. Electronic conductivity and transition point in magnetite. *Physics* 8, 979–82.

Vilks, G., J. M. Hall and D. J. W. Piper 1977. The natural remanent magnetism of sediment cores from the Beaufort sea. *J. Earth Sci.* 14, 2007–12.

Vine, F. J. and D. H. Matthews 1963. Magnetic anomalies over oceanic ridges. *Nature* 199 (4897), 947–9.

Vitorello, I. and R. Van der Voo 1977. Magnetic stratigraphy of Lake Michigan sediments obtained from cores of lacustrine clay. *Quat. Res.* 7, 398–412.

Vlasov, A. Y., G. V. Kovalenko and V. A. Chikhacher 1967. The superparamagnetism of a-FeOOH. Izv. *Earth Physics* 7, 64–69. English translation, 1968. *Physics of the solid Earth*, 460–4.

Walcott, C., J. L. Gould and J. L. Kirschvink 1979. Pigeons have magnets. *Science* 205, 1027–9.

Walcott, C. and R. P. Green 1974. Orientation of homing pigeons altered by a change in the direction of an applied magnetic field. *Science* 184, 180–2.

Walling, D. E., M. R. Peart, F. Oldfield and R. Thompson 1979. Suspended sediment sources identified by magnetic measurements. *Nature* 281, 110–13.

Wasilewski, P. J. 1973. Magnetic hysteresis in natural materials. *Earth Planet. Sci. Letters* 20, 67–72.

Watkins, N. D. 1968. Short period geomagnetic polarity events in deep-sea sedimentary cores. *Earth Planet. Sci. Letters* 4, 341–9.

Watkins, N. D. 1971. Geomagnetic polarity events and the problem of 'The Reinforcement Syndrome'. *Comments Earth Sci. Geophys.* 2, 36–43.

Watkins, N. D. 1972. Review of the development of the geomagnetic polarity time scale and discussion of prospects for its finer definition. *Geol. Soc. Am. Bull.* 83, 551–74.

Watson, G. S. 1970. Orientation statistics in the Earth sciences. *Bull. Geol. Instn. Univ. Upsala N.S.* 2 (9), 78–89.

Weiss, P. 1907. L'hypothese du champ moleculaire et la propriete ferromagnetique. *J. Phys.* 6, 661.

West, G. F. and D. J. Dunlop 1971. An improved ballistic magnetometer for rock magnetic experiments. *J. Phys.* E 4, 37–40.

Whitby, K. T. and B. Cantrell 1975. *Atmospheric aerosols – characterisation and measurements.* Proceeding of the International Conference on Environmental Sensing and Assessment, Las Vagas, Nevada.

Williamson, S. J. and L. Kaufman 1981. Biomagnetism. *J. Magnetism Magnetic Materials* 22 (2), 129–202.

Wilson, M. N. 1983. *Superconducting magnets.* Oxford: Clarendon Press.

Wilson, R. L. 1962. The palaeomagnetism of baked contact rocks and reversals of the Earth's magnetic field. *Geophys. J. R. Astr. Soc.* 7, 194–202.

Wiltschko, R., D. Nohr and W. Wiltschko 1981. Pigeons with a deficient sun compass use the magnetic compass. *Science* 214, 343–5.

Windom, H. L. 1969. Atmospheric dust records in permanent snowfields: implications to marine sedimentation. *Geol. Soc. Am. Bull.* 80, 761–82.

Winterer, E. L. 1973. Sedimentary facies and plate tectonics of equatorial Pacific. *Am. Assoc. Petrol. Geol. Bull.* 57, 2, 265–82.

Wise, S. M. 1979. *Magnetic and radiometric studies of erosion.* Unpubl. MS of paper presented to UK Geophysical Assembly, Southampton.

Wollin, G., W. B. F. Ryan, D. B. Ericson and J. H. Foster 1977. Palaeoclimate, palaeomagnetism and the eccentricity of the Earth's orbit. *Geophys. Res. Lett.* 44, 267.

Worsley, A. T. (nee O'Garra) 1983. *A palaeoecological study of recent environmental change in the highlands of Papua New Guinea.* Unpubl. PhD thesis. University of Liverpool.

Yanak, F. and I. Uman 1967. The distribution function of magnetic susceptibilities. *Izv. Earth physics.* 6, 137–43. English translation, 1968. *Physics of the solid Earth*, 416–20.

Yukutake, T. 1979. Review of the geomagnetic secular variations on the historical time scale. *Phys. Earth Planet. Interiors* 20, 83–95.

Yukatake, T. and H. Tachinaka 1968. The non-dipole part of the Earth's magnetic field. *Bull. Earthquake Res. Inst.* 46, 1027–62.

Zoeger, J., J. R. Dunn and M. Fuller 1981. Magnetic material in the head of the common Pacific Dolphin. *Science* 213, 892–4.

Index